Mathematical Solutions for Complex Engineering Systems

RIVER PUBLISHERS SERIES IN MATHEMATICAL, STATISTICAL AND COMPUTATIONAL MODELLING FOR ENGINEERING

Series Editor:

MANGEY RAM
Graphic Era University, India

Applied mathematical techniques along with statistical and computational data analysis has become vital skills across the physical sciences. The purpose of this book series is to present novel applications of numerical and computational modelling and data analysis across the applied sciences. We encourage applied mathematicians, statisticians, data scientists and computing engineers working in a comprehensive range of research fields to showcase different techniques and skills, such as differential equations, finite element method, algorithms, discrete mathematics, numerical simulation, machine learning, probability and statistics, fuzzy theory, etc.

Books published in the series include professional research monographs, edited volumes, conference proceedings, handbooks and textbooks, which provide new insights for researchers, specialists in industry, and graduate students.

Topics included in this series are as follows:

- Discrete mathematics and computation
- Fault diagnosis and fault tolerance
- Finite element method (FEM) modeling/simulation
- Fuzzy and possibility theory
- Fuzzy logic and neuro-fuzzy systems for relevant engineering applications
- Game Theory
- Mathematical concepts and applications
- Modelling in engineering applications
- Numerical simulations
- Optimization and algorithms
- Queueing systems
- Resilience
- Stochastic modelling and statistical inference
- Stochastic Processes
- Structural Mechanics
- Theoretical and applied mechanics

For a list of other books in this series, visit www.riverpublishers.com

Mathematical Solutions for Complex Engineering Systems

Editors

Satyvir Singh

RWTH Aachen University, Germany

Mukesh Kumar Awasthi

Babasaheb Bhimrao Ambedkar University, Lucknow, India

River Publishers

Routledge
Taylor & Francis Group
NEW YORK AND LONDON

Published 2026 by River Publishers
River Publishers
Broagervej 10, 9260 Gistrup, Denmark
www.riverpublishers.com

Distributed exclusively by Routledge
605 Third Avenue, New York, NY 10017, USA
4 Park Square, Milton Park, Abingdon, Oxon OX14 4RN

Mathematical Solutions for Complex Engineering Systems / by Satyvir Singh, Mukesh Kumar Awasthi.

Routledge is an imprint of the Taylor & Francis Group, an informa business

ISBN 978-87-4380-668-4 (hardback)
ISBN 978-87-4381-312-5 (paperback)
ISBN 978-87-4380-670-7 (online)
ISBN 978-87-4380-669-1 (master ebook)
DOI: 10.1201/9788743806691

Contents

Preface

Mathematical Solutions for Complex Engineering Systems

Engineering systems are inherently complex, requiring sophisticated mathematical tools to analyze, model, and optimize their performance. As technology advances, the need for mathematical frameworks that bridge theoretical principles with real-world engineering challenges has become more critical than ever. This book, *Mathematical Solutions for Complex Engineering Systems*, offers a comprehensive collection of mathematical methodologies to address multifaceted problems across diverse engineering disciplines. By integrating deterministic and stochastic modeling, numerical techniques, and computational simulations, this book serves as a valuable resource for researchers, engineers, and academicians seeking innovative solutions to contemporary engineering problems.

The book begins with **Chapter 1**, which introduces the fundamental role of mathematical modeling in engineering applications, covering deterministic, stochastic, linear, and nonlinear models, alongside modern computational tools such as finite element methods and data-driven approaches. **Chapter 2** builds on this with stochastic techniques for reliability analysis, focusing on critical infrastructure systems.

Nonlinear dynamics take center stage in **Chapter 3**, where epidemic models reveal chaos and stability in spreading phenomena. **Chapter 4** extends this to ecological systems, analyzing plant-herbivore interactions with time-delayed dynamics. **Chapter 5** shifts to agriculture, modeling pesticide effects on crop yields through logistic growth frameworks. **Chapter 6** transitions to tumor dynamics, integrating biological insights with mathematical precision. **Chapter 7** employs high-order computational methods to dissect shock-induced instabilities in gas interfaces, highlighting Richtmyer-Meshkov instability.

Fluid dynamics and transport phenomena dominate **Chapters 8–12**. **Chapter 8** compares hydrodynamic models for rarefied gases, while **Chapter 9** solves non-equilibrium flow problems using moment equations. **Chapter 10** derives unified formulas for turbulent boundary layers over rotating surfaces. **Chapter 11** optimizes heat transfer and fluid flow, and **Chapter 12** adapts meshfree methods to nonlinear PDEs.

Advanced numerical schemes are explored in **Chapters 13–16**. **Chapter 13** leverages fractional calculus for variational problems, and **Chapter 14** devises a fourth-order scheme for wave equations. **Chapter 15** tackles the Sine-Gordon equation via B-spline methods, while **Chapter 16** evaluates high-resolution CFD schemes for convective flows.

Chapter 17 offers a forward-looking perspective on *Trends, Challenges, and Future Directions in Mathematical Solutions for Complex Engineering Systems*. It synthesizes emerging themes across the chapters, identifies current limitations, and proposes promising directions for future research, including data-driven modeling, multiscale methods, and cross-disciplinary integration.

This book offers a well-rounded perspective on mathematical modeling techniques applicable to engineering systems, blending theoretical rigor with computational efficiency. Each chapter presents innovative methodologies, case studies, and real-world applications that demonstrate the power of mathematical solutions in addressing engineering complexities. We hope this book serves as an invaluable reference for researchers, professionals, and students interested in advancing their understanding of mathematical approaches to engineering problems.

We extend our gratitude to the contributors whose expertise has shaped this book, and we hope that it inspires further research and innovation in the field of mathematical modeling for complex engineering systems.

Editors

Satyvir Singh
Mukesh Kumar Awasthi

List of Figures

List of Tables

List of Contributors

Alsaeed, Salman Saud: *Department of Mathematics, College of Science, Jouf University, Saudi Arabia*

Anand, Monika: *Department of Sciences, Quantum University, Roorkee–247667, Uttarakhand, India*

Awasthi, Mukesh Kumar: *Department of Mathematics, Babasaheb Bhimrao Ambedkar University, India*

Bazyar, Pourya: *Department of Mechanical Engineering and Production Management, Hamburg University of Applied Science, Germany*

Bhattacharjee, Rahul: *Department of Mathematics, Birla Institute of Technology and Science Pilani, Pilani Campus, India*

Chourushi, Tushar: *Department of Aerospace Engineering, MIT Art, Design and Technology University, India*

Dhiman, Neeraj: *Department of Mathematics, Graphic Era Deemed to be University, India*

Dingare, Sunil: *Department of Aerospace Engineering, MIT Art, Design and Technology University, India*

Dipesh: *Department of Mathematics, SR University, India*

Farkya, Ankit: *Department of Mathematics, Birla Institute of Technology and Science Pilani, Pilani Campus, India*

Fatima, Nahid: *Department of Mathematics and Sciences, Prince Sultan University, Saudi Arabia*

Gupta, Mukesh: *Department of Mathematics, Govt. Degree College, India; Department of Mathematics, Babasaheb Bhimrao Ambedkar University, India*

Gupta, Surbhi: *Department of Applied Mathematics, Amity University, India*

Kanwar, Shefali: *Department of Applied Mathematics, Amity University, India*

Kashyap, Manju: *Department of Applied Mathematics, Galgotias College of Engineering and Technology, India*

Kishore, Rajeev: *Department of Mathematics, Galgotias College of Engineering & Technology, India*

Kumar, Anil: *Department of Mathematics, Birla Institute of Technology and Science Pilani, Pilani Campus, India*

Kumar, Deepak: *Department of Mathematics, SRPS College, B.R.A. Bihar University, India*

Kumar, Pankaj: *Department of Mathematics, Lovely Professional University, India*

Malik, Pradeep: *Department of Mathematics, Faculty of Applied and Basic Sciences, SGT University, India*

Mandal, Saptarshi: *Department of Aerospace Engineering, Amity University Mumbai, India*

Msmali, Ahmed Hussein: *Department of Mathematics, College of Science, Jazan University, Kingdom of Saudi Arabia*

Rana, Anirudh Singh: *Department of Mathematics, Birla Institute of Technology and Science Pilani, Pilani Campus, India*

Sanap, Sudarshan B: *Department of Aerospace Engineering, MIT Art, Design and Technology University, India*

Sheidaee, Ehsan: *Biosystems Engineering Department, Tarbiat Modares University, Iran*

Shvydkyi, Oleh G: *Independent researcher, Ukraine*

Singh, Brajesh Kumar: *Department of Mathematics, Babasaheb Bhimrao Ambedkar University, India*

Singh, Satyvir: *Institute for Applied and Computational Mathematics, RWTH Aachen University, Germany; Department of Mathematics, Graphic Era Deemed to be University, India*

Sonawane, Chandrakant: *Symbiosis Institute of Technology Pune, Symbiosis International Deemed University, India*

Song, Hang: *Institute of Applied and Computational Mathematics, RWTH Aachen University, Germany*

Srivastava, Vineet Kishore: *Department of Mathematics, School of Life and Applied Sciences, Uttaranchal University, India*

Tamsir, Mohammed: *Department of Mathematics, College of Science, Jazan University, Jazan, Saudi Arabia*

Torrilhon, Manuel: *Institute of Applied and Computational Mathematics, RWTH Aachen University, Germany*

Verma, Rupali: *Department of Mathematics, Lovely Professional University, India*

Yadav, Bhawana: *Department of Mathematics, Faculty of Applied and Basic Sciences, SGT University, India*

Yadav, Manisha: *Department of Mathematics, Faculty of Applied and Basic Sciences, SGT University, India*

List of Abbreviations

2T	Two-temperature
ACO	Ant colony optimization
AE	Absolute error
AI	Artificial intelligence
AM	Additive manufacturing
ANN	Artificial neural network
ARMA	Autoregressive moving average
BAIR	Boundedness, accuracy, and interpolative reasonableness
BC	Boundary condition
BEM	Boundary element method
BRN	Basic reproduction number
BTE	Boltzmann transport equation
BVP	Boundary value problem
CBC	Convection boundedness criteria
CCR	Coupled constitutive relation
CFD	Computational fluid dynamics
CHT	Conjugate heat transfer
CTCM	Continuous-time Markov chain
DDM	Discrete direct method
DG	Discontinuous Galerkin
DGSEM	Discontinuous Galerkin spectral element method
DIC	Diagonal incomplete-Cholesky
DQM	Differential quadrature method
DRM	Dual reciprocity method
DSMC	Direct Simulation Monte Carlo
DTMP	Discrete-time Markov process
EA	Evolutionary algorithm
ECM	Extracellular matrix
EHD	Electrohydrodynamic

ELE	Euler-Lagrange equation
EWM	Equilibrium wall model
ExCBCM	Exponential cubic b-spline collocation method
FDE	Fractional differential equation
FEM	Finite element method
FOTW	Fractional order Taylor wavelet
FVI	Fractional variational integrator
FVM	Finite volume method
FVP	Fractional variational problem
GA	Genetic algorithms
HE	Heat exchanger
HO	Higher-order
HRS	High-resolution scheme
HVAC	Heating, ventilation, and air conditioning
IBVP	Initial boundary value problem
IC	Initial condition
IMQ	Inverse multiquadrics
IS	Incident shock
LGL	Legendre-Gauss-Lobatto
LoW	Law of the wall
LSM	Least square method
MD	Molecular dynamics
MDE	Matrix-degrading enzyme
MFS	Method of fundamental solution
ML	Machine learning
MMP	Matrix metalloproteinases
MQ	Multiquadrics
MRM	Multiple reciprocity method
MSS	Multi-state system
NBC	Neumann boundary condition
NPP	Nuclear power plant
NSF	Navier–Stokes–Fourier
NVF	Normalized variable formulation
ODE	Ordinary differential equation
PCG	Preconditioned conjugate gradient
PDE	Partial differential equation
PINN	Physics-informed neural networks
PM	Preventive maintenance
PSO	Particle swarm optimization

QFT	Quantum field theory
RBF	Radial basis function
RBS	Rayleigh–Brillouin scattering
RCBC	Refined convective boundedness criterion
RME	Realistic mathematics education
RMI	Richtmyer–Meshkov instability
RMS	Root mean square
RNSF	Refined Navier–Stokes–Fourier
SBP	Summation-by-parts
SDE	Stochastic differential equation
SEIR	Susceptible exposed infectious recovered
SOLP	Shifted orthonormal Legendre polynomial
TO	Topology optimization
TPS	Thin-plate splines
TVD	Total variation diminishing
uPA	Urokinase plasminogen activator
VSS	Variable soft sphere
VTE	Vorticity transport equation
XAI	Explainable artificial intelligence

1

Recent Innovations and Advances in Mathematical Modeling for Engineering Systems

Mukesh Kumar Awasthi[1] **and Satyvir Singh**[2,3]

[1]Department of Mathematics, Babasaheb Bhimrao Ambedkar University, India
[2]Institute for Applied and Computational Mathematics, RWTH Aachen University, Germany
[3]Department of Mathematics, Graphic Era Deemed to be University, India
E-mail: mukeshiitr.kumar@gmail.com; singh@acom.rwth-aachen.de

Abstract

Mathematical modeling is a powerful tool for analyzing and optimizing complex engineering systems, enabling the translation of real-world challenges into solvable frameworks. This chapter provides a comprehensive introduction to the foundational principles of mathematical modeling in the context of engineering systems. It highlights the critical role of modeling in translating complex real-world phenomena into solvable mathematical formulations, enabling analysis, prediction, optimization, and informed decision-making. The chapter categorizes engineering systems based on key characteristics such as linearity, dynamics, stochasticity, and control, while illustrating each with relevant mathematical expressions. It underscores the importance of computational tools in solving complex equations, simulating physical systems, and optimizing performance through numerical methods and data-driven approaches. Furthermore, it addresses the major challenges in modeling complex systems, including nonlinearity, high dimensionality, uncertainty, multiscale behavior, and validation issues. By bridging theory with practical applications, this chapter lays the groundwork for advanced

DOI: 10.1201/9788743806691-1

modeling techniques and sets the stage for interdisciplinary innovation in modern engineering.

Keywords: Mathematical modeling, engineering systems, computational tools, system classification, modeling challenges

1.1 Importance of Mathematical Modeling

Mathematical modeling stands at the heart of modern engineering and scientific practice. It plays a vital role in helping engineers, scientists, and decision-makers represent, interpret, and solve real-world problems through a structured and logical framework. As engineering challenges grow more complex – encompassing multidisciplinary phenomena, large-scale systems, and high-precision requirements – mathematical models provide an indispensable means for understanding and innovation. This section elaborates on why mathematical modeling is not only important but essential for modern engineering and technological progress. It explores the key reasons for modeling, including analysis, prediction, optimization, and decision-making, supported by illustrative applications across engineering domains.

1.1.1 Modeling for analysis and design

- *Foundation for problem-solving:* Mathematical modeling serves as the foundation for problem-solving in engineering. It allows engineers to represent real-world systems in a simplified yet accurate manner, enabling the analysis of system behavior under various conditions. For example, in civil engineering, models are used to predict the stress and strain on structures like bridges and buildings [1]. This predictive capability is essential for ensuring safety and reliability.
- *Enabling innovation and design optimization:* Mathematical models are indispensable in the design and optimization of engineering systems. By simulating different scenarios, engineers can identify the most efficient designs and minimize costs. In aerospace engineering, for instance, computational fluid dynamics (CFD) models are used to optimize aircraft shapes for reduced drag and improved fuel efficiency [2].
- *Enabling virtual prototyping:* Virtual prototyping, made possible by mathematical modeling, allows engineers to test and refine designs without the need for physical prototypes. This is particularly valuable in automotive engineering, where crash simulations are used to improve vehicle safety [3].

- *Ensuring scalability and adaptability:* Mathematical models enable engineers to scale solutions and adapt them to different contexts. In industrial engineering, models are used to design production systems that can be easily scaled up or down to meet changing demands [4].

1.1.2 Modeling for decision-making and prediction

- *Supporting decision-making:* Engineers rely on mathematical models to make data-driven decisions. Models provide insights into system performance, helping engineers choose the best course of action. For example, in environmental engineering, models are used to assess the impact of pollution and develop mitigation strategies [5].
- *Enhancing predictive capabilities:* One of the most significant advantages of mathematical modeling is its ability to predict future outcomes. In mechanical engineering, models are used to predict the lifespan of machinery and plan maintenance schedules, reducing downtime and costs [6].
- *Improving resource management:* Mathematical models play a crucial role in resource management. In chemical engineering, for example, models are used to optimize the use of raw materials and energy in manufacturing processes, reducing waste and improving sustainability [7].
- *Supporting risk assessment and mitigation:* Engineers use mathematical models to assess risks and develop mitigation strategies. In nuclear engineering, models are used to simulate reactor behavior and ensure safety under extreme conditions [8].
- *Supporting policy and regulation:* Mathematical models are used to inform policy and regulation in engineering. In transportation engineering, for example, models are used to evaluate the impact of infrastructure projects and develop traffic management strategies [9].
- *Addressing global challenges:* Mathematical modeling is essential for addressing global challenges such as climate change, energy sustainability, and public health. In renewable energy engineering, models are used to optimize the design and operation of solar panels and wind turbines [10].

1.1.3 Modeling for innovation and communication

- *Bridging theory and practice:* Mathematical modeling bridges the gap between theoretical concepts and practical applications. It allows engineers to test hypotheses and validate theories in a controlled environment before implementing them in real-world systems. This is particularly important in fields like electrical engineering, where circuit models are used to design and test electronic devices [11].

- *Facilitating interdisciplinary collaboration:* Mathematical models provide a common language for interdisciplinary collaboration. Engineers, scientists, and mathematicians can work together to develop and refine models, leading to innovative solutions. For example, in biomedical engineering, models are used to simulate physiological systems and develop medical devices [12].
- *Driving technological advancements:* Mathematical modeling has been instrumental in driving technological advancements. In robotics, models are used to design control systems that enable robots to perform complex tasks with precision [13].
- *Supporting education and training:* Mathematical models are widely used in engineering education to teach fundamental concepts and principles. They provide students with hands-on experience in solving real-world problems, preparing them for professional practice [14].
- *Enhancing communication and visualization:* Mathematical models help engineers communicate complex ideas and visualize system behavior. In software engineering, models are used to represent system architectures and workflows, facilitating collaboration among team members [15].

The process of developing a mathematical model involves a series of structured steps that guide engineers from problem definition to real-world application. Figure 1.1 illustrates a typical modeling workflow, highlighting key phases such as abstraction, mathematical formulation, solution strategies, validation, and deployment. This systematic approach ensures that models are not only mathematically sound but also aligned with the practical goals of engineering design and analysis.

In order to fully leverage the power of mathematical models in these diverse applications, it is essential to understand how engineering systems differ in their nature and behavior. The following section classifies these systems to build a deeper insight into modeling strategies.

1.2 Types of Engineering Systems

Engineering systems encompass a wide range of applications, from mechanical and electrical systems to chemical and biological processes. These

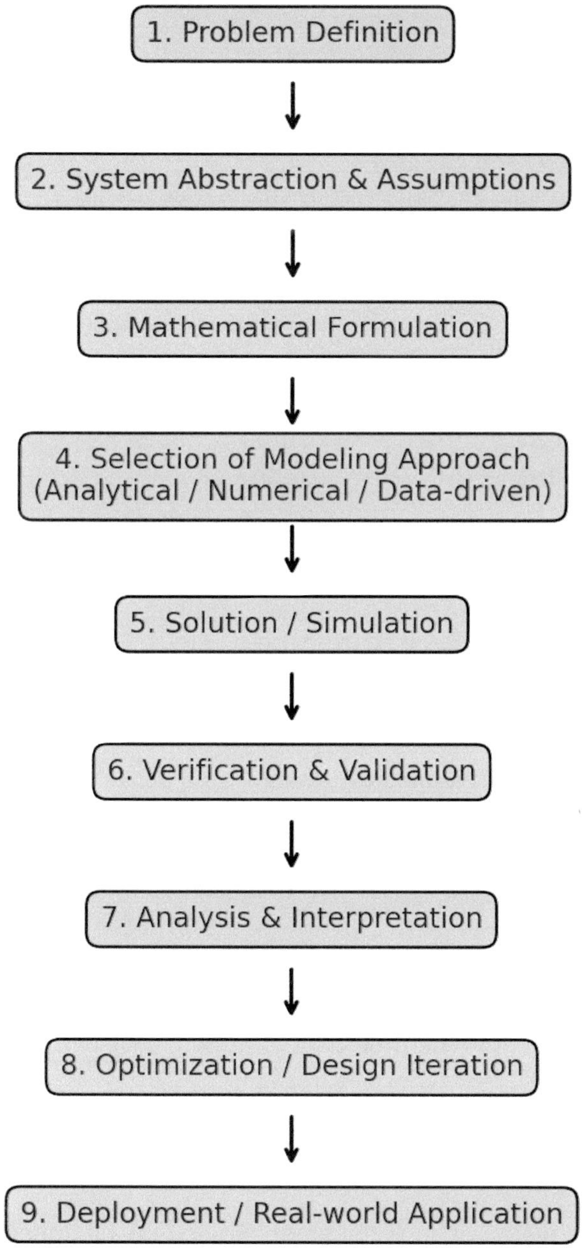

Figure 1.1 Flowchart of the mathematical modeling process.

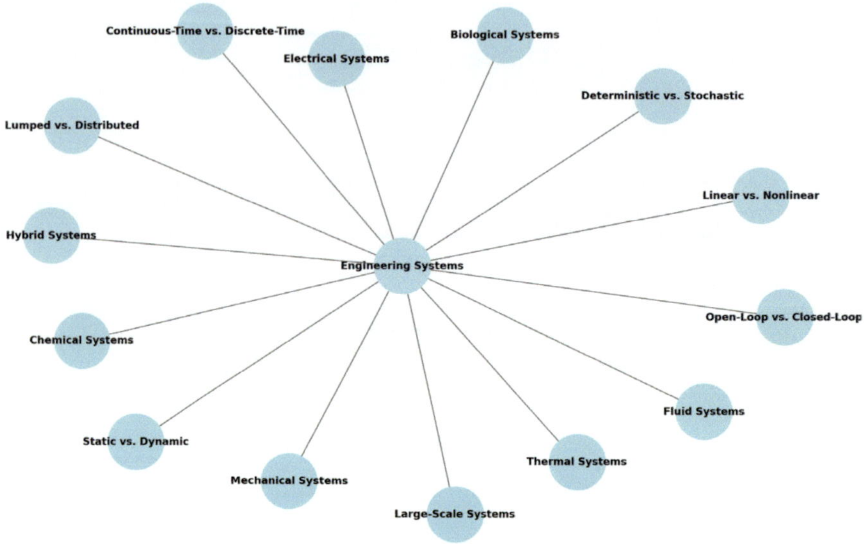

Figure 1.2 Classification of engineering systems.

systems can be classified based on their complexity, behavior, and the mathematical frameworks used to model them. Below, we discuss the primary types of engineering systems, their characteristics, and the mathematical tools used to analyze them.

Figure 1.2 organizes engineering systems into behavioral, structural, and domain-specific categories. It visually distinguishes between how systems behave, how they are structured, and the fields they apply to – providing a holistic view that supports the modeling frameworks discussed in the chapter.

1.2.1 Linear versus nonlinear systems

Linear systems are characterized by their superposition property, where the output is directly proportional to the input. Mathematically, a linear system can be represented as

$$y(t) = \int_{-\infty}^{\infty} h(t-\tau)x(\tau)d\tau, \tag{1.1}$$

where $y(t)$ is the output, $x(t)$ is the input, and $h(t)$ is the impulse response of the system [16]. Nonlinear systems, on the other hand, do not obey superposition and often exhibit complex behaviors such as chaos and bifurcations.

For example, the Lorenz system, a classic nonlinear model, is described by:

$$\frac{dx}{dt} = \sigma(y - x), \quad \frac{dy}{dt} = x(\rho - z) - y, \quad \frac{dz}{dt} = xy - \beta z, \tag{1.2}$$

where σ, ρ, and β are parameters [17].

1.2.2 Static versus dynamic systems

Static systems are memory-less, meaning their output depends only on the current input. For example, Ohm's law in electrical systems is a static relationship:

$$V = IR, \tag{1.3}$$

where V is voltage, I is current, and R is resistance [18]. Dynamic systems, however, depend on both current and past inputs, often described by differential equations. For instance, the motion of a spring-mass-damper system is governed by

$$m\frac{d^2x}{dt^2} + c\frac{dx}{dt} + kx = F(t), \tag{1.4}$$

where m is mass, c is damping, k is stiffness, and $F(t)$ is the external force [19].

1.2.3 Lumped-parameter versus distributed-parameter systems

Lumped-parameter systems assume that system properties are concentrated at discrete points. For example, an electrical circuit with resistors, capacitors, and inductors is a lumped-parameter system [20]. Distributed-parameter systems, such as heat conduction in a rod, require partial differential equations (PDEs) to describe spatial and temporal variations:

$$\frac{\partial T}{\partial t} = \alpha\frac{\partial^2 T}{\partial x^2}, \tag{1.5}$$

where $T(x, t)$ is temperature, and α is thermal diffusivity [21].

1.2.4 Deterministic versus stochastic systems

Deterministic systems have predictable outputs for given inputs, such as the trajectory of a projectile modeled by

$$y(t) = v_0 t \sin(\theta) - \frac{1}{2}gt^2, \tag{1.6}$$

where v_0 is initial velocity, θ is launch angle, and g is gravitational acceleration [22].

Stochastic systems, however, incorporate randomness, often modeled using probability distributions. For example, the Langevin equation describes Brownian motion:

$$\frac{dx}{dt} = -\gamma x + \eta(t), \tag{1.7}$$

where γ is a damping coefficient, and $\eta(t)$ is a random noise term [23].

1.2.5 Continuous-time versus discrete-time systems

Continuous-time systems are described by differential equations, such as the R-C circuit:

$$RC\frac{dV}{dt} + V = V_{\mathrm{in}}(t), \tag{1.8}$$

where $V_{\mathrm{in}}(t)$ is the input voltage [24].

Discrete-time systems, on the other hand, are modeled using difference equations, such as the logistic map:

$$x_{\mathrm{n}+1} = rx_{\mathrm{n}}(1 - x_{\mathrm{n}}), \tag{1.9}$$

where r is a parameter, and x_{n} represents the state at time step n [25].

1.2.6 Open-loop versus closed-loop systems

In open-loop systems, there is no feedback, meaning the system's output does not influence the control action. For example, a toaster heats bread for a fixed time [26]. In such a case, the controller acts solely based on the input reference or command. The mathematical model of an open-loop control system is typically represented as

$$y(t) = G(s) \cdot u(t), \tag{1.10}$$

where $y(t)$ is the output of the system, $u(t)$ is the input or control signal, and $G(s)$ is the transfer function of the plant/system in the Laplace domain.

Closed-loop systems, like a thermostat-controlled heater, use feedback to adjust performance:

$$u(t) = K_{\mathrm{p}}e(t) + K_{\mathrm{i}}\int e(t)dt + K_{\mathrm{d}}\frac{de(t)}{dt}, \tag{1.11}$$

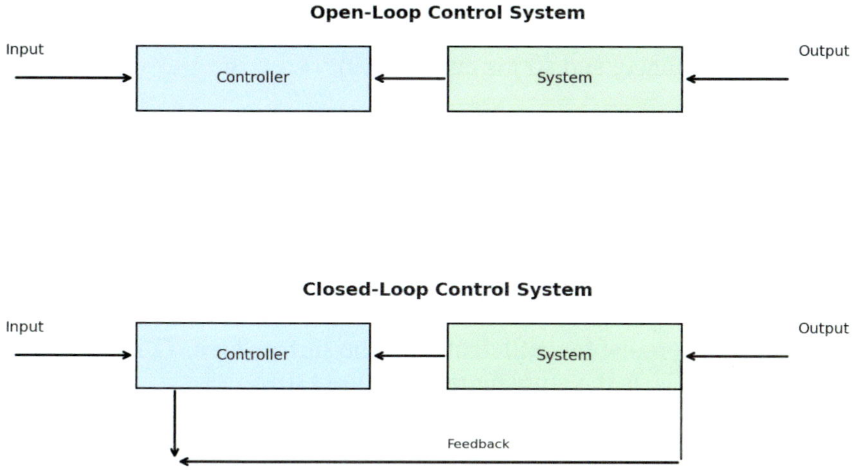

Figure 1.3 Comparison of open-loop and closed-loop control systems.

where $u(t)$ is the control input, $e(t)$ is the error, and K_p, K_i, and K_d are PID controller gains [27].

Figure 1.3 illustrates the structural difference between open-loop and closed-loop control systems, highlighting the presence of feedback in the closed-loop configuration.

1.2.7 Mechanical systems

Mechanical systems, such as robotic arms, are often modeled using Newtonian mechanics. For example, the dynamics of a pendulum are given by

$$\frac{d^2\theta}{dt^2} + \frac{g}{L}\sin(\theta) = 0, \tag{1.12}$$

where θ is the angular displacement, g is gravity, and L is the pendulum length [28].

1.2.8 Electrical systems

Electrical systems, such as RLC circuits, are modeled using Kirchhoff's laws. The voltage across a capacitor in an RLC circuit is:

$$V_C(t) = \frac{1}{C}\int i(t)dt, \tag{1.13}$$

where C is capacitance, and $i(t)$ is current [29].

1.2.9 Thermal systems

Thermal systems, such as heat exchangers, are modeled using energy balance equations. For example, the heat transfer rate is:

$$Q = hA(T_s - T_\infty), \tag{1.14}$$

where h is the heat transfer coefficient, A is the surface area, T_s is the surface temperature, and T_∞ is the ambient temperature [30].

1.2.10 Fluid systems

Fluid systems, such as pipelines, are modeled using the Navier–Stokes equations:

$$\rho\left(\frac{\partial u}{\partial t} + (u \cdot \nabla)u\right) = -\nabla p + \mu\nabla^2 u + f, \tag{1.15}$$

where ρ is density, u is velocity, p is pressure, μ is viscosity, and f is body force [31].

1.2.11 Chemical systems

Chemical systems, such as reactors, are modeled using mass and energy balances. For example, the rate of reaction in a CSTR is:

$$\frac{dC_A}{dt} = \frac{F}{V}(C_{A0} - C_A) - kC_A, \tag{1.16}$$

where C_A is concentration, F is flow rate, V is volume, and k is the reaction rate constant [32].

1.2.12 Biological systems

Biological systems, such as population dynamics, are often modeled using differential equations. The Lotka–Volterra equations describe predator-prey interactions:

$$\frac{dx}{dt} = \alpha x - \beta xy, \quad \frac{dy}{dt} = \delta xy - \gamma y, \tag{1.17}$$

where x and y are prey and predator populations, respectively, and α, β, δ, and γ are parameters [33].

1.2.13 Hybrid systems

Hybrid systems combine continuous and discrete dynamics, such as a bouncing ball:

$$\frac{d^2y}{dt^2} = -g, \quad \text{with reset conditions} \quad y(t^+) = -ey(t^-), \tag{1.18}$$

where e is the coefficient of restitution [34].

1.2.14 Large-scale systems

Large-scale systems, such as power grids, require decomposition techniques for analysis. For example, the power flow equations are:

$$P_\mathrm{i} = \sum_{j=1}^{N} |V_\mathrm{i}||V_\mathrm{j}| \left(G_\mathrm{ij}\cos\theta_\mathrm{ij} + B_\mathrm{ij}\sin\theta_\mathrm{ij}\right), \tag{1.19}$$

where P_i is power, V_i is voltage, G_ij and B_ij are conductance and susceptance, and θ_ij is the phase difference [35].

1.3 Role of Computational Tools in Modern Modeling

Computational tools have become indispensable in the field of mathematical modeling for engineering systems. These tools enable engineers to simulate, analyze, and optimize complex systems that are often impossible to solve analytically. From solving differential equations to performing large-scale simulations, computational tools bridge the gap between theoretical models and real-world applications.

1.3.1 Solving complex equations

Many engineering systems are governed by differential equations, integral equations, or systems of nonlinear equations. Analytical solutions to these equations are often intractable, especially for systems with multiple variables or nonlinearities. Computational tools, such as numerical methods, provide approximate solutions with high accuracy. For example, the finite difference method can be used to solve partial differential equations (PDEs) like the heat equation:

$$\frac{\partial T}{\partial t} = \alpha \frac{\partial^2 T}{\partial x^2}, \tag{1.20}$$

where $T(x, t)$ represents the temperature distribution, and α is thermal diffusivity. Numerical solvers like MATLAB or Python's SciPy library can discretize this equation and compute solutions efficiently [36].

1.3.2 Simulation of engineering systems

Computational tools enable the simulation of engineering systems under various conditions. For instance, computational fluid dynamics (CFD) tools like ANSYS Fluent or COMSOL Multiphysics simulate fluid flow, heat transfer, and other phenomena. These tools solve the Navier–Stokes equations:

$$\rho \left(\frac{\partial u}{\partial t} + (u \cdot \nabla) u \right) = -\nabla p + \mu \nabla^2 u + f, \tag{1.21}$$

where ρ is fluid density, u is velocity, p is pressure, μ is viscosity, and f is body force. Such simulations are critical in designing aircraft, automobiles, and HVAC systems [37]. One prominent application of fluid modeling is in Formula 1 (F1) racing, where CFD simulations are used to optimize the aerodynamics of race cars. Engineers model airflow over the car's body to minimize drag and maximize downforce, which are critical for speed and handling. High-fidelity simulations based on the Navier–Stokes equations help fine-tune components like the front wing, rear diffuser, and side pods. This process drastically reduces the need for expensive wind tunnel testing while accelerating design iterations.

1.3.3 Optimization of system performance

Optimization is a key aspect of engineering design, and computational tools like MATLAB's Optimization Toolbox or Python's Pyomo library facilitate this process. For example, in structural engineering, the goal might be to minimize the weight of a structure while ensuring it meets strength requirements. This can be formulated as a constrained optimization problem:

$$\text{Minimize } f(x) \quad \text{subject to } g_{\mathrm{i}}(x) \leq 0, \quad i = 1, 2, \ldots, m, \tag{1.22}$$

where $f(x)$ is the objective function (e.g., weight), and $g_{\mathrm{i}}(x)$ are the constraints (e.g., stress limits). Computational tools use algorithms like gradient descent or genetic algorithms to find optimal solutions [38].

1.3.4 Handling large-scale data

Modern engineering systems often generate vast amounts of data, such as sensor readings or experimental measurements. Computational tools like Python's Pandas library or MATLAB's Data Analytics Toolbox help process and analyze this data. For example, machine learning algorithms can be used to identify patterns in data and improve system performance. A simple linear regression model can be expressed as

$$y = \beta_0 + \beta_1 x + \varepsilon, \tag{1.23}$$

where y is the dependent variable, x is the independent variable, β_0 and β_1 are coefficients, and ε is the error term. Such models are widely used in predictive maintenance and quality control [39]. A notable example of data-driven modeling is the use of machine learning in predictive maintenance for industrial machinery. By analyzing vibration, temperature, and operational data collected from sensors, ML models such as decision trees and recurrent neural networks can detect patterns that precede equipment failure. Libraries like scikit-learn and TensorFlow are employed to train these models, significantly reducing unplanned downtime and maintenance costs while improving operational reliability.

1.3.5 Visualization of results

Visualization is a critical component of modeling, as it helps engineers interpret results and communicate findings. Tools like Tecplot, Paraview, MATLAB, Python's Matplotlib, or Tableau create plots, graphs, and 3D visualizations. For example, a contour plot of temperature distribution in a heat exchanger can provide insights into thermal performance.

1.3.6 Integration with real-time systems

Computational tools are increasingly being integrated with real-time systems for monitoring and control. For example, in a smart grid, real-time data from sensors is processed using computational algorithms to optimize power distribution. The power flow equations in a grid can be expressed as

$$P_{\mathrm{i}} = \sum_{j=1}^{N} |V_{\mathrm{i}}||V_{\mathrm{j}}| \left(G_{\mathrm{ij}} \cos\theta_{\mathrm{ij}} + B_{\mathrm{ij}} \sin\theta_{\mathrm{ij}}\right), \tag{1.24}$$

where P_{i} is the power injected at bus i, V_{i} and V_{j} are voltage magnitudes, G_{ij} and B_{ij} are conductance and susceptance, and θ_{ij} is the phase angle difference. Real-time computational tools ensure efficient and reliable grid operation [40].

1.3.7 Challenges and limitations in the computation

Despite their advantages, computational tools face challenges such as high computational costs, numerical instability, and the need for accurate input data. For example, in molecular dynamics simulations, the computational cost increases exponentially with the number of particles. Additionally, numerical methods like the Runge–Kutta method for solving ordinary differential equations (ODEs) may suffer from instability if the step size is not chosen carefully [41].

1.3.8 Modern computational tools and techniques

Recent advances in computational modeling have introduced powerful new tools that enhance model accuracy and flexibility. Python libraries such as SymPy (symbolic mathematics), TensorFlow (deep learning), and scikit-learn (machine learning) allow for seamless integration of analytical and data-driven methods. Furthermore, digital twin technology – virtual replicas of physical systems – enables real-time simulation and monitoring, particularly in manufacturing, aerospace, and smart cities. Another growing area is the use of ML-based surrogate models and reduced-order models, which approximate complex simulations with significantly reduced computational effort, making them ideal for real-time optimization and control.

1.4 Challenges in Modeling Complex Engineering Systems

Mathematical modeling is a cornerstone of engineering, enabling the analysis, design, and optimization of systems. However, as engineering systems grow in complexity, so do the challenges associated with modeling them. Complex systems often involve nonlinear dynamics, multiple interacting components, and uncertainties that are difficult to capture using traditional modeling approaches. This section explores the key challenges in modeling complex engineering systems, supported by mathematical formulations and examples.

1.4.1 Nonlinear dynamics and chaos

Many engineering systems exhibit nonlinear behavior, where the output is not directly proportional to the input. Nonlinear dynamics can lead to chaotic behavior, making long-term predictions nearly impossible. For example, the Lorenz system, a set of nonlinear differential equations, is often used to model atmospheric convection and exhibits chaotic behavior:

$$\frac{dx}{dt} = \sigma(y - x), \quad \frac{dy}{dt} = x(\rho - z) - y, \quad \frac{dz}{dt} = xy - \beta z, \tag{1.25}$$

where σ, ρ, and β are parameters. Small changes in initial conditions can lead to vastly different outcomes, a phenomenon known as the "butterfly effect" [17].

1.4.2 High dimensionality

Complex systems often involve numerous variables and parameters, leading to high-dimensional models. For instance, in computational fluid dynamics (CFD), the Navier–Stokes equations describe fluid flow:

$$\rho\left(\frac{\partial u}{\partial t} + (u \cdot \nabla)u\right) = -\nabla p + \mu \nabla^2 u + f, \tag{1.26}$$

where ρ is fluid density, u is velocity, p is pressure, μ is viscosity, and f is body force. Solving these equations for real-world systems requires discretization, leading to millions of degrees of freedom and significant computational costs [42].

1.4.3 Uncertainty and stochasticity

Uncertainty is inherent in complex systems due to measurement errors, environmental variability, and incomplete knowledge. Stochastic differential equations (SDEs) are often used to model such systems. For example, the Langevin equation describes the motion of a particle under random forces:

$$m\frac{d^2x}{dt^2} = -\gamma\frac{dx}{dt} + \eta(t), \tag{1.27}$$

where m is mass, γ is a damping coefficient, and $\eta(t)$ represents Gaussian white noise. Quantifying and managing uncertainty is a major challenge in modeling [43].

1.4.4 Multiscale phenomena

Complex systems often operate across multiple spatial and temporal scales. For example, in materials science, the behavior of a material at the macroscale depends on its microstructure. Bridging these scales requires multiscale modeling techniques, such as homogenization or coarse-graining. A common approach is to use coupled equations, such as

$$\frac{\partial u}{\partial t} = D\nabla^2 u + f(u, v), \quad \frac{\partial v}{\partial t} = g(u, v), \tag{1.28}$$

where u and v represent variables at different scales, and f and g are coupling functions [44].

1.4.5 Data-driven modeling challenges

With the rise of big data, data-driven modeling techniques like machine learning are increasingly used. However, these methods face challenges such as overfitting, interpretability, and the need for large datasets. For example, a neural network model can be represented as

$$y = f(\mathbf{x}; \mathbf{W}) \tag{1.29}$$

where $\mathbf{x}$ is the input, $\mathbf{W}$ are the weights, and f is a nonlinear activation function. Ensuring that such models generalize well to new data is a significant challenge [45].

1.4.6 Emerging trends in complex modeling

The increasing integration of AI with physics-based modeling has led to new paradigms such as physics-informed neural networks (PINNs), which embed differential equations into neural networks to solve PDEs more efficiently. These models are particularly useful for problems where limited data is available but governing equations are known. Another important development is data assimilation, which blends observational data with model predictions to improve accuracy in real time – widely used in weather forecasting and hydrology. Moreover, the need for explainable AI (XAI) in engineering has grown, ensuring that AI-driven decisions are transparent, interpretable, and trustworthy, especially in safety-critical domains like healthcare and transportation.

1.4.7 Computational complexity

The computational cost of solving complex models can be prohibitive. For example, simulating the behavior of a large-scale power grid involves solving thousands of coupled differential equations. Parallel computing and reduced-order modeling are often employed to mitigate these challenges.

1.4.8 Validation and verification

Validating and verifying complex models is a critical but challenging task. Models must be tested against experimental data, but obtaining accurate data for complex systems can be difficult. Additionally, discrepancies between model predictions and observations may arise due to oversimplifications or incorrect assumptions.

1.4.9 Interdisciplinary nature

Complex engineering systems often span multiple disciplines, requiring expertise in areas such as physics, biology, and computer science. Integrating knowledge from these domains into a cohesive model is a significant challenge.

1.4.10 Scalability, ethical, and societal implications

As systems grow in size and complexity, ensuring that models remain scalable is crucial. For example, modeling a smart city involves integrating transportation, energy, and communication networks, each with its own complexities. Finally, modeling complex systems raises ethical and societal questions. For instance, models used in autonomous vehicles or healthcare must be transparent and fair, ensuring that they do not perpetuate biases or harm individuals.

1.5 Concluding Remarks

This chapter has laid the groundwork for understanding the fundamental role of mathematical modeling in engineering systems. It began by highlighting the importance of modeling as a bridge between theory and real-world application, enabling engineers to analyze, predict, and optimize system performance across diverse domains such as mechanical, electrical, chemical, thermal, and biological systems.

Through a classification of engineering systems-ranging from linear and nonlinear to deterministic, stochastic, static, dynamic, lumped, and distributed-the chapter demonstrated how mathematical frameworks are tailored to system behavior and complexity. The role of computational tools was underscored, illustrating how numerical methods, simulation software, optimization algorithms, and data-driven techniques enhance the modeling and analysis of modern engineering challenges.

Furthermore, the chapter addressed the multifaceted challenges of modeling complex systems, including nonlinearity, high dimensionality, uncertainty, multiscale phenomena, data-driven limitations, and computational constraints. Each challenge reflects the evolving landscape of engineering, where integration with real-time systems, ethical considerations, and scalability demand robust and adaptive modeling strategies.

In conclusion, mathematical modeling is not only a foundational pillar of engineering practice but also a continuously evolving discipline. It empowers engineers to convert complex physical phenomena into solvable mathematical representations, leading to better design, deeper insights, and informed decision-making. As the chapters ahead delve into specialized methods and applications, the concepts introduced here will serve as essential tools for advancing knowledge and practice in engineering modeling.

References

[1] J. Smith. *Structural Analysis and Design: Principles and Applications*. Cambridge University Press, 2020.

[2] J. D. Anderson. *Computational Fluid Dynamics: The Basics with Applications*. McGraw-Hill Education, 2019.

[3] T. D. Gillespie. *Fundamentals of Vehicle Dynamics*. SAE International, 1992.

[4] M. P. Groover. *Automation, Production Systems, and Computer-Integrated Manufacturing*. Pearson, 2020.

[5] S. C. Chapra. *Surface Water-Quality Modeling*. Waveland Press, 2018.

[6] R. L. Norton. *Machine Design: An Integrated Approach*. Pearson, 2020.

[7] H. S. Fogler. *Elements of Chemical Reaction Engineering*. Prentice Hall, 2016.

[8] J. R. Lamarsh and A. J. Baratta. *Introduction to Nuclear Engineering*. Pearson, 2018.

[9] J. D. Ortúzar and L. G. Willumsen. *Modelling Transport*. Pearson, 2011.

[10] J. Twidell and T. Weir. *Renewable Energy Resources*. Routledge, 2015.

[11] J. W. Nilsson and S. A Riedel. *Electric Circuits*. Pearson, 2019.

[12] W. M. Saltzman. *Biomedical Engineering: Bridging Medicine and Technology*. Cambridge University Press, 2015.

[13] B. Siciliano and O. Khatib. *Springer Handbook of Robotics*. Springer, 2016.
[14] E. Kreyszig. *Advanced Engineering Mathematics*. Wiley, 2018.
[15] I. Sommerville. *Software Engineering*. Pearson, 2016.
[16] A. V. Oppenheim and A. S. Willsky. *Signals and Systems*. Prentice Hall, 1997.
[17] E. N. Lorenz. 'Deterministic Nonperiodic Flow'. *J. Atmos. Sci.*, 20:130—141, 1963.
[18] W. H. Hayt and J. E. Kemmerly. *Engineering Circuit Analysis*. McGraw-Hill, 2012.
[19] D. J. Inman. *Engineering Vibration*. Pearson, 2014.
[20] R. C. Dorf and J. A. Svoboda. *Introduction to Electric Circuits*. Wiley, 2010.
[21] F. P. Incropera and D. P. DeWitt. *Fundamentals of Heat and Mass Transfer*. Wiley, 2002.
[22] J. B. Marion and S. T. Thornton. *Classical Dynamics of Particles and Systems*. Brooks/Cole, 1995.
[23] N. G. Van Kampen. *Stochastic Processes in Physics and Chemistry*. Elsevier, 2007.
[24] J. W. Nilsson and S. A. Riedel. *Electric Circuits*. Pearson, 2015.
[25] S. H. Strogatz. *Nonlinear Dynamics and Chaos*. Westview Press, 1994.
[26] K. Ogata. *Modern Control Engineering*. Pearson, 2010.
[27] K. J. Åström and T. Hägglund. *Advanced PID Control*. ISA, 2006.
[28] H. Goldstein, C. Poole, and J. Safko. *Classical Mechanics*. Addison-Wesley, 2002.
[29] C. K. Alexander and M. N. O. Sadiku. *Fundamentals of Electric Circuits*. McGraw-Hill, 2016.
[30] T. L. Bergman, A.S. Lavine, F. P. Incropera, and D. P. DeWitt. *Introduction to Heat Transfer*. Wiley, 2011.
[31] F. M. White. *Fluid Mechanics*. McGraw-Hill, 2011.
[32] H. S. Fogler. *Essentials of chemical reaction engineering*. Pearson education, 2010.
[33] J. D. Murray. *Mathematical biology: I. An introduction*. Springer Science & Business Media, 2007.
[34] R. Goebel, R. G. Sanfelice, and A. R. Teel. *Hybrid Dynamical Systems*. Princeton University Press, 2012.
[35] J. J. Grainger and W. D. Stevenson. *Power System Analysis*. McGraw-Hill Press, 1994.

[36] J. A. Smith and B. C. Johnson. *Numerical Methods for Engineers*. Cambridge University Press, 2020.

[37] J. D. Anderson. *Computational Fluid Dynamics: The Basics with Applications*. McGraw-Hill Education, 2019.

[38] J. Nocedal and S. J. Wright. *Numerical Optimization*. Springer, 2006.

[39] G. James, D. Witten, T. Hastie, and R. Tibshirani. *An Introduction to Statistical Learning*. Springer, 2013.

[40] P. Kundur. *Power System Stability and Control*. McGraw-Hill Education, 1994.

[41] E. Hairer, S. P. Nørsett, and G. Wanner. *Solving Ordinary Differential Equations I: Nonstiff Problems*. Springer, 1993.

[42] J. H. Ferziger and M. Perić. *Computational Methods for Fluid Dynamics*. Springer, 2002.

[43] C. W. Gardiner. *Stochastic Methods: A Handbook for the Natural and Social Sciences*. Springer, 2009.

[44] E. Weinan and B. Engquist. 'The Heterogeneous Multiscale Methods'. *Commun. Math. Sci.*, 1(1):87–132, 2003.

[45] I. Goodfellow, Y. Bengio, and A. Courville. *Deep Learning*. MIT Press, 2016.

2

Stochastic Modeling in Engineering Systems: A Study of a Nuclear Power Plant System

Surbhi Gupta[1], Manju Kashyap[2], Shefali Kanwar[1], and Nahid Fatima[3]

[1]Department of Applied Mathematics, Amity University, India
[2]Department of Applied Mathematics, Galgotias College of Engineering and Technology, India
[3]Department of Mathematics and Sciences, Prince Sultan University, Saudi Arabia
E-mail: sgupta11@amity.edu; drmanju.kashyap@galgotiacollege.edu; skanwar1@amity.edu; drnahidfatima@gmail.com

Abstract

This chapter presents a stochastic modeling approach for reliability analysis of nuclear power plant systems, with a particular focus on multi-state systems that reflect varying levels of subsystem performance. Unlike traditional binary-state models, which classify systems as either fully operational or failed, the multi-state framework provides a more nuanced and realistic assessment of operational degradation. The complexity inherent in analyzing such systems is addressed using stochastic processes, particularly through the application of artificial neural networks (ANNs) to estimate time-dependent state probabilities. The model investigates subsystem behavior under both constant (useful life) and time-dependent (wear-out) failure rates, modeled via exponential and Weibull distributions, respectively. Profitability is evaluated across different preventive maintenance strategies – no maintenance, minimum, and maximum repair policies – based on changes in upstate probabilities. Results demonstrate the effectiveness of the ANN-based model in predicting system reliability and optimizing profit. The findings underscore

DOI: 10.1201/9788743806691-2

that, despite higher initial costs, maximum preventive maintenance yields the highest long-term profitability by minimizing downtime and sustaining system performance.

Keywords: Stochastic process, transition diagram, multi-state systems, nuclear power plant, reliability analysis

2.1 Introduction

Nuclear power plants (NPPs) represent one of the most complex and safety-critical components of modern energy infrastructure. Ensuring their reliable and efficient operation is paramount, necessitating the use of advanced reliability modeling techniques. Traditional binary-state reliability models – which classify system behavior strictly as "operational" or "failed" – are often insufficient to capture the multifaceted operational states inherent in NPPs. In reality, nuclear systems may experience various levels of performance degradation, operating at full capacity, reduced efficiency, or in complete failure. Neglecting these intermediate states can lead to oversimplified and potentially misleading assessments of system reliability.

To overcome this limitation, multi-state system (MSS) reliability models have emerged as a more realistic alternative. These models allow for a continuum of performance levels, enabling the representation of systems that function in degraded but still operational modes. For instance, a reactor cooling system might operate at 100%, 75%, or 50% capacity, each associated with distinct operational implications. MSS models accommodate such variability and thereby offer a more granular and accurate evaluation of system performance and failure risk.

Analyzing such models requires sophisticated mathematical tools. Stochastic processes provide a powerful theoretical and computational framework for modeling systems subject to inherent uncertainties. These processes are essential for analyzing dynamic systems whose evolution is influenced by random events. Their applicability spans numerous domains – finance, biology, environmental science, and engineering – where uncertainty is a defining factor [1]. In engineering systems, stochastic models – incorporating random variables and probabilistic transitions – are instrumental in forecasting and managing system behavior under uncertainty [2].

Fundamental concepts in stochastic modeling, such as stationarity, the Markov property, independence, and correlation, enable detailed characterization of system behavior over time [3]. Several classes of stochastic processes-including Poisson processes, renewal processes, and Markov chains-have been developed to describe various random phenomena in system operations [4]. In reliability engineering, these models are indispensable for accurately estimating and improving the dependability of complex systems [5].

While traditional reliability models rely on binary logic, multi-state reliability analysis captures the nuanced operational degradation that real systems undergo [6]. This distinction is especially critical for NPPs, where gradual performance deterioration, rather than abrupt failure, often dominates system dynamics. Markovian models, particularly continuous-time markov chains (CTMCs), are commonly used in this context to represent probabilistic transitions among multiple states [7]. However, their applicability is often limited by the state space explosion problem, where system complexity leads to an exponential increase in the number of states, posing significant computational challenges [8].

The long-term reliability of infrastructure systems like NPPs is essential not only for public safety but also for economic stability and energy security [9]. As these infrastructures age, their performance declines, increasing the likelihood of failures [10]. Therefore, comprehensive aging management programs are required to sustain system safety and reliability throughout their operational and extended lifetimes. A critical challenge in this domain is dealing with uncertainty in the deterioration processes and accurately predicting component lifespans. While traditional lifetime distribution models are effective for time-based maintenance strategies, they are less suitable for condition-based maintenance, where real-time system condition data is used to optimize intervention timing [11]. To address these challenges, advanced stochastic models are needed to improve condition assessment accuracy, enhance remaining life predictions, and support more effective life-cycle management strategies [12].

In this chapter, we present a stochastic modeling approach for analyzing the reliability of a nuclear power plant system using an artificial neural network (ANN)-based model. The study focuses on short-term state probability behaviors during two critical operational periods: the useful life phase and the wear-out phase. The economic implications of these phases are evaluated under two preventive maintenance (PM) strategies – minimal and maximal repairs/replacements. System profitability is assessed by comparing the profit per unit time in scenarios with no PM and with each PM strategy, based on changes in upstate probability.

2.2 Stochastic Processes in Reliability

Stochastic processes in reliability refer to the application of probability theory to model and analyze systems whose behavior evolves randomly over time. This framework is particularly useful in evaluating how uncertainties – such as random component failures, repairs, and maintenance actions – affect the overall reliability and performance of engineering systems. By leveraging advanced mathematical tools, stochastic reliability modeling enables the prediction of system behavior under uncertainty and supports the development of strategies to enhance system dependability across various operational contexts.

2.2.1 Poisson process

Consider a system with one unit subject to repair or replacement as and when required, assuming the repair or replacement time is negligible. After each repair or replacement, the system is restored to a like-new condition. If the unit fails at a constant rate λ in each infinitesimal time interval $(t, t + \mathrm{d}t)$ – regardless of the time the system has already been in operation – then the failure process can be modeled as a Poisson process with intensity λ.

Mathematically, a Poisson process $N(t)$ with intensity λ is a stochastic counting process where $N(t) = 0, 1, 2, \ldots$ and satisfies the following properties:

1. $N(0) = 0$
2. The increments of $N(t)$ are independent
3. The increments of $N(t)$ are stationary
4. The probability that x events occur in time interval t is given by

$$\Pr(x) = \frac{e^{-\lambda t}(\lambda t)^x}{x!}.$$

Such a process is known as a Poisson process. In practical scenarios, when a failed system is repaired, it may not be fully restored to its original state. This implies that although the failure rate in the post-repair interval may remain constant, it could still be higher than what was expected prior to failure. As a result, the increments of the Poisson process $N(t)$ may no longer remain stationary and may instead exhibit a monotonic trend, such as a decreasing rate of occurrence over time. When condition (iii) – stationarity of increments – is violated, the process is referred to as a non-homogeneous Poisson process [13].

2.2.2 Markov process

The Markov property, also referred to as the *memoryless property*, states that the probability of transitioning from one state to another depends solely on the current state of the system and is independent of any past states. For instance, consider a single-unit system that undergoes repair after failure and is restored to an operational condition – functioning as good as new. The future transition of this system to a failed state depends only on its current (post-repair) state and is completely independent of its previous states, including its condition before the repair.

A stochastic process that satisfies the Markov property at all times is called a Markov process. A homogeneous Poisson process is an example of a Markov process. If the states of the system are considered at discrete time intervals, the process is known as a discrete-time markov process (DTMP) [14]. Conversely, if time is treated as a continuous variable, the process becomes a continuous-time markov process (CTMP). In the context of continuous time, the specific time points at which the system probabilistically restarts are referred to as *regenerative points*. The mean sojourn time in a given state refers to the expected amount of time the system remains in that state before transitioning to another.

2.2.3 Semi-Markov process

A stochastic process that employs a Markov chain to model state transitions, but where the time between transitions is governed by a general probability distribution, is known as a semi-Markov process. In such a process, the time interval between two consecutive transitions is modeled by a random variable, whose distribution may depend on both the previous and subsequent states [15, 16]. Many real-world systems, particularly those in reliability engineering, do not strictly satisfy the memoryless property of the Markov process. Instead, they adhere to the Markov property only at the discrete time instants when transitions occur from one state to another. Such systems are aptly modeled by semi-Markov processes [17].

2.2.4 Markov-renewal process

The Markov-renewal process is conceptually equivalent to the semi-Markov process. The primary distinction lies in the state specification: in a semi-Markov process, the system state is defined for every point in time, whereas in a Markov-renewal process, it is defined only at transition epochs [18, 19].

2.2.5 Lifetime distributions and failure rates

Various statistical distributions are used in reliability engineering to model the lifetime distribution of systems. The reliability life cycle of a hardware system typically consists of three phases:

1. *Infant mortality phase*: Characterized by a high initial failure rate that gradually decreases with time. This phase often reflects early-life defects and production flaws.
2. *Useful life phase*: The failure rate stabilizes and remains approximately constant. Systems are considered to be in their most reliable operating period.
3. *Wear-out phase*: Due to aging and degradation of components, the system experiences an increasing failure rate over time.

To accurately model these phases, appropriate statistical distributions – such as the Weibull, Exponential, or Lognormal distributions – are commonly employed.

2.2.6 Weibull distribution

When a component or system experiences wear-out (i.e., deterioration) during its final life phase or undergoes infant mortality during its initial phase, the two-parameter Weibull distribution is widely used for reliability modeling. The probability density function (pdf) is given by

$$f(t) = \frac{\beta}{\eta^\beta} t^{\beta-1} e^{-(t/\eta)^\beta}.$$

The associated reliability functions are

- Failure rate function: $\lambda(t) = \dfrac{\beta t^{\beta-1}}{\eta^\beta}.$
- Survivor function: $R(t) = e^{-(t/\eta)^\beta}.$

Here, β denotes the shape parameter and η denotes the scale parameter. If $\beta < 1$, the distribution characterizes the infant mortality phase. If $\beta > 1$, it represents the wear-out phase, and when $\beta = 1$, it reduces to the exponential distribution, indicating a constant failure rate.

2.3 Case Studies

2.3.1 System description

Nuclear power plants generate electricity by harnessing the energy released during nuclear fission. At the core of the plant lies the nuclear reactor, where neutrons bombard fissionable material, typically uranium-235, causing atomic nuclei to split. This fission process produces a substantial amount of heat along with additional neutrons, thus initiating a controlled chain reaction. The chain reaction is moderated and regulated using two critical components: control rods, which absorb excess neutrons to manage the reaction rate, and a moderator, usually water, which slows down neutrons to enhance the probability of further fission.

The reactor core is enclosed within a robust reactor pressure vessel, designed to withstand high pressure, facilitate coolant flow, and provide essential radiation shielding. The immense heat generated in the reactor is transferred to the secondary water loop via a heat exchanger or steam generator. In this process, radioactive water from the primary loop transfers heat to clean water in the secondary loop through sealed tubes, producing high-pressure steam.

This high-pressure steam drives a turbine, converting thermal energy into mechanical energy. The spinning turbine shaft is coupled to an alternator or generator, which then transforms the mechanical energy into electrical energy through electromagnetic induction.

After passing through the turbine, the steam is cooled and condensed back into water in the condenser, creating a vacuum that enhances turbine efficiency. The condensed water is then recycled into the steam generator, forming a closed-loop system. This continuous cycle ensures a reliable and controlled supply of electrical energy through sustained nuclear fission.

The operation of a nuclear power plant can be broadly categorized into the following key subsystems:

1. Nuclear reactor: The reactor is the central component where controlled nuclear fission occurs. Neutrons bombard fuel rods containing uranium-235, causing fission and releasing significant thermal energy along with additional neutrons. Control rods absorb the excess neutrons to maintain a steady, regulated chain reaction [20].
2. Heat exchangersteam generator: The primary function of the heat exchanger is to transfer heat from the radioactive primary coolant to a clean secondary water loop, generating high-pressure steam. Importantly, the heat exchanger prevents radioactive contamination by isolating the primary and secondary fluids. Failure of this component may result in plant shutdown [21, 22].
3. Steam turbine: The high-pressure steam generated in the secondary loop is directed to turbine blades. The steam's thermal energy is converted into mechanical energy as it rotates the blades. The turbine is typically

multi-staged to extract maximum energy. The rotating shaft of the turbine is mechanically coupled to the generator [23].

4. Condenser: The condenser enhances both water conservation and plant efficiency. It cools and condenses the spent steam from the turbine back into liquid water, which is then recycled. The condensation process also generates a vacuum that increases turbine efficiency [24].
5. Alternatorgenerator: The generator, connected to the turbine shaft, consists of a rotor equipped with strong magnets and a stator with wire coils. As the rotor spins, it induces a varying magnetic field, producing alternating current through electromagnetic induction. This process converts mechanical energy into electrical energy for transmission to the grid [25].

To evaluate the reliability of a nuclear power plant (NPP), its core operations can be effectively represented as a series of interconnected subsystems. Figure 2.1 illustrates a schematic of the working procedure of an NPP. This diagram clearly depicts the sequence of operations and the functional dependencies among critical components involved in the power generation process. The flow begins at the nuclear reactor, where controlled fission

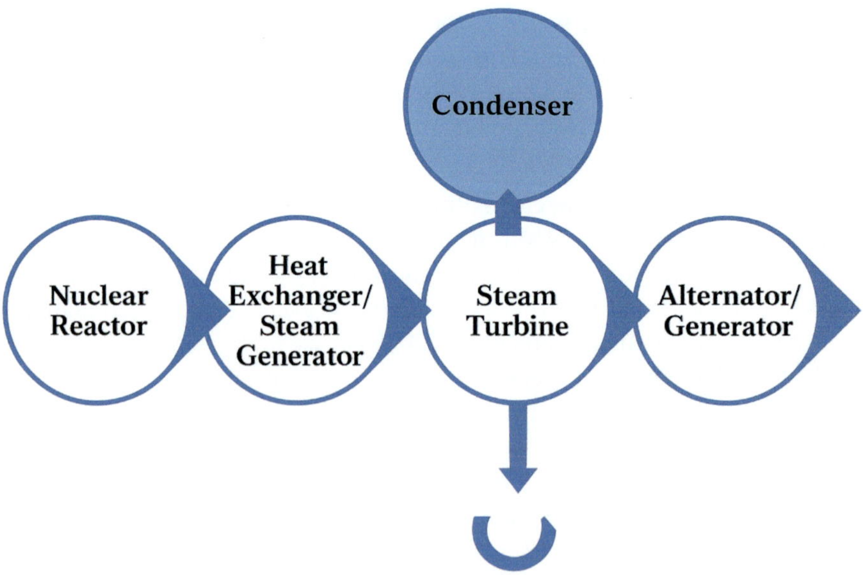

Figure 2.1 Working procedure of nuclear power plant.

reactions generate heat. This thermal energy is then transferred via the heat exchanger/steam generator to produce high-pressure steam in the secondary loop. The steam drives the steam turbine, converting thermal energy into mechanical energy, which is subsequently transformed into electrical energy by the alternator/generator. Finally, the steam is cooled and condensed in the condenser and recycled, completing the closed-loop system. This process flow model provides a structural basis for analyzing the system's reliability, where each component can be treated as an independent yet interacting subsystem.

Figure 2.2 illustrates the series-parallel configuration of these subsystems that is crucial for successful plant operation. This diagram visually represents the logical dependencies between the subsystems. It clearly demonstrates which subsystems must function in series (failure of any one leads to system failure) and which subsystems operate in parallel (redundancy, where multiple subsystems can perform the same function, increasing reliability). Understanding this configuration is essential for identifying critical paths and potential single points of failure. The diagram highlights how the successful operation of these interconnected subsystems contributes to the overall

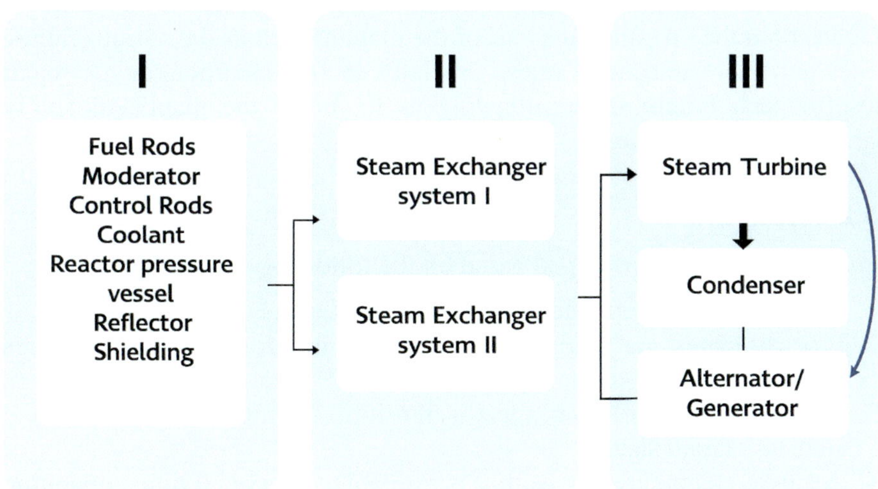

Figure 2.2 Block diagram of nuclear power plant.

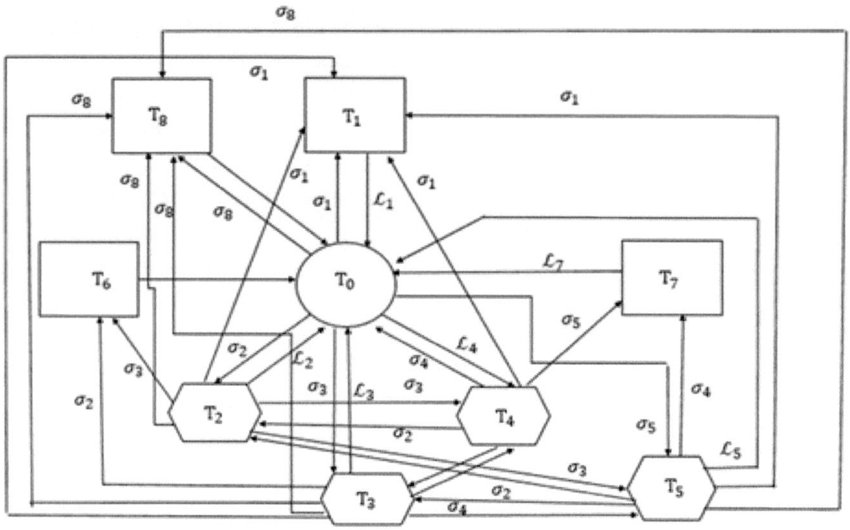

Figure 2.3 Block diagram of nuclear power plant.

success of the NPP. Figure 2.3 presents a transition diagram of the NPP, which provides a dynamic view of the plant's operational states. This diagram includes both the operational states of the plant as well as the various failure modes and their associated rates. Crucially, it also incorporates the repair rates for each failure mode, allowing us to model the plant's reliability characteristics over time.

2.3.2 Assumption

The system model is developed based on the following assumptions:

1. All units initially operate at their normal efficiency levels.
2. Process states are statistically independent throughout the system's operation.
3. A unit's partial failure triggers a transition from the normal operating state to a failed state.
4. A single-channel repair facility is available to serve all units, operating with a constant repair rate.
5. Repaired units are restored to their original, fully functional condition.
6. Corrective maintenance is implemented unless explicitly stated otherwise.
7. Inter-process transitions occur without introducing defects or errors.

2.3.3 Notations

State probabilities are defined as follows

- T_0 = System is in a fully working state.
- T_1 = Subsystem I failure
- T_2 = Subsystem IIA failure
- T_3 = Subsystem IIB failure
- T_4 = Subsystem IIIA failure
- T_5 = Subsystem IIIB failure
- T_6 = Subsystems IIA and IIB failure
- T_7 = Subsystems IIIA and IIIB failure
- T_8 = Human error/environmental failure

Failure rates are denoted by

- σ_1 = Rate of failure of subsystem I
- σ_2 = Rate of failure of subsystem IIA
- σ_3 = Rate of failure of subsystem IIB
- σ_4 = Rate of failure of subsystem IIIA
- σ_5 = Rate of failure of subsystem IIIB
- σ_6 = Rate of failure of subsystems IIA and IIB
- σ_7 = Rate of failure of subsystems IIIA and IIIB
- σ_8 = Rate of failure due to human error and environmental factors

Repair rates are denoted by

- $\mathcal{L}_1$ = Subsystem I repair rate
- $\mathcal{L}_2$ = Subsystem IIA repair rate
- $\mathcal{L}_3$ = Subsystem IIB repair rate
- $\mathcal{L}_4$ = Subsystem IIIA repair rate
- $\mathcal{L}_5$ = Subsystem IIIB repair rate
- $\mathcal{L}_6$ = Subsystems IIA and IIB repair rate
- $\mathcal{L}_7$ = Subsystems IIIA and IIIB repair rate

2.4 Mathematical Model

2.4.1 Formulation of ANN model of system

The learning processes in the human brain serve as the model for artificial neural networks (ANNs). They learn by being presented with input data and corresponding target output data. Through an iterative training process involving multiple epochs, the model adjusts its internal parameters to minimize the difference between its predictions and the target outputs.

This process aims to optimize the model's ability to provide accurate output estimations. Three layers make up the ANN model described here: an input layer, a hidden layer, and an output layer.

Input layer:

The inputs are defined as shown in eqn (2.1)

$$X_{\mathrm{i}} = P_{\mathrm{i}}(t), \quad i = 1, \ldots, 7. \tag{2.1}$$

Neurons:

The number of states shown in the transition diagram (Figure 2.3) corresponds to the number of neurons in the input layer. As shown below, the matching failure and repair rates for each state determine the connection weights assigned to its neuron.

The weights of the neural network are:

$$w_{01} = w_{21} = w_{31} = w_{41} = w_{51} = \sigma_1 \Delta t, \tag{2.2}$$
$$w_{02} = w_{34} = w_{35} = w_{36} = \sigma_2 \Delta t, \tag{2.3}$$
$$w_{03} = w_{24} = w_{26} = \sigma_3 \Delta t, \tag{2.4}$$
$$w_{04} = w_{52} = w_{53} = w_{57} = \sigma_4 \Delta t, \tag{2.5}$$
$$w_{05} = w_{25} = w_{42} = w_{43} = w_{47} = \sigma_5 \Delta t, \tag{2.6}$$
$$w_{06} = \sigma_6 \Delta t, \tag{2.7}$$
$$w_{07} = \sigma_7 \Delta t, \tag{2.8}$$
$$w_{08} = w_{28} = w_{38} = w_{48} = w_{58} = \sigma_8 \Delta t. \tag{2.9}$$

Repair weights are defined as

$$w_{10} = \mathcal{L}_1 \Delta t, \tag{2.10}$$
$$w_{20} = \mathcal{L}_2 \Delta t, \tag{2.11}$$
$$w_{30} = \mathcal{L}_3 \Delta t, \tag{2.12}$$
$$w_{40} = \mathcal{L}_4 \Delta t, \tag{2.13}$$
$$w_{50} = \mathcal{L}_5 \Delta t, \tag{2.14}$$
$$w_{60} = \mathcal{L}_6 \Delta t, \tag{2.15}$$
$$w_{70} = \mathcal{L}_7 \Delta t, \tag{2.16}$$
$$w_{80} = \mathcal{L}_8 \Delta t. \tag{2.17}$$

Normalization constraints are expressed as

$$w_{00} = 1 - w_{01} - w_{02} - w_{03} - w_{04} - w_{05} - w_{06} - w_{07} - w_{08}, \quad (2.18)$$
$$w_{11} = 1 - w_{10}, \quad (2.19)$$
$$w_{22} = 1 - w_{21} - w_{24} - w_{25} - w_{26} - w_{20}, \quad (2.20)$$
$$w_{33} = 1 - w_{31} - w_{34} - w_{35} - w_{36} - w_{30}, \quad (2.21)$$
$$w_{44} = 1 - w_{41} - w_{42} - w_{43} - w_{47} - w_{40}, \quad (2.22)$$
$$w_{55} = 1 - w_{51} - w_{52} - w_{53} - w_{57} - w_{50}, \quad (2.23)$$
$$w_{66} = 1 - w_{60}, \quad (2.24)$$
$$w_{77} = 1 - w_{70}, \quad (2.25)$$
$$w_{88} = 1 - w_{28} - w_{38} - w_{48} - w_{58}. \quad (2.26)$$

Hidden layer:
This layer employs an activation function, which is defined as

$$f(\mathcal{Z}) = \mathcal{Z} \quad (2.27)$$

Output layer:
The outputs of the ANN are represented by:

$$Y_{\mathrm{i}} = P_{\mathrm{i}}(t + \Delta t), \quad \text{where } i \text{ ranges from 1 to 7.} \quad (2.28)$$

The overall structure of the ANN, as shown in Figure 2.4, is defined as

$$Y_{\mathrm{j}} = f\left(\sum w_{\mathrm{ij}} T_{\mathrm{i}} + \ell_{\mathrm{j}}\right), \quad \text{where } i, j = 0 \text{ to } 7, \quad (2.29)$$

where w_{ij} represents the weights connecting neurons i and j, and ℓ_{j} represents the bias for neuron j. A linear activation function, as defined in eqn (2.27), is used within the network.

2.4.2 Solved ANN model of system

The artificial neural network (ANN) model is analytically solved by substituting the weight and bias expressions defined in eqn (2.2) through (2.26) into the general ANN output structure given in eqn (2.29). The corresponding outputs are subsequently computed using the activation function defined in eqn (2.27) and the ANN output eqn (2.28).

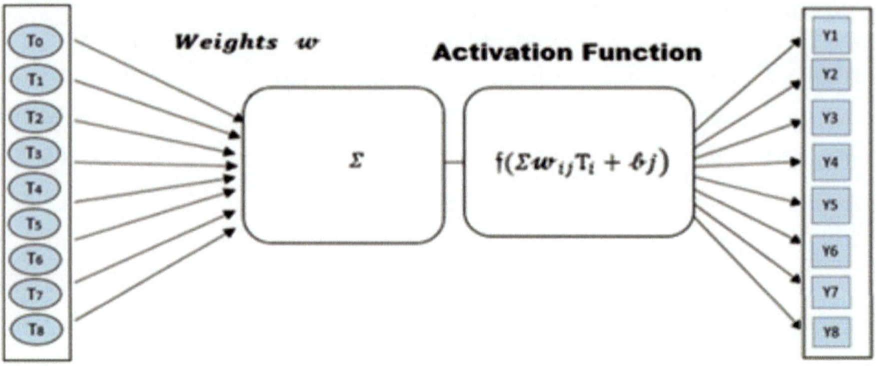

Figure 2.4 The system model of artificial neural networks (ANN).

The individual output neurons Y_0 through Y_8 are evaluated as follows:

$$Y_0 = w_{00}X_0 + w_{10}X_1 + w_{20}X_2 + w_{30}X_3 + w_{40}X_4 + w_{50}X_5 + w_{60}X_6, + w_{70}X_7 + w_{80}X_8 + \ell_0, \quad (2.30)$$

$$Y_1 = w_{01}T_0 + w_{11}T_1 + w_{21}T_2 + w_{31}T_3 + w_{41}T_4 + w_{51}T_5 + \ell_1, \quad (2.31)$$

$$Y_2 = w_{02}T_0 + w_{22}T_2 + w_{42}T_4 + w_{52}T_5 + \ell_2, \quad (2.32)$$

$$Y_3 = w_{03}T_0 + w_{33}T_3 + w_{43}T_4 + w_{53}T_5 + \ell_3, \quad (2.33)$$

$$Y_4 = w_{04}T_0 + w_{24}T_2 + w_{34}T_3 + w_{44}T_4 + \ell_4, \quad (2.34)$$

$$Y_5 = w_{05}T_0 + w_{25}T_2 + w_{35}T_3 + w_{55}T_5 + \ell_5, \quad (2.35)$$

$$Y_6 = w_{06}T_0 + w_{26}T_2 + w_{36}T_6 + \ell_6, \quad (2.36)$$

$$Y_7 = w_{07}T_0 + w_{47}T_4 + w_{57}T_5 + w_{77}T_7 + \ell_7, \quad (2.37)$$

$$Y_8 = w_{08}T_0 + w_{28}T_2 + w_{38}T_3 + w_{48}T_4 + w_{58}T_5 + \ell_8. \quad (2.38)$$

Based on the above outputs, the total probability of the system being in the upstate or downstate can be computed as:

$$P_{\text{upstate}} = Y_0 + Y_2 + Y_3 + Y_4 + Y_5, \quad (2.39)$$

$$P_{\text{downstate}} = Y_1 + Y_6 + Y_7 + Y_8. \quad (2.40)$$

Equation (2.1) defines the state probabilities at time t (serving as the ANN input), whereas eqn (2.27) and (2.28) characterize the variations in these state probabilities at a future time step $(t + \Delta t)$ (serving as the ANN output).

2.5 Results and Discussion

2.5.1 Estimated variations in state probabilities over time

To facilitate the numerical evaluation and comparison of state probabilities, a continuously operating system was analyzed over a defined time period (in months). The initial state probabilities at time t are presented in Table 2.1, under the assumption of uninterrupted system operation. Details regarding the continuous repair rates employed in the analysis are provided in Table 2.2. Furthermore, two distinct failure rate scenarios were considered, as outlined in Table 2.3.

1. *Useful life (constant failure rate):* This phase represents the period during which each subsystem maintains a constant failure rate. The reliability of each subsystem in this phase is modeled using an exponentially distributed survival function.
2. *Wear-out phase (time-dependent failure rate):* As the system ages, its subsystems transition into a wear-out phase characterized by degradation and increasing failure rates. The reliability of each subsystem during this phase follows a Weibull distributed survival function, with

Table 2.1 Subsystem initial state probabilities (ANN input values).

$T_0(t) = T_0$	$T_1(t) = T_1$	$T_2(t) = T_2$	$T_3(t) = T_3$	$T_4(t) = T_4$	$T_5(t) = T_5$	$T_6(t) = T_6$	$T_7(t) = T_7$	$T_8(t) = T_8$
0.35	0.1	0.1	0.1	0.1	0.1	0.05	0.05	0.05

Table 2.2 Subsystems repair rate.

Repair rates	$\mathcal{L}_1$	$\mathcal{L}_2$	$\mathcal{L}_3$	$\mathcal{L}_4$	$\mathcal{L}_5$	$\mathcal{L}_6$	$\mathcal{L}_7$	$\mathcal{L}_8$
Values	0.02	0.01	0.01	0.01	0.01	0.02	0.01	0.01

Table 2.3 Subsystem failure rates under time-dependent and constant failure hypotheses.

Failure rates (per month)	σ_1	σ_2	σ_3	σ_4	σ_5	σ_6	σ_7	σ_8
Scale parameter	0.65	1.55	2.46	3.10	3.5	3.5	1.36	1.5
Shape parameter	2	2	2	2	2	2	2	2
Failure rate at $t = 0$ **(Weibull)**	0.0497	0.0102	0.005	0.0031	0.0011	0.0022	0.015	0.010
Failure rate (constant)	0.05	0.01	0.005	0.003	0.001	0.02	0.01	0.005

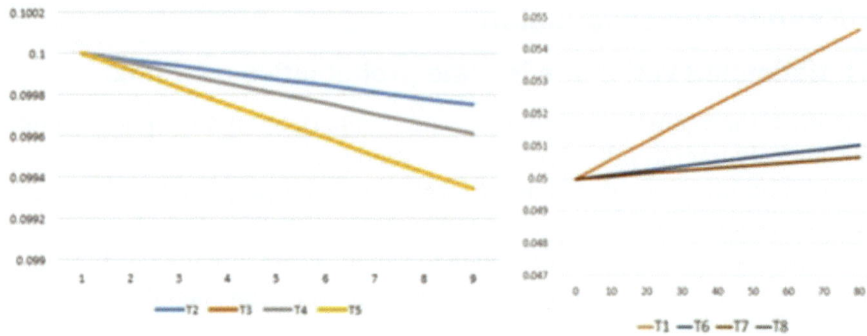

Figure 2.5 Variation in state probabilities for subsystems with constant failure rates. (a) Time evolution of T_2, T_3, T_4, and T_5. (b) Time evolution of T_1, T_6, T_7, and T_8.

the corresponding failure rate given by

$$\lambda(t) = \beta t^{\beta-1}/\eta^{\beta}. \tag{2.41}$$

In eqn (2.41), β represents the shape parameter and η the scale parameter of the Weibull distribution. These parameters are chosen at $t = 0$ such that they align with the initial failure rates of the corresponding subsystems, thereby ensuring consistency between the failure rates in the wear-out phase and the constant rates assumed in the useful life phase.

In the artificial neural network (ANN) model proposed in this study, a bias term is typically included in the activation function to improve the alignment of estimated state probabilities with observed values. However, because observed state probabilities are not utilized in this computational framework, the bias term is set to zero.

Time, measured in months, serves as a key factor when applying the ANN model to the initial state probabilities (Table 2.1). The repair and failure rates defined in eqn (2.2) through (2.26) are used to allocate the connection weights. Subsequently, eqn (2.30) through (2.38) are employed to compute the resulting changes in state probabilities over a 10-hour increment.

Assuming constant failure rates, the expected variation in state probabilities during the "useful life" period is illustrated in Figure 2.4. In contrast, Figure 2.5 displays the corresponding results for the "wear-out phase," which incorporates time-dependent failure rates derived from a Weibull distribution.

2.5.2 Comparison of useful life and wear-out phase

A strong similarity was observed in the estimated decrease of state probabilities between the "useful life" and "wear-out" phases for the state pairs T_3

versus T_5 (degraded efficiencies) and T_2 versus T_4. This similarity can be attributed to two main factors:

1. Weight influence in the ANN model: In the ANN model, the summation of the product of weights and input states is largely influenced by the weights associated with specific states. For the mentioned state pairs, these dominant weights minimize the influence of other weights and states.
2. Similarity in initial state probabilities: The initial state probabilities at time $t = 0$ for the compared state pairs (T_2 vs. T_4 and T_3 vs. T_5) are both 0.1. Since each pair represents a single, comparable failure mechanism, this justifies the observed resemblance in their behaviors.
3. Variation in state T_1: As shown in Figure 2.4, the increase in the probability of failure state T_1 was relatively minor under constant failure rates, rising from 5% to 5.47% over 80 hours. Conversely, Figure 2.5 demonstrates a more significant increase in T_1, from 5% to 7%, under time-dependent failure rates during the same period. This difference is due to the nature of subsystem I, which comprises eight highly loaded processes with rapidly deteriorating equipment. Without proper maintenance, this deterioration activates the failure mode represented by T_1. The absence of maintenance impacts the Weibull distribution's shape parameter (β), increasing degradation. Consequently, the combination of subsystem I's complexity and its time-varying failure rates explains the sharper rise in T_1 during the wear-out phase.
4. Effect on upstate and downstate probabilities: Using eqn (2.39) and (2.40), Figure 2.6 illustrates the change in upstate and downstate probabilities. The upstate probability is estimated to decrease from 86% to 86.34% during the useful life, and from 86% to 82.87% during the wear-out phase. This larger decrease during wear-out is primarily caused by the increase in failure probability of T_1, leading to a more significant drop in upstate probability and a corresponding rise in downstate probability. As expected, this outcome aligns with the system's aging and wear progression.

2.5.3 Profit analysis considering preventive maintenance (PM)

Profit analysis with preventive maintenance (PM) balances upfront PM costs against long-term gains. It considers reduced downtime, extended asset lifespan, and fewer costly repairs. Effective PM minimizes production losses, boosts efficiency, and stabilizes operational expenses, ultimately enhancing overall profitability.

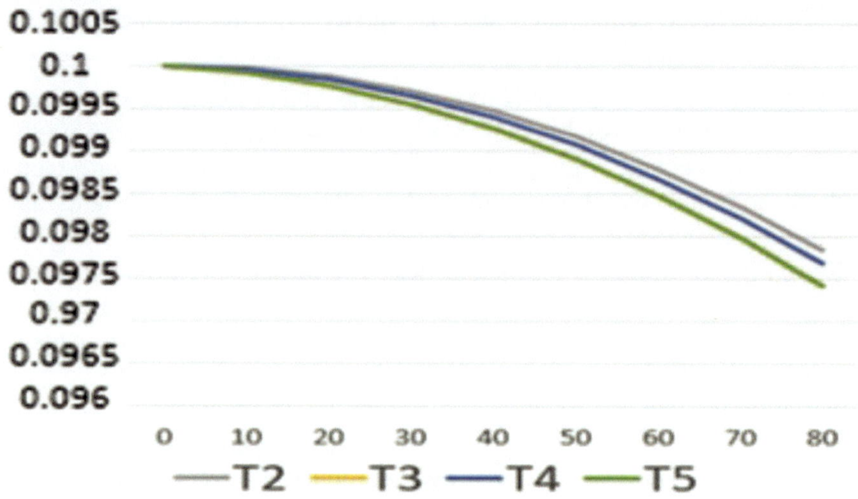

Figure 2.6 Change in state probabilities for subsystems with time-dependent failure rates: time evolution of subsystems T_2, T_3, T_4, and T_5.

A key goal in assessing the reliability of an industrial process is cost optimization for maximum profit. While achieving over 90% uptime seems desirable, it often drives maintenance costs up significantly. Various maintenance performance measures exist to minimize costs, but these can still impact profitability. Consequently, researchers have used reliability parameters to model profit in various industrial processes.

In nuclear power plants, total cost comprises numerous elements, including material, labor, engineering (including machine running costs), maintenance, and overhead. The profit function, or expected total profit per unit time, is calculated as follows:

$$\text{Profit per unit time} = K_I \mathcal{A}_0(t) - C_T - C_{PM}, \tag{2.42}$$

where

- K_I = Revenue per unit uptime
- $\mathcal{A}_0(t)$ = Steady-state availability of the system
- C_T = Total cost per unit time
- C_{PM} = Total cost of preventive maintenance (PM) per unit time

The total cost of PM per unit time is further decomposed as

$$C_{PM} = C_{PRS} + C_{DT}, \tag{2.43}$$

where

- C_{PRS} = Preventive replacement and service cost
- C_{DT} = Downtime cost

2.5.4 Numerical computation

The key numerical values used in the profit analysis are as follows: $K_I =$ 14,000 USD (revenue per unit uptime), $C_T = 8400$ USD (total cost per unit time), and $\Delta t = 10$ hours.

Three preventive maintenance (PM) strategies are evaluated:

- Maximum repairs and replacements: This PM strategy incurs the highest total cost, with $C_{PM} = 10{,}000$ USD and a corresponding cost of 46.29 USD per 10-hour cycle. However, it enables the longest PM cycle interval (2160 hours), thereby minimizing downtime and maximizing operational continuity.
- No preventive maintenance (No PM): This scenario eliminates scheduled maintenance costs altogether ($C_{PM} = 0$ USD). While it avoids immediate expenses, it significantly increases the risk of downtime and equipment failure, potentially resulting in substantial economic losses.
- Minimum repairs and replacements: This strategy provides a cost-effective middle ground, with $C_{PM} = 2000$ USD and a 10-hour cost of 27.77 USD. It requires more frequent maintenance intervals (720 hours), offering a trade-off between maintenance frequency and cost.

The choice of an appropriate PM strategy depends on various factors, including initial cost tolerance, long-term system reliability, and risk management preferences, particularly in high-stakes environments such as nuclear power plants.

Figure 2.7 illustrates the profit variation over time for the three strategies, computed using eqn (2.42) (with unit time $= 10$ hours) and state probabilities generated by an artificial neural network (ANN) model.

- *Case 1: No preventive maintenance (No PM)* - In the absence of PM, the system's profit declines sharply after a certain period due to accelerated equipment degradation and increased downtime. This scenario highlights the essential role of PM in sustaining profitability.
- *Case 2: Minimum repairs and replacements* - This strategy minimizes per-unit-time maintenance costs, making it economically favorable for systems where moderate reliability is acceptable.

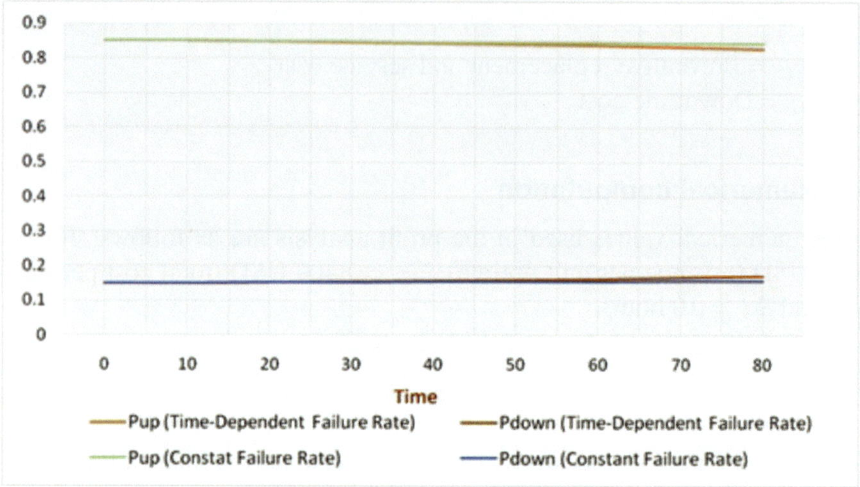

Figure 2.7 Variation of upstate and downstate probabilities over time under constant and time-dependent failure rate assumptions. The figure compares P_{up} and P_{down} for both constant failure rates and time-dependent failure rates.

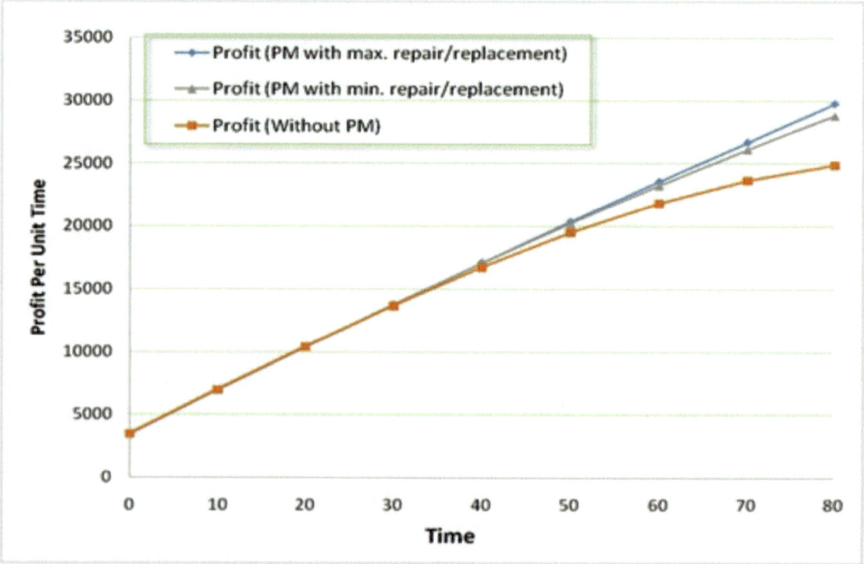

Figure 2.8 Variation in profit over time for a system under three different maintenance strategies: (i) Preventive maintenance with maximum repair/replacement, (ii) PM with minimum repairs, and (iii) no preventive maintenance.

- *Case 3: Maximum repairs and replacements* - Although the upfront maintenance cost is higher (approximately double that of the minimum strategy), this approach yields the highest long-term profit, as illustrated in Figure 2.8. The extended maintenance cycle and reduced downtime demonstrate the substantial economic advantage of a robust PM policy.

The study concludes that, although the preventive maintenance strategy involving maximum repairs and replacements entails higher upfront costs, it yields the highest long-term profitability.

2.6 Conclusion

This chapter presents a comprehensive analysis of the reliability and profitability of nuclear power plant operations using a stochastic modeling approach integrated with artificial neural networks (ANNs). The model captures the transition of subsystems across multiple states – ranging from full functionality to partial and complete failure – thereby offering a more realistic representation than conventional binary-state reliability models.

Two critical operational phases of the nuclear power plant – namely, the useful life (with constant failure rates) and the wear-out phase (with time-dependent failure rates) – were examined. The ANN-based simulation revealed sharper increases in the probability of failure states during the wear-out phase, particularly for heavily utilized subsystems such as Subsystem I. This trend is attributed to the higher degradation rate modeled by the Weibull distribution.

The study further incorporated economic evaluation by computing profit per unit time under three preventive maintenance (PM) strategies: no PM, minimum PM, and maximum PM. While the no-PM approach minimized short-term costs, it led to rapid degradation and a decline in profitability. Conversely, the maximum PM strategy, despite its higher initial cost, resulted in superior long-term performance and profitability due to extended maintenance intervals and reduced downtime.

In conclusion, the proposed ANN-based stochastic model effectively demonstrates that integrating system degradation patterns with maintenance strategies enables more informed and economically sound decision-making. Among the strategies evaluated, the maximum preventive maintenance policy proves to be the most beneficial in sustaining high reliability and maximizing long-term profit in nuclear power plant systems.

References

[1] Toshio Nakagawa. *Stochastic processes: With applications to reliability theory*. Springer Science & Business Media, 2011.

[2] Samuel Karlin and Howard E Taylor. *A second course in stochastic processes*. Elsevier, 1981.

[3] Erhan Cinlar. Exceptional paper – markov renewal theory: A survey. *Management Science*, 21(7):727–752, 1975.

[4] GR Grimmett and DR Strizaker. Probability and random processes. ed. ke-3, 2001.

[5] Marvin Rausand. høyland a: System reliability theory: Models. *Statistical Methods, and Applications, 2nd Edition, John Wiley and Sons, New York*, 2004.

[6] Anatoly Lisnianski and Gregory Levitin. *Multi-state system reliability: assessment, optimization and applications*. World scientific, 2003.

[7] Kishor S Trivedi. Sharpe 2002: Symbolic hierarchical automated reliability and performance evaluator. In *Proceedings international conference on dependable systems and networks*, page 544. IEEE, 2002.

[8] William J Stewart. Introduction to the numerical solution of markov chains. 2021.

[9] Richard G Little. Toward more robust infrastructure: observations on improving the resilience and reliability of critical systems. In *36th Annual Hawaii International Conference on System Sciences, 2003. Proceedings of the*, pages 9–pp. IEEE, 2003.

[10] Richard E Brown and H Lee Willis. The economics of aging infrastructure. *IEEE Power and Energy Magazine*, 4(3):36–43, 2006.

[11] Xiao-Jing Wang. Probabilistic decision making by slow reverberation in cortical circuits. *Neuron*, 36(5):955–968, 2002.

[12] Jan M Van Noortwijk. A survey of the application of gamma processes in maintenance. *Reliability Engineering & System Safety*, 94(1):2–21, 2009.

[13] Balbir S Dhillon. *Human reliability, error, and human factors in power generation*. Springer, 2014.

[14] Toshio Nakagawa. *Stochastic processes: With applications to reliability theory*. Springer, 2011.

[15] P. Levy. Proceesus semi-markoviens. *Proc. Int. Cong. Math. Amsterdam*, pages 416–426, 1954.

[16] Walter L Smith. Regenerative stochastic processes. *Proceedings of the Royal Society of London. Series A. Mathematical and Physical Sciences*, 232(1188):6–31, 1955.

[17] Franciszek Grabski. *Semi-Markov processes: applications in system reliability and maintenance*. Elsevier, 2014.

[18] Yutian Chen, Xianyun Meng, and Shengqiang Chen. Reliability analysis of a cold standby system with imperfect repair and under poisson shocks. *Mathematical Problems in Engineering*, 2014(1):507846, 2014.

[19] Shu-Zhen Yu, Wen-Bin Niu, Su-Li Wu, Wei Ma, and Shu-Fen Zhang. Robust and flexible thermal-plasticizing 3d shaped composite films with invariable and brilliant structural color. *Journal of Materials Chemistry C*, 6(47):12814–12821, 2018.

[20] Saeed A Alameri and Ahmed K Alkaabi. Fundamentals of nuclear reactors. In *Nuclear Reactor Technology Development and Utilization*, pages 27–60. Elsevier, 2020.

[21] KAA Ramadan and KV Slyusarskiy. Effect of heat recovery steam generator on characteristics of combined cycle npp with gas-cooled reactors. *Progress in Nuclear Energy*, 164:104879, 2023.

[22] Peiwei Sun, Zi Wang, Jianmin Zhang, and Guanghui Su. Thermal–hydraulic design and analysis code development for steam generator of cfr600. *Annals of Nuclear Energy*, 90:256–263, 2016.

[23] Ali Chaibakhsh and Ali Ghaffari. Steam turbine model. *Simulation Modelling Practice and Theory*, 16(9):1145–1162, 2008.

[24] K Harby, Doaa R Gebaly, Nader S Koura, and Mohamed S Hassan. Performance improvement of vapor compression cooling systems using evaporative condenser: An overview. *Renewable and sustainable energy reviews*, 58:347–360, 2016.

[25] Paweł Madejski, Krzysztof Banasiak, Paweł Ziółkowski, Dariusz Mikielewicz, Jarosław Mikielewicz, Tomasz Kuś, Michał Karch, Piotr Michalak, Milad Amiri, Paweł Dąbrowski, et al. Development of a spray-ejector condenser for the use in a negative co2 emission gas power plant. *Energy*, 283:129163, 2023.

3

Nonlinear Dynamics and Chaos in Engineering Systems: Through an Epidemic Model

Rajeev Kishore[1], Deepak Kumar[2], and Ahmed Hussein Msmali [3]

[1]Department of Mathematics, Galgotias College of Engineering & Technology, India
[2]Department of Mathematics, SRPS College, B.R.A. Bihar University Muzaffarpur, India
[3]Department of Mathematics, College of Science, Jazan University, Kingdom of Saudi Arabia
E-mail: dr.rajeevkishorepandey@gmail.com; dkriitr@gmail.com; amsmali@jazanu.edu.sa

Abstract

Nonlinear dynamics and chaos play a crucial role in understanding and predicting the behavior of complex engineering and social systems. In this study, we develop a novel SEIR-based epidemic model to examine the academic behavior of slow learners in classroom settings. By drawing parallels between infectious disease transmission and the spread of motivation and academic engagement, the model offers a dynamic framework to simulate behavioral transitions among learners. The proposed model categorizes students into four compartments – slow learners, learners with abnormal behavior, those involved in non-academic activities, and improved learners – and captures the nonlinear interactions among them. Key factors such as awareness, motivation, social interaction, and teaching innovation are embedded in the model to reflect real-world educational dynamics. Stability analysis, reproduction number (R_0), and equilibrium points are computed to assess the impact of interventions. Numerical simulations validate the model's predictions, demonstrating its ability to represent both stable and

DOI: 10.1201/9788743806691-3

endemic behavioral outcomes. The results emphasize the value of integrating mathematical modeling with educational strategies to support and transform the learning behavior of slow learners.

Keywords: Nonlinear dynamics, SEIR model, slow learners, academic motivation, behavioral modeling, and epidemic framework

3.1 Introduction

Understanding and addressing the academic challenges of slow learners has become a critical area of focus in modern education systems. These learners often lag behind their peers due to a variety of cognitive, emotional, and environmental factors. This study explores innovative approaches to improve their learning outcomes using tools from nonlinear dynamics and epidemic modeling. We propose the use of a modified SEIR (Susceptible Exposed Infectious Recovered) mathematical model to capture and analyze the behavioral transitions of slow learners under the influence of social and academic interactions. The model allows us to simulate how motivation, awareness, and external stimuli spread across a population of students in a manner analogous to infectious diseases.

Social and educational behaviors are significantly influenced by the learner's surroundings, including family, friends, peers, and institutional environments. Academic success depends not only on intelligence but also on motivation, discipline, and teacher engagement. Moreira-Morales and García-Loor [1] emphasized that motivation serves as a major driver of academic performance. This is supported by Abdulrahman et al. [2], who found a strong correlation between academic motivation and success among medical learners. Similarly, Formento-Torres et al. [3] demonstrated that metacognitive regulation is a key outcome of motivational strategies and contributes significantly to students' academic achievement.

Efforts to enhance learning in students who exhibit slow progress have led to the adoption of contextual and adaptive teaching techniques. Listiawati et al. [4] showed that incorporating principles of realistic mathematics education (RME) can improve mathematical skills among slow learners. Meanwhile, Al-Osaimi and Fawaz [5] highlighted the positive influence of blended learning and motivational strategies in improving academic outcomes. Extending this idea, Johar et al. [6] presented a mathematical model capturing the effects of social interaction and motivation on learning behavior, underscoring the value of modeling in educational settings.

Mathematical models originally developed in epidemiology have proven useful in non-biological systems such as rumor propagation, social behavior, and learning environments. Kishore et al. [7] proposed an epidemic model to simulate rumor dynamics in e-commerce, and a follow-up study by Kishore et al. [8] extended this work to control rumor spreading in online social networks using SEIR-type models. These models reflect interaction-driven dynamics and can be adapted to represent student transitions between academic states based on motivation and behavioral feedback loops.

The significance of motivation and teacher intervention in promoting academic performance is further supported by several educational investigations. Setyawan et al. [9] emphasized the role of rigorous mathematical thinking in enhancing problem-solving among slow learners. Nieto-Márquez et al. [10] confirmed that motivation is positively correlated with metacognitive awareness and performance in primary education students. Korikana [11] emphasized that when teachers identify and develop students' inherent talents, especially through real-life situations, slow learners can flourish academically. Abdelrahman [12] and Hidayat et al. [13] also highlighted the importance of meta-cognitive behavior and its mediation through performance goals.

Assessment approaches must be tailored to the needs of slow learners, as discussed by Khaira and Herman [14]. Often, slow learners are not intellectually limited but rather disengaged or uninterested due to rigid educational structures. Numerous studies suggest that slow learning may result from factors such as ineffective instruction, lack of individual attention, or social distractions [15–18]. Addressing these through innovative teaching, personalized support, and peer engagement has been shown to significantly improve learner outcomes [19, 20].

This study builds upon these foundations and introduces a computational SEIR-based model tailored to track slow learners' academic progression. Learners are classified into four compartments: (i) X – slow learners; (ii) Y – those with abnormal or disengaged academic behavior; (iii) Z – learners involved in non-academic and social activities; and (iv) R – improved learners who responded positively to academic stimuli. The classification is used to simulate transitions over time and assess the stability of different behavioral states. The model structure is consistent with the classical epidemiological frameworks developed by Brauer and Castillo-Chávez [21], which are widely used in the modeling of dynamic population systems.

Ultimately, this research highlights how a nonlinear epidemic model, when applied to educational behavior, can provide valuable insights into learning dynamics and suggest practical interventions for slow learners. Through classroom observations and mathematical analysis, the study supports peer learning, group activities, and individualized teaching as effective strategies. The proposed model provides a robust tool for designing targeted interventions, fostering motivation, and ultimately guiding slow learners toward academic success.

3.2 Model Construction

In the proposed model, the learner population is categorized into four distinct behavioral compartments. Let $X(t)$ denote the number of slow learners who exhibit limited academic progress due to various cognitive or motivational challenges. The variable $Y(t)$ represents learners who do not fall under the traditional definition of slow learners but display abnormal or disengaged academic behavior. The compartment $Z(t)$ includes learners who are actively involved in non-academic pursuits and social interactions, potentially diverting attention from academic tasks. Finally, $R(t)$ represents the group of learners who have shown academic improvement over time, primarily as a result of effective teaching strategies, peer support, and motivational interventions. The conceptual framework and dynamic interactions among these compartments are illustrated schematically in Figure 3.1. The key model variables and parameters, including transmission and transition rates between learner compartments, are summarized in Table 3.1. This table

Table 3.1 The mathematical model's epidemiological coefficients.

Symbol	Description
X	Learners who are slow learners
Y	Learners who are not slow learners but engage in abnormal behavior
Z	Learners who actively participate in non-academic pursuits and social interactions
R	Learners who, during academic activities and motivation at time t, have improved themselves
T	Constant admitted learners
β_1	Transmission rate between learners X and Z
β_2	Transmission rate between learners X and Y
γ	Rate at which Y learners join the Z learners
γ_1	Rate of improvement for Y learners
γ_2	Rate of improvement for Z learners
σ	Quit rate of Z learners through the slow learning process
μ	Natural quit rate of the learning process
N	Total strength

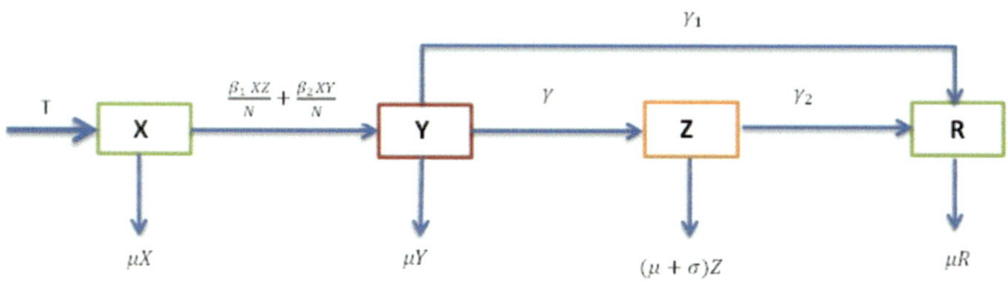

Figure 3.1 Mathematical flowchart of slow learners. The diagram illustrates the dynamic transitions among four behavioral compartments in the SEIR-based model: X represents slow learners; Y represents learners exhibiting abnormal academic behavior despite not being categorized as slow learners; Z represents students engaged in non-academic pursuits and social interactions; and R denotes learners who have shown academic improvement due to teaching motivation and academic engagement at time t.

serves as a reference for interpreting the interactions among different learner groups in the proposed SEIR-based framework. The following system of nonlinear differential equations defines the computational epidemic model formulated based on our assumptions concerning the behavioral dynamics of slow learners:

$$\begin{cases} \frac{dX}{dt} = T - \frac{\beta_1 XZ}{N} - \frac{\beta_2 XY}{N} - \mu X \\ \frac{dY}{dt} = \frac{\beta_1 XZ}{N} + \frac{\beta_2 XY}{N} - \gamma Y - \gamma_1 Y - \mu Y \\ \frac{dZ}{dt} = \gamma Y - \sigma Z - \gamma_2 Z - \mu Z \\ \frac{dR}{dt} = \gamma_1 Y + \gamma_2 Z - \mu R. \end{cases} \tag{3.1}$$

The feasible region corresponding to system (4.1) is defined as

$$\varphi = \left\{ (X, Y, Z, R) \in \mathbb{R}^4 : X \geq 0,\ Y \geq 0,\ Z \geq 0,\ R \geq 0,\ X + Y + Z + R \leq 1 \right\}.$$

This region φ is positively invariant under the dynamics of system (4.1), implying that any solution initiating within φ remains in φ for all future time. Hence, the model adheres to both epidemiological consistency and mathematical rigor, ensuring boundedness and non-negativity of all state variables throughout the simulation.

3.3 Basic Reproductive Number R_0

The basic reproduction number R_0 is a critical threshold parameter that determines the stability of the model and the long-term behavior of the learning dynamics. It is computed using the *Next Generation Matrix Method*, a widely accepted technique in epidemic modeling. The model is considered locally asymptotically stable in the feasible region φ if $R_0 \leq 1$, indicating that undesirable learning behaviors will eventually diminish. Conversely, the model becomes unstable within φ when $R_0 > 1$, signifying the persistence or spread of abnormal learning behavior.

The basic reproduction number R_0 for the system (3.1) is derived by focusing on the dynamics of learners in compartments Y (abnormal behavior) and Z (non-academic activities). The subsystem representing their interactions is given as

$$\begin{cases} \frac{dY}{dt} = \frac{\beta_1 XZ}{N} + \frac{\beta_2 XY}{N} - \gamma Y - \gamma_1 Y - \mu Y, \\ \frac{dZ}{dt} = \gamma Y - \sigma Z - \gamma_2 Z - \mu Z. \end{cases} \tag{3.2}$$

Based on this subsystem, the transmission matrix F and the transition matrix V are formulated to compute the next-generation matrix. These matrices characterize the rate of appearance of new "infections" (abnormal learning behaviors) and the rate of transition or recovery from them. The dominant eigenvalue (spectral radius) of the matrix product FV^{-1} yields the basic reproduction number R_0, which governs the system's stability and the long-term impact of academic interventions.

We obtain the transmission matrix F and the transition matrix V using the dynamics of the Y and Z compartments as follows:

$$F = \begin{bmatrix} \beta_2 & \beta_1 \\ 0 & 0 \end{bmatrix}, \quad V = \begin{bmatrix} \gamma + \gamma_1 + \mu & 0 \\ -\gamma & \sigma + \gamma_2 + \mu \end{bmatrix}.$$

Determining V^{-1}:

$$V^{-1} = \frac{1}{(\gamma + \gamma_1 + \mu)(\sigma + \gamma_2 + \mu)} \begin{bmatrix} \sigma + \gamma_2 + \mu & 0 \\ \gamma & \gamma + \gamma_1 + \mu \end{bmatrix}.$$

Computing FV^{-1}:

$$FV^{-1} = \frac{1}{(\gamma + \gamma_1 + \mu)(\sigma + \gamma_2 + \mu)} \begin{bmatrix} \beta_2 & \beta_1 \\ 0 & 0 \end{bmatrix} \begin{bmatrix} \sigma + \gamma_2 + \mu & 0 \\ \gamma & \gamma + \gamma_1 + \mu \end{bmatrix},$$

$$FV^{-1} = \frac{1}{(\gamma + \gamma_1 + \mu)(\sigma + \gamma_2 + \mu)} \begin{bmatrix} \beta_2(\sigma + \gamma_2 + \mu) + \beta_1 \gamma & \beta_1(\gamma + \gamma_1 + \mu) \\ 0 & 0 \end{bmatrix}.$$

The basic reproduction number R_0 is the spectral radius (dominant eigenvalue) of the matrix FV^{-1}. Hence, it is given by

$$R_0 = \frac{\beta_2(\gamma_2 + \sigma + \mu) + \beta_1 \gamma}{(\gamma_2 + \sigma + \mu)(\gamma + \gamma_1 + \mu)}.$$

Equivalently, it can be expressed as

$$R_0 = \frac{\beta_2}{\gamma + \gamma_1 + \mu} + \frac{\beta_1 \gamma}{(\gamma_2 + \sigma + \mu)(\gamma + \gamma_1 + \mu)}.$$

3.4 Analysis of Stability in Equilibrium

We now determine the equilibrium points of system (3.1) and examine their stability characteristics. The equilibrium states are derived by setting the system of differential equations to zero as follows:

$$\begin{aligned} T - \frac{\beta_1 XZ}{N} - \frac{\beta_2 XY}{N} - \mu X = 0, \\ \frac{\beta_1 XZ}{N} + \frac{\beta_2 XY}{N} - \gamma Y - \gamma_1 Y - \mu Y = 0, \\ \gamma Y - \sigma Z - \gamma_2 Z - \mu Z = 0 \\ \gamma_1 Y + \gamma_2 Z - \mu R = 0. \end{aligned} \tag{3.3}$$

(i) Equilibrium points free from abnormal learning behavior

The equilibrium point in the absence of any abnormal learning behavior is given by

$$E_0(X_0, Y_0, Z_0, R_0) = \left(\frac{T}{\mu}, 0, 0, 0\right).$$

This point represents a steady-state scenario where no learners exhibit disengaged or non-academic behavior, indicating that the entire population remains in the compartment of slow learners without influence from behavioral transitions.

(ii) Endemic equilibrium point (abnormal learning behavior persists with a defined threshold)

The endemic equilibrium $E^*(X_0, Y_0, Z_0, R_0) = (X^*, Y^*, Z^*, R^*)$ corresponds to a state where abnormal learning behavior exists consistently but within a certain controlled range. This occurs when a subset of learners is influenced by disengaged behavior yet not entirely dominated by it. The endemic equilibrium reflects the ongoing social interactions and motivational influences within the learning population.

The expressions for the endemic equilibrium components are given as

$$X^* = \frac{T(\beta_1\alpha + \beta_2 k_2) - (T\beta_1\alpha + T\beta_2 k_2 - N\mu k_1 k_2)}{\mu(\beta_1\alpha + \beta_2 k_2)},$$

$$Y^* = \frac{T}{k_1}\left[\frac{T\beta_1\alpha + T\beta_2 k_2 - N\mu k_1 k_2}{\beta_1\alpha + \beta_2 k_2}\right],$$

$$Z^* = \frac{\alpha T}{k_1 k_2}\left[\frac{T\beta_1\alpha + T\beta_2 k_2 - N\mu k_1 k_2}{\beta_1\alpha + \beta_2 k_2}\right],$$

$$R^* = \frac{T}{k_1\mu}\left[\gamma_1 + \frac{\gamma_2\alpha}{k_2}\right]\left[\frac{T\beta_1\alpha + T\beta_2 k_2 - N\mu k_1 k_2}{\beta_1\alpha + \beta_2 k_2}\right],$$

where $k_1 = \gamma + \gamma_1 + \mu$, and $k_2 = \sigma + \gamma_2 + \mu$.

3.4.1 Equilibrium point free from abnormal learning behavior and its stability

Lemma 1. The equilibrium point is free from learners' abnormal learning behavior if $R_0 \leq 1$. In this case, the equilibrium is locally asymptotically stable within the feasible region, indicating that the abnormal learning behavior of the learners does not influence others. Conversely, if $R_0 > 1$, the equilibrium point becomes unstable, and the learning behavior tends to propagate within the population.

Proof: To analyze the local stability of the equilibrium point, we compute the Jacobian matrix of system (4.1) as

$$J = \begin{bmatrix} -\frac{\beta_1 Z}{N} - \frac{\beta_2 Y}{N} - \mu & -\frac{\beta_2 X}{N} & -\frac{\beta_1 X}{N} & 0 \\ \frac{\beta_1 Z}{N} + \frac{\beta_2 Y}{N} & -(\gamma + \gamma_1 + \mu) & \frac{\beta_1 X}{N} & 0 \\ 0 & \gamma & -(\sigma + \gamma_2 + \mu) & 0 \\ 0 & \gamma_1 & \gamma_2 & -\mu \end{bmatrix}.$$

Evaluating the Jacobian at the behavior-free equilibrium $E_0 = \left(\frac{T}{\mu}, 0, 0, 0\right)$, we obtain:

$$J = \begin{bmatrix} -\mu & -\frac{\beta_2}{N}\cdot\frac{T}{\mu} & -\frac{\beta_1}{N}\cdot\frac{T}{\mu} & 0 \\ 0 & -(\gamma+\gamma_1+\mu) & \frac{\beta_1}{N}\cdot\frac{T}{\mu} & 0 \\ 0 & \gamma & -(\sigma+\gamma_2+\mu) & 0 \\ 0 & \gamma_1 & \gamma_2 & -\mu \end{bmatrix}.$$

The characteristic equation of this Jacobian yields the following eigenvalues:

$$\lambda_1 = -\mu, \quad \lambda_2 = -(\gamma+\gamma_1+\mu),$$
$$\lambda_3 = -(\sigma+\gamma_2+\mu) + \frac{\gamma}{\gamma+\gamma_1+\mu}\cdot\frac{\beta_1 T}{N\mu}, \qquad \lambda_4 = -\mu$$

To simplify notation, let $A = \dfrac{T}{\mu}$. Then the critical eigenvalue λ_3 can be re-expressed as

$$\lambda_3 = \beta\cdot\frac{T}{\mu}\left(e^{-\mu\delta_1} + e^{-\mu\delta_2}\right) - (\mu+\gamma+\eta).$$

This eigenvalue λ_3 is negative if:

$$\mu(\mu+\gamma_2+\eta) > \beta T(e^{-\mu\delta_1} + e^{-\mu\delta_2}).$$

Therefore, the equilibrium point $E_0 = \left(\dfrac{T}{\mu}, 0, 0, 0\right)$ is locally asymptotically stable in the feasible region φ if $R_0 \leq 1$ and the above inequality holds. If $R_0 > 1$, the fourth eigenvalue becomes positive, rendering the equilibrium unstable. This confirms that the persistence or elimination of abnormal learning behavior is fundamentally governed by the basic reproduction number R_0.

3.4.2 Endemic equilibrium points and their stability

Lemma 2. The unique endemic equilibrium point

$$E^*(X_0, Y_0, Z_0, R_0) = (X^*, Y^*, Z^*, R^*)$$

is locally asymptotically stable in the feasible region φ if $R_0 > 1$, and unstable if $R_0 < 1$.

Proof: To investigate the stability of the endemic equilibrium, we analyze the Jacobian matrix evaluated at $E^* = (X^*, Y^*, Z^*, R^*)$. The characteristic matrix is given by

$$J = \begin{bmatrix} -\mu + \beta R^* & 0 & 0 & -\beta X^* + \varepsilon e^{-\mu\delta_3} \\ \beta R^*(P - e^{-\mu\delta_1}) & -\mu & 0 & \beta X^*(P - e^{-\mu\delta_1}) \\ \beta R^*(1 - P - e^{-\mu\delta_2}) & 0 & -\mu & \beta X^*(1 - P - e^{-\mu\delta_2}) \\ \beta R^*(e^{-\mu\delta_1} + e^{-\mu\delta_2}) & 0 & 0 & \beta X^*(e^{-\mu\delta_1} + e^{-\mu\delta_2}) - (\mu + \gamma + \eta) \end{bmatrix}.$$

From this matrix, we observe that the eigenvalues

$$\lambda_2 = -\mu, \quad \lambda_3 = -\mu$$

are strictly negative. The remaining eigenvalues are also negative for $R_0 > 1$, confirming that the endemic equilibrium is locally asymptotically stable when the basic reproduction number exceeds unity.

3.5 Results and Discussion

In this study, the proposed epidemic mathematical model is simulated for two scenarios: $R_0 < 1$ and $R_0 > 1$, to explore the dynamics of student learning behavior.

Example 1: This example illustrates the local stability of the behavior-free equilibrium point E_0 under the condition $R_0 < 1$. Figures 3.2–3.4 provide graphical validation of the theoretical analysis. The simulation is performed using the following initial conditions and parameter values:

$$X = 0.4, \quad Y = 0.2, \quad Z = 0.2, \quad R = 0.2, \quad T = 0.8, \quad \mu = 0.4,$$

$$\gamma = 0.1, \quad \beta_1 = 0.15, \quad \beta_2 = 0.18, \quad \gamma_1 = 0.3, \quad \gamma_2 = 0.3, \quad \sigma = 0.1$$

The corresponding basic reproduction number is computed as $R_0 = 0.248$, which is less than 1. This confirms the theoretical prediction that the behavior-free equilibrium is locally asymptotically stable.

Figure 3.2 shows the temporal evolution of each compartment. Over time, the proportion of slow learners (X) increases, while the number of learners in abnormal behavior (Y), non-academic activity (Z), and even the improved group (R) decreases, illustrating the system's return to the equilibrium state. Figures 3.3 and 3.4 provide the phase portraits of the system. These visualizations show the dynamic interplay among the learner groups. The convergence of trajectories in these phase plots confirms the system's tendency toward

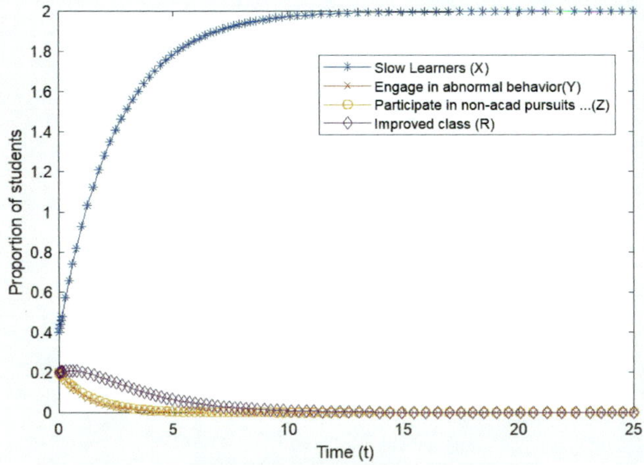

Figure 3.2 Local stability of the equilibrium point E_0 when $R_0 < 1$, showing the temporal dynamics of different learner categories. The proportion of slow learners (X) increases over time, while those engaging in abnormal behavior (Y), participating in non-academic activities (Z), and improved learners (R) decrease, indicating the system's convergence to a behavior-free equilibrium.

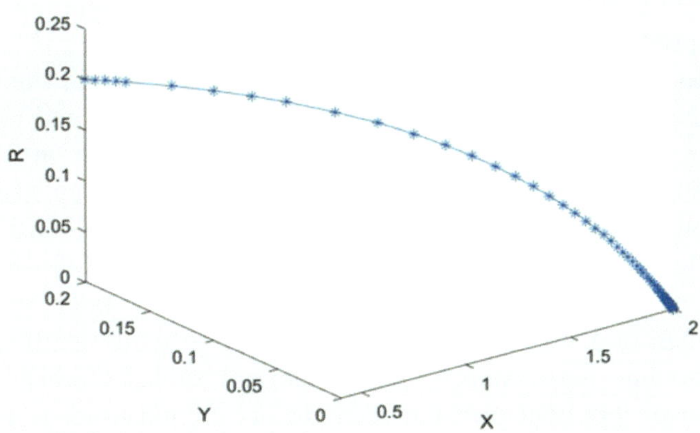

Figure 3.3 Phase portrait of the system (3.1) under the condition $R_0 < 1$. The trajectory illustrates the convergence of the system from initial states toward the equilibrium point, showing the interaction between slow learners (X), learners engaging in abnormal behavior (Y), and improved learners (R). The system exhibits stable behavior as the variables evolve toward a steady state.

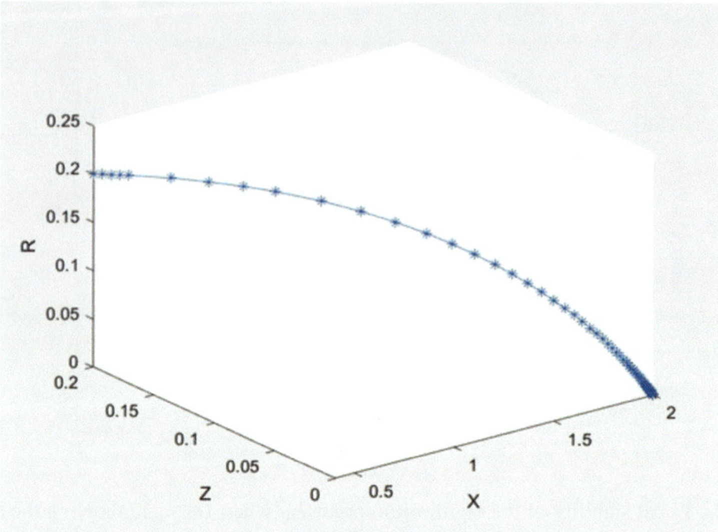

Figure 3.4 Phase portrait of system (1) when $R_0 < 1$, illustrating the interaction between slow learners (X), participants in non-academic pursuits (Z), and improved learners (R). The trajectory demonstrates a stable progression where learners gradually transition from disengaged categories toward academic improvement, confirming the system's convergence to the stable equilibrium point.

a stable equilibrium in the presence of educational interventions and low behavioral reproduction numbers.

From a pedagogical perspective, this behavior-free equilibrium reflects a scenario where effective classroom strategies, motivation, and peer interactions suppress abnormal learning behaviors. The decline in Y and Z populations suggests a successful redirection of learners away from disengaged and non-academic behaviors. Overall, this example demonstrates that when the influence of abnormal behavior is weak ($R_0 < 1$), the system self-corrects over time. This highlights the value of early interventions and the implementation of targeted teaching strategies to guide learners toward improved academic performance. The numerical values corresponding to the simulation of this behavior-free scenario are summarized in Table 3.2, further validating the convergence of each learner group toward the stable equilibrium.

Example 2: This example illustrates the local stability of the endemic equilibrium point $E^* = (X^*, Y^*, Z^*, R^*)$ under the condition $R_0 > 1$. Numerical

Table 3.2 Slow learners' distribution of individual classes against time for abnormal learning behavior and non-academic pursuits and social interactions ($R_0 < 1$).

Parameter values	t	X	Y	Z	R
$X = 0.4$	2.2483	1.3490	0.0331	0.0405	0.1572
$Y = 0.2$	4.2142	1.7035	0.0069	0.0098	0.0859
$Z = 0.2$	5.4434	1.8186	0.0026	0.0040	0.0555
$R = 0.2$	7.9039	1.9322	0.0004	0.0006	0.0220
$T = 0.8$	8.1515	1.9386	0.0003	0.0005	0.0199
$\mu = 0.4$	11.0819	1.9810	0.0000	0.0001	0.0063
$\gamma = 0.1$	13.0519	1.9914	0.0000	0.0000	0.0029
$\beta_1 = 0.15$	17.4822	1.9985	0.0000	0.0000	0.0005
$\beta_2 = 0.18$	20.5959	1.9996	0.0000	0.0000	0.0001
$\gamma_1 = 0.3$	24.5130	1.9999	0.0000	0.0000	0.0000
$\gamma_2 = 0.3$	25.0000	1.9999	0.0000	0.0000	0.0000
$\sigma = 0.1$	2.2483	1.3490	0.0331	0.0405	0.1572

simulations are conducted using the same initial conditions as in Example 1:

$$X = 0.4, \quad Y = 0.2, \quad Z = 0.2, \quad R = 0.2,$$

with the parameter values:

$$T = 0.8, \quad \mu = 0.4, \quad \gamma = 0.1, \quad \beta_1 = 0.72,$$

$$\beta_2 = 0.7, \quad \gamma_1 = 0.3, \quad \gamma_2 = 0.3, \quad \sigma = 0.1.$$

Based on these parameters, the basic reproduction number is computed as $R_0 = 1.146$, indicating that the system is within the threshold where persistent abnormal learning behavior is expected.

Figures 3.5–3.7 depict the numerical results, confirming the stability of the endemic equilibrium. As shown in Figure 3.5, the proportions of learners in the abnormal behavior (Y) and non-academic activity (Z) compartments remain at non-zero levels over time. This persistence confirms that the system stabilizes in an endemic state, where behavioral deviations continue to exist despite academic interventions.

Moreover, there is a gradual increase in the proportion of improved learners (R), suggesting that while the learning environment is not free from abnormal behavior, effective teaching strategies and peer influences contribute to partial academic recovery. Figures 3.6 and 3.7 further reinforce these findings through phase portraits that reveal stable trajectories converging toward the endemic equilibrium. This example emphasizes that when $R_0 > 1$, the system does not revert to a behavior-free state. Instead, it

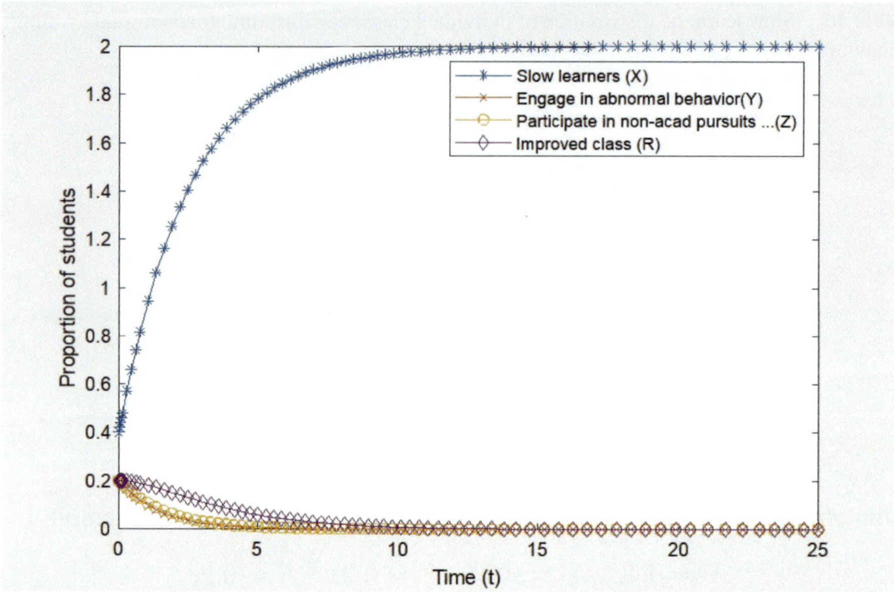

Figure 3.5 Local stability of the endemic equilibrium point when $R_0 > 1$. The plot shows the temporal dynamics of slow learners (X), learners engaging in abnormal behavior (Y), those involved in non-academic pursuits (Z), and improved learners (R). As time progresses, the proportions of Y and Z persist at non-zero levels, confirming the stability of the endemic equilibrium. The gradual increase in R indicates partial recovery, despite the persistence of abnormal behavior in the learning environment.

settles into a state where abnormal behaviors coexist with academic progress, highlighting the necessity of sustained and dynamic interventions to manage and mitigate non-academic tendencies among slow learners.

These dynamics are depicted clearly in Figures. 3.5–3.7, which validates the stability of the endemic equilibrium. The corresponding simulated values are summarized in Table 3.3, which provides a detailed account of the temporal evolution of all four learner compartments. As shown in Table 3.3, the proportion of slow learners (X) increases steadily and eventually saturates near a stable upper bound, while the proportions of learners in the abnormal behavior (Y) and non-academic activity (Z) categories decrease to negligible levels over time. Simultaneously, the number of improved learners (R) gradually diminishes after a brief initial rise, stabilizing near zero. This numerical trend highlights the persistence of endemic abnormal behavior in the population, even in the presence of moderate academic recovery. The data reinforce the theoretical outcome that, for $R_0 > 1$, the system evolves

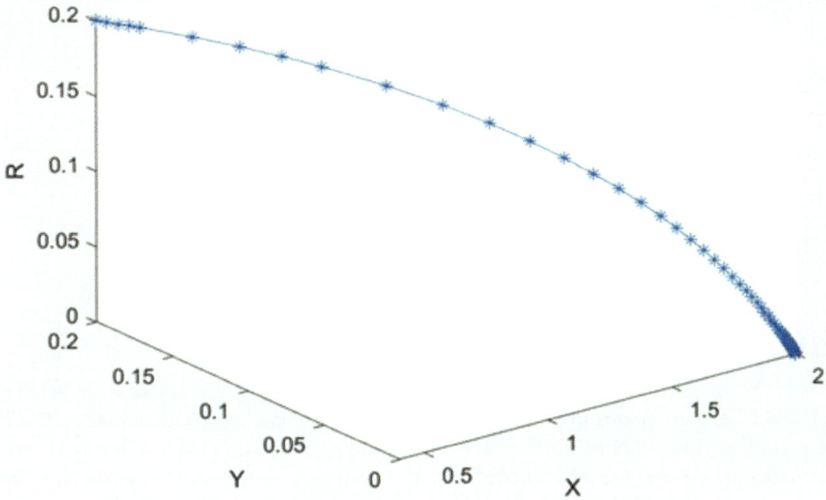

Figure 3.6 Phase portrait of system (3.1) when $R_0 > 1$, illustrating the interaction between slow learners (X), learners engaging in abnormal behavior (Y), and improved learners (R). The trajectory demonstrates the persistence of abnormal behavior within the learner population, with convergence toward a non-zero endemic equilibrium. Despite the presence of interventions, the system stabilizes with a portion of learners remaining in abnormal or partially improved states.

Table 3.3 Slow learners' distribution of individual classes against time for an abnormal learning behavior and non-academic pursuits and social interactions ($R_0 > 1$).

Parameter values	t	X	Y	Z	R
$X = 0.4$	0	0.4	0.2	0.2	0.2
$Y = 0.2$	1.0418	0.9452	0.0965	0.1065	0.1802
$Z = 0.2$	3.3103	1.5742	0.0198	0.0262	0.101
$R = 0.2$	5.8425	1.8454	0.0034	0.0053	0.0429
$T = 0.8$	7.8131	1.9297	0.0009	0.0015	0.0207
$\mu = 0.4$	9.7844	1.968	0.0002	0.0004	0.0097
$\gamma = 0.1$	11.0654	1.9809	0.0001	0.0002	0.0059
$\beta_1 = 0.72$	12.9512	1.991	0	0.0001	0.0028
$\beta_2 = 0.7$	15.2363	1.9964	0	0	0.0011
$\gamma_1 = 0.2$	19.8983	1.9994	0	0	0.0002
$\gamma_2 = 0.2$	24.4748	1.9999	0	0	0
$\sigma = 0.1$	25	1.9999	0	0	0

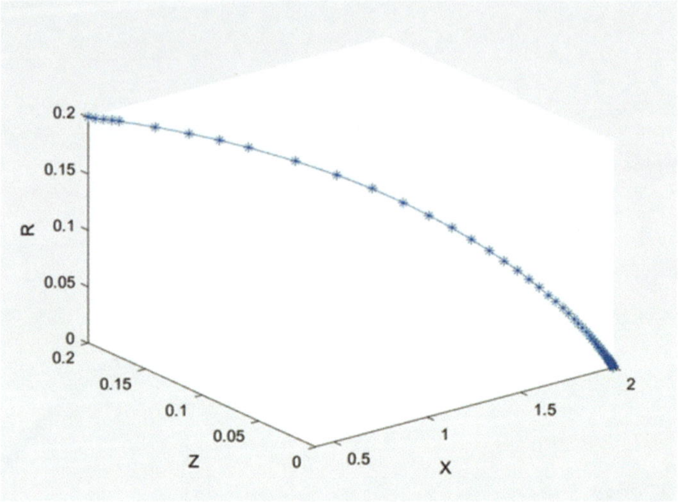

Figure 3.7 Phase portrait of system (3.1) when $R_0 > 1$, showing the relationship between slow learners (X), participants in non-academic pursuits (Z), and improved learners (R). The trajectory confirms the stability of the endemic equilibrium, with persistent levels of non-academic engagement and partial improvement. The system does not return to a behavior-free state, highlighting the sustained impact of abnormal learning behavior.

toward a non-zero endemic equilibrium, emphasizing the need for sustained and more intensive educational interventions.

3.6 Validation of the Model

The validation of the proposed SEIR-based behavioral model is substantiated through a series of numerical simulations and graphical representations (Figures. 3.2–3.7), which confirm the theoretical findings discussed in the previous sections. The model equations were numerically solved using MATLAB, implementing the classical fourth-order Runge–Kutta method to achieve precise and stable results over time. Two distinct scenarios, namely $R_0 < 1$ and $R_0 > 1$, were considered to examine the system's dynamic behavior under varying behavioral reproduction numbers. Under the condition $R_0 < 1$, the simulation results reveal that the proportion of slow learners (X) increases steadily while the proportions of students with abnormal behavior (Y), those engaged in non-academic activities (Z), and improved learners (R) diminish over time. This dynamic behavior signifies the system's convergence toward a behavior-free equilibrium – a state where motivational strategies and academic discipline suppress the spread of disengaged learning behavior. Conversely, for $R_0 > 1$, the model predicts a non-zero endemic equilibrium. The simulations depict persistent proportions of Y and Z learners despite academic interventions, emphasizing the necessity of sustained, multi-layered teaching strategies. The improved learners (R) increase initially but stabilize, highlighting partial recovery amidst continued behavioral deviations.

In addition to simulation-based validation, the model's relevance is further strengthened through empirical data analysis. The study examined the learning profiles of 1695 engineering students, classified into four distinct categories: slow learners (X), students with abnormal academic behavior (Y), those involved in non-academic and social activities (Z), and learners showing academic improvement (R). Figure 3.8 presents the categorical distribution of learners based on learning pace, showing that out of the total students surveyed, 123 were fast learners, 880 were average, and 692 were slow learners. This high percentage of slow learners reinforces the need for targeted academic support and justifies the use of behavioral modeling to understand learning patterns. Figure 3.9 illustrates the behavioral classification of slow learners at a specific time t, dividing them among the four defined

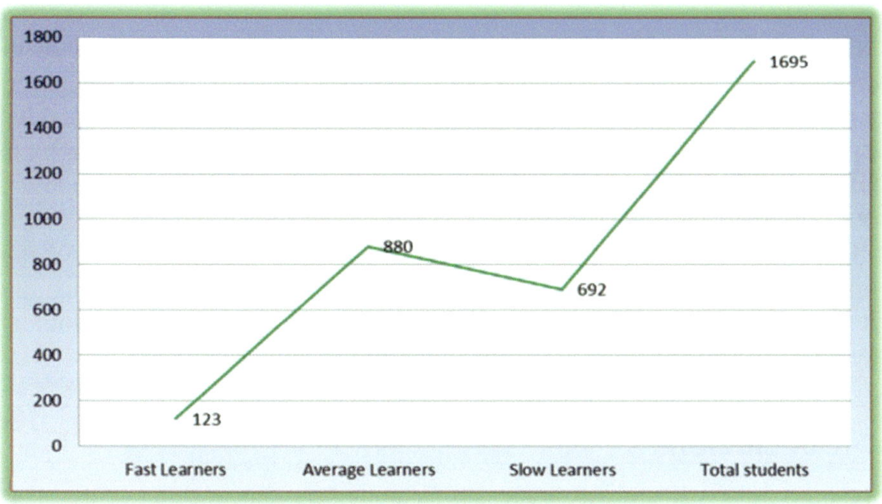

Figure 3.8 Classification of students based on learning pace. The figure illustrates the distribution of learners categorized as fast, average, and slow, along with the total number of students surveyed. Out of 1695 students, 123 are fast learners, 880 are average learners, and 692 are identified as slow learners. This classification highlights the significant proportion of students requiring academic support.

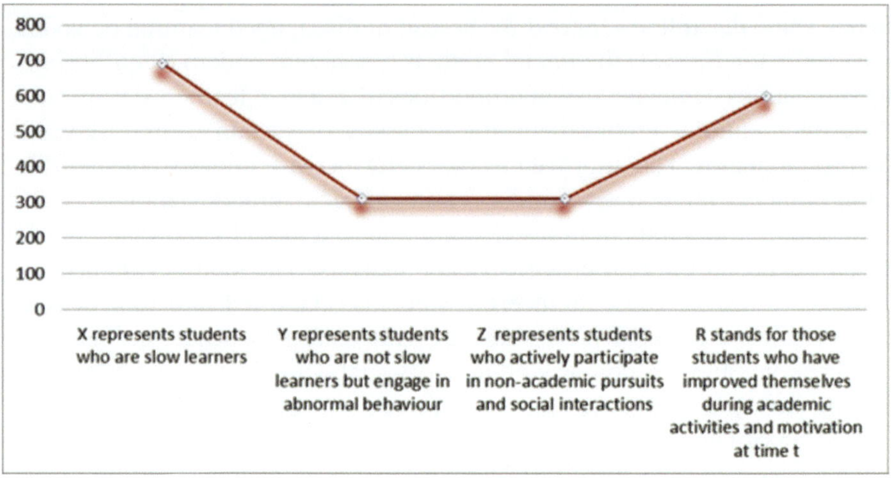

Figure 3.9 Improved chart of slow learners based on behavioral categories. The chart visualizes the distribution of learners in four key groups: X represents slow learners; Y denotes students exhibiting abnormal academic behavior despite not being slow learners; Z includes those engaged in non-academic pursuits and social interactions; and R corresponds to students who have shown academic improvement due to motivational and educational interventions at a given time t.

groups: X, Y, Z, and R. The chart provides a snapshot of the current state of learners within the model's framework, demonstrating how real-world learners are distributed across compartments based on behavioral patterns. This supports the predictive capability of the model, which accurately captures not only the dynamics but also the transitional pathways between academic disengagement and improvement.

Overall, both simulation and empirical data affirm the robustness of the proposed model. It effectively mirrors real-world academic transitions and offers a powerful quantitative tool for evaluating and guiding pedagogical interventions aimed at improving the outcomes of slow learners.

3.7 Conclusions

This study presents a novel approach to analyzing and improving the academic behavior of slow learners using a nonlinear epidemic modeling framework. By adapting the SEIR model, commonly used in epidemiology, we simulate the spread and transition of learning behaviors within a student population. The model captures how various factors-such as motivation, social interaction, innovative teaching practices, and environmental influences-impact the learning trajectories of slow learners.

Through the classification of students into four behavioral compartments (slow learners, learners exhibiting abnormal behavior, those engaged in non-academic activities, and improved learners), the model provides a structured method for understanding educational dynamics. The mathematical analysis, including the derivation of equilibrium points and the basic reproduction number (R0), enables the identification of conditions under which academic improvement or degradation occurs. The model's stability analysis reveals critical thresholds that determine whether positive learning behavior becomes dominant or fades over time. Numerical simulations further validate the theoretical results, illustrating how targeted teaching interventions and supportive learning environments can convert disengaged or under-performing learners into academically improved students. The model also highlights the importance of timely motivation, awareness-building, and personalized teaching in influencing long-term behavioral outcomes.

Overall, the integration of nonlinear dynamics and epidemic modeling offers a powerful framework for designing adaptive educational strategies. The findings of this study can inform educators, academic counselors, and policymakers in developing data-driven approaches to support slow learners. Future work may involve refining the model by incorporating real-time classroom data, psychological variables, or extending it to other learning populations and educational systems.

Acknowledgment

Rajeev Kishore is thankful to the Chairman of Galgotias Group of Institutions, Greater Noida, for providing the necessary infrastructure, and Dr. Deepak Kumar is appreciative to the Principal, SRPS College, Jaintpur, constituent unit of B.R.A. Bihar University, Muzaffarpur, Bihar, India, for promoting the research facilities.

References

[1] D. B. Moreira-Morales and M. I. García-Loor. Motivation in academic performance. *International Research Journal of Management, IT and Social Sciences*, 11(1):30–38, 2024.

[2] Khalid A Bin Abdulrahman, Abdulrahman S Alshehri, Khalid M Alkhalifah, Ahmed Alasiri, Mohammad S Aldayel, Faisal S Alahmari,

Abdulrahman M Alothman, Mohammed A Alfadhel, Khalid Alkhalifah Sr, and Mohammad Aldayel. The relationship between motivation and academic performance among medical students in riyadh. *Cureus*, 15(10), 2023.

[3] Motivation and academic performance in adolescence: a systematic meta-analytic review/motivaciã^{3}n y rendimiento academico en la adolescencia: una revisiã^{3}n sistematica meta-analãtica/motivaã §.

[4] Nur Listiawati, Simon Sili Sabon, Slamet Wibowo, Bambang Riyanto, et al. Analysis of implementing realistic mathematics education principles to enhance mathematics competence of slow learner students. *Journal on Mathematics Education*, 14(4):683–700, 2023.

[5] Dalyal Nader Al-Osaimi and Mirna Fawaz. Nursing students' perceptions on motivation strategies to enhance academic achievement through blended learning: a qualitative study. *Heliyon*, 8(7), 2022.

[6] Rahmah Johar, Marwan Ramli, et al. Mathematical model of student learning behavior with the effect of learning motivation and student social interaction. *Journal on Mathematics Education*, 13(3):415–436, 2022.

[7] Rajeev Kishore, Indu Tyagi, Yerra Shankar Rao, and Deepak Kumar. Epidemic model on rumor propagation in e-commerce. *Materials Today: Proceedings*, 57:2056–2060, 2022.

[8] Rajeev Kishore, Indu Tyagi, Deepak Kumar, and Yerra Shankar Rao. Control of rumor spreading in online social network through an epidemic model. In *2022 4th International Conference on Advances in Computing, Communication Control and Networking (ICAC3N)*, pages 2036–2043. IEEE, 2022.

[9] F Setyawan, Tri Kinasih Handayani, Koesoemo Ratih, Anam Sutopo, Tri Indah Rusli, Nur Rizky Alfiany, et al. Rigorous thinking in mathematics modelling for slow learners. In *Journal of Physics: Conference Series*, volume 1720, page 012005. IOP Publishing, 2021.

[10] Natalia Lara Nieto-Márquez, Sandra García-Sinausía, Miguel Ángel Pérez Nieto, et al. Relaciones de la motivación con la metacognición y el desempeño en el rendimiento cognitivo en estudiantes de educación primaria. *Anales de Psicología/Annals of Psychology*, 37(1):51–60, 2021.

[11] Appaji Korikana. Slow learners-a universal problem and providing educational opportunities to them to be a successful learner. *PEOPLE: International Journal of Social Sciences*, 6(1):29–42, 2020.

[12] Rasha M Abdelrahman. Metacognitive awareness and academic motivation and their impact on academic achievement of ajman university students. *Heliyon*, 6(9), 2020.

[13] Riyan Hidayat, Sharifah Norul Akmar Syed Zamri, Hutkemri Zulnaidi, and Putri Yuanita. Meta-cognitive behaviour and mathematical modelling competency: Mediating effect of performance goals. *Heliyon*, 6(4), 2020.

[14] U Khaira and T Herman. Assessment processes for slow learners in mathematics learning. In *Journal of Physics: Conference Series*, volume 1521, page 032097. IOP Publishing, 2020.

[15] Eshthih Fithriyana and Hidayah Hidayah. Game therapy based on local wisdom in cognitive development of slow learner children. In *Annual Conference on Islamic Early Childhood Education (ACIECE)*, volume 4, pages 129–138, 2019.

[16] A Mary and AUS Jebaseelan. Student learning behavior and academic achievement-unraveling its relationship. *Indian journal of applied research*, 4(12):57–59, 2014.

[17] K Dasaradhi, Ch Sri Raja Rajeswari, Head Mistress, and PVS Badarinath. 30 methods to improve learning capability in slow learners. *International Journal of English Language, Literature and Humanities*, 6(2):556–570, 2016.

[18] Panida Varachanon. Study of learning behaviors of nursing student at the royal thai navy college of nursing. *Procedia-Social and Behavioral Sciences*, 197:1043–1047, 2015.

[19] Vijay Pratap Singh. *Education of the slow learners*, volume 2. Sarup & Sons, 2004.

[20] Stan Maria Magdalena. The relationship of learning styles, learning behaviour and learning outcomes at the romanian students. *Procedia-Social and Behavioral Sciences*, 180:1667–1672, 2015.

[21] Fred Brauer, Carlos Castillo-Chavez, and Carlos Castillo-Chavez. *Mathematical models in population biology and epidemiology*, volume 2. Springer, 2012.

4

Eco-epidemiological Analysis of Plant–Herbivore Interactions: Allee Effect and Time-dependent Dynamics

Dipesh[1], Pankaj Kumar[2], and Rupali Verma[2]

[1]Department of Mathematics, SR University, India
[2]Department of Mathematics, Lovely Professional University, India
E-mail: Dipeshdalal2411@gmail.com; pankaj.kumar1@lpu.co.in; rupaliverma171317@gmail.com

Abstract

The Allee effect plays a significant role in plant–herbivore dynamics, occurring when the fitness of individuals in a sparse population decreases in proportion to population density. To study this phenomenon, an eco epidemiological model is proposed, incorporating a weak Allee effect and a Holling type-I functional response. Herbivores consume only infectious plants, which are generated by the infection of susceptible plants. The stability is examined at non-zero equilibrium point. The system shows asymptotic stability and destabilizes when the time lag is below and above the threshold point. Hopf-bifurcation is observed when the time lag exceeds the threshold point. Sensitivity analysis is conducted with respect to the herbivore gain from consuming infectious plants and the herbivore mortality rate. MATLAB software is used to graphically represent the results of the analysis.

Keywords: Eco-epidemiological model, functional response, Hopf - bifurcation, Allee effect, Holling type-I

4.1 Introduction

The study of infectious disease is known as ecological epidemiology. Eco-epidemiological models are models where the predator-prey population

DOI: 10.1201/9788743806691-4

suffers from an infectious disease [1]. Lotka and Volterra [2, 3] are the first who started to create realistic models mathematically, which include two or more than two interacting species. Initially, the eco-epidemiological model is established by Hadeler and Freedman [4]. Thereafter, many researchers worked on eco-epidemiology models. Different researchers focused on the spread of disease in different populations such as prey population only, predator population only, or both prey and predator populations [5, 6]. There are some epidemic models present in literature where authors supposed that the predators consume infectious prey, some supposed that the predator population does not consume the infectious prey population, and some authors also supposed that the predator population consumes both the susceptible and infectious plants [7–9]. Some researchers incorporated the Allee effect accompanied by infectious disease in the population [10, 11].

Initially, the biologist, Allee [12], defined the term Allee effect, which describes the correspondence between population size and mean individual fitness. The Allee effect occurs in natural populations such as plants [13], birds [14], insects [15], and animals [16]. There are many mechanisms that cause the Allee effect. The key mechanism is mate limitation, which causes the Allee effect in both plant populations and animals. The other mechanisms include environmental mechanisms, inbreeding depression, demographic genetics, etc. [17, 18]. Plants can suffer from the Allee effect if the density of plants is low and sparsely distributed. The Allee effect falls into two categories. The strong Allee effect signifies that the population shrinks when the concentration of the population is low below the Allee threshold. Adversely, the weak Allee effect means the increment of the growth rate with the increment of population size, but there is no threshold value [19, 20]. The large effects of the extinction of species are documented due to the diseases [21]. For instance, if predators consumed infectious prey, then there is negative growth of the predators [1]. When predators eat infectious plants, there is a necessary time lag for the gestation period. Therefore, delay differential equations have a significant role in realistic models [22–24].

The use of the weak Allee effect with infectious disease in plants and time lag is very rare in eco-epidemiological models. Therefore, the purpose of this contribution is to study an eco-epidemiological model accompanied by a weak Allee effect. The susceptible, as well as infectious plants, are incorporated with herbivores. Additionally, delay differential equations are used as there is an essential time lag for the gestation of the infectious plants by herbivores.

4.2 Mathematical Model and Analysis of the System

4.2.1 Assumptions of the model

There are assumptions for the Allee effect, as well as for plants:

- Reproduction is not possible without mates.
- The increment in population density causes a decrement of the Allee effect.
- When population density is large, that is if population density approaches to infinity, then the Allee function approaches to one.
- The predation of susceptible plants by herbivores is ignored in this study.
- The assumption is that the infectious plants are generated by the infection of susceptible plants. Furthermore, infectious plants cannot grow, recover, or reproduce. Therefore, $\gamma > 0$ is the mortality rate of the infectious plant.
- The supposition for the weak infectious plant is that the handling time of herbivores for the infectious plant is zero, so there is a functional response with a linear mass-action βP_{I} where $\beta > 0$ is the attack rate of herbivores.

4.2.2 Mathematical formulation

The plant–herbivore model is given by the following system of delay differential equations:

$$\frac{dP_{\mathrm{S}}}{dt} = P_{\mathrm{S}}\{[1-(P_{\mathrm{S}}+P_{\mathrm{I}})](P_{\mathrm{S}}-b)-\alpha P_{\mathrm{I}}\}, \tag{4.1}$$

$$\frac{dP_{\mathrm{I}}}{dt} = P_{\mathrm{I}}(\alpha P_{\mathrm{S}}-\beta H-\gamma), \tag{4.2}$$

$$\frac{dH}{dt} = \delta H(t-\tau)P_{\mathrm{I}}(t-\tau)-\mu H, \tag{4.3}$$

where $P_{\mathrm{S}}(t) > 0$, $P_{\mathrm{I}}(t) > 0$, $H(t) > 0\ \forall t \geq 0$ and $b \leq 0$ for weak Allee effect and $P(t-\tau) = \varepsilon$, constant for all $t \in [0, \tau]$. Also, all the parameters $b, \alpha, \beta, \gamma, \delta$, and μ are positive constants.

4.3 Non-zero Equilibrium of the Model

There exist eight equilibrium points in the system, each satisfying distinct non-negativity criteria as discussed in [25]. For a more comprehensive analysis, we focus on the non-zero equilibrium point, which characterizes the simultaneous existence of all state variables $(P_\mathrm{S}^*, P_\mathrm{I}^*, H^*)$, where none of them is zero. This is particularly relevant since the stability of the system is unaffected by the delay term at the remaining equilibrium points.

To determine the non-zero equilibrium, we set the right-hand sides of the model eqn (4.1)–(4.3) to zero. At the equilibrium state $E^*(P_\mathrm{S}^*, P_\mathrm{I}^*, H^*)$, we assume that the delay terms approximate their current values, that is

$$P_\mathrm{I}^*(t-\tau) \approx P_\mathrm{I}^*(t), \quad H^*(t-\tau) \approx H^*(t).$$

From the second equation, the equilibrium condition yields:

$$\frac{dP_\mathrm{I}^*}{dt} = 0 \quad \Rightarrow \quad H^* = \frac{\alpha P_\mathrm{S}^* - \gamma}{\beta}, \quad \text{provided that } P_\mathrm{S}^* > \frac{\gamma}{\alpha}.$$

From the third equation:

$$\frac{dH^*}{dt} = 0 \quad \Rightarrow \quad P_\mathrm{I}^* = \frac{\mu}{\delta}, \quad \text{provided that } \mu > 0.$$

Substituting these into the first equation and setting the derivative to zero gives:

$$\frac{dP_\mathrm{S}^*}{dt} = 0 \quad \Rightarrow \quad P_\mathrm{S}^{*2} - \left(1 + b - \frac{\mu}{\delta}\right) P_\mathrm{S}^* + \left[b + (\alpha - b)\frac{\mu}{\delta}\right] = 0.$$

The equilibrium value P_S^* has real and distinct roots if the discriminant of the quadratic equation is positive. This yields the condition:

$$\left(1 + b - \frac{\mu}{\delta}\right)^2 > 4\left[b + (\alpha - b)\frac{\mu}{\delta}\right].$$

4.4 Stability Analysis at Non-zero Equilibrium $E^*(P_S^*, P_I^*, H^*)$

The original system of eqn (4.1)–(4.3) can be reformulated at the non-zero equilibrium point $E^*(P_\mathrm{S}^*, P_\mathrm{I}^*, H^*)$ as follows:

$$\frac{dP_\mathrm{S}^*}{dt} = P_\mathrm{S}^* \left\{\left[1 - (P_\mathrm{S}^* + P_\mathrm{I}^*)\right](P_\mathrm{S}^* - b) - \alpha P_\mathrm{I}^*\right\}, \tag{4.4}$$

$$\frac{dP_\mathrm{I}^*}{dt} = P_\mathrm{I}^*(\alpha P_\mathrm{S}^* - \beta H^* - \gamma), \tag{4.5}$$

$$\frac{dH^*}{dt} = \delta H^*(t-\tau)P_{\mathrm{I}}^*(t-\tau) - \mu H^*. \tag{4.6}$$

The exponential characteristic equation at the equilibrium $E^*(P_{\mathrm{S}}^*, P_{\mathrm{I}}^*, H^*)$ is derived from the Jacobian matrix as

$$\begin{vmatrix} \lambda - m_1 & -m_2 & -m_3 \\ -m_4 & \lambda - m_5 & -m_6 \\ -m_7 & -m_8 & \lambda - m_9 \end{vmatrix} = 0,$$

where the coefficients m_{i} are defined as

$$\begin{aligned}
m_1 &= -3P_{\mathrm{S}}^{*2} + 2(1+b)P_{\mathrm{S}}^* - 2P_{\mathrm{S}}^*P_{\mathrm{I}}^* + (b-\alpha)P_{\mathrm{I}}^* - b, \\
m_2 &= \alpha P_{\mathrm{I}}^*, \quad m_3 = 0, \\
m_4 &= -P_{\mathrm{S}}^{*2} + bP_{\mathrm{S}}^* - \alpha P_{\mathrm{S}}^*, \quad m_5 = \alpha P_{\mathrm{S}}^* - \beta H^* - \gamma, \\
m_6 &= \delta H^* e^{-\lambda\tau}, \quad m_7 = 0, \quad m_8 = -\beta P_{\mathrm{S}}^*, \\
m_9 &= \delta P_{\mathrm{I}}^* e^{-\lambda\tau} - \mu.
\end{aligned}$$

This determinant yields the exponential characteristic equation:

$$\left(\lambda^3 + A_1\lambda^2 + A_2\lambda + A_3\right) + \left(B_1\lambda^2 + B_2\lambda + B_3\right)e^{-\lambda\tau} = 0, \tag{4.7}$$

where the coefficients A_{i} and B_{i} are:

$$\begin{aligned}
A_1 &= -m_1 - m_5 + \mu, \\
A_2 &= m_1m_5 - m_2m_4 - \mu(m_1 + m_5), \\
A_3 &= \mu(m_1m_5 - m_2m_4), \\
B_1 &= -\delta P_{\mathrm{I}}^*, \\
B_2 &= (m_1 + m_5)\delta P_{\mathrm{I}}^* - m_6m_8, \\
B_3 &= (-m_1m_5 + m_2m_4)\delta P_{\mathrm{I}}^* + m_1m_6m_8.
\end{aligned}$$

To analyze stability, assume $\lambda = i\omega$ is a solution of eqn (4.7). Substituting gives:

$$(i\omega)^3 + A_1(i\omega)^2 + A_2(i\omega) + A_3 + \left(B_1(i\omega)^2 + B_2(i\omega) + B_3\right)e^{-i\omega\tau} = 0. \tag{4.8}$$

Separating into real and imaginary parts yields:

$$A_3 - A_1\omega^2 + (B_3 - B_1\omega^2)\cos(\omega\tau) + B_2\omega\sin(\omega\tau) = 0, \tag{4.9}$$

$$A_2\omega - \omega^3 + B_2\omega\cos(\omega\tau) - (B_3 - B_1\omega^2)\sin(\omega\tau) = 0. \tag{4.10}$$

Combining eqn (4.9) and (4.10), we obtain the sixth-degree algebraic equation:

$$\omega^6+(A_1^2-B_1^2-2A_2)\omega^4+(A_2^2-B_2^2+2B_1B_3-2A_1A_3)\omega^2+(A_3^2-B_3^2)=0. \tag{4.11}$$

Let us define:

$$\begin{aligned} c_1 &= A_1^2 - B_1^2 - 2A_2, \\ c_2 &= A_2^2 - B_2^2 + 2B_1B_3 - 2A_1A_3, \\ c_3 &= A_3^2 - B_3^2. \end{aligned}$$

Substituting $\omega^2 = x$, eqn (4.11) becomes a cubic polynomial:

$$x^3 + c_1x^2 + c_2x + c_3 = 0. \tag{4.12}$$

Conjecture 1. If $c_3 < 0$, then eqn (4.12) has at least one positive real root.

Proof. Let

$$l(x) = x^3 + c_1x^2 + c_2x + c_3.$$

Since $l(0) = c_3 < 0$ and $\lim_{x\to\infty} l(x) = \infty$, by the intermediate value theorem, there exists $x_0 \in (0,\infty)$ such that $l(x_0) = 0$.

Conjecture 2. For $c_3 \geq 0$, a necessary condition for eqn (4.12) to possess at least one positive real root is that the discriminant $D = c_1^2 - 3c_2$ satisfies $D \geq 0$.

Proof. Let us consider the cubic function

$$l(x) = x^3 + c_1x^2 + c_2x + c_3.$$

To analyze the turning points and behavior of $l(x)$, we differentiate:

$$l'(x) = \frac{d}{dx}l(x) = 3x^2 + 2c_1x + c_2.$$

To locate critical points, we solve the equation:

$$l'(x) = 0 \quad \Rightarrow \quad 3x^2 + 2c_1x + c_2 = 0. \tag{4.13}$$

The solutions to the quadratic eqn (4.13) are given by:

$$x_{1,2} = \frac{-2c_1 \pm \sqrt{4c_1^2 - 12c_2}}{6} = \frac{-c_1 \pm \sqrt{D}}{3}, \quad \text{where } D = c_1^2 - 3c_2. \tag{4.14}$$

Case 1: If $D < 0$, then eqn (4.13) has no real solution. This implies that $l'(x) > 0$ for all $x \in \mathbb{R}$, meaning that $l(x)$ is a strictly increasing function. Given that $l(0) = c_3 \geq 0$, it follows that:

$$l(x) > 0 \quad \text{for all } x > 0.$$

Hence, eqn (4.12) does not admit any positive real root.

Case 2: If $D \geq 0$, then $l(x)$ has two real critical points. In particular, the point:

$$x_1 = \frac{-c_1 + \sqrt{D}}{3}$$

is a local minimum of $l(x)$. The possibility of a positive root for eqn (4.12) then depends on whether $l(x_1) < 0$. This establishes that $D \geq 0$ is a necessary (but not sufficient) condition for the existence of positive real roots when $c_3 \geq 0$.

Conjecture 3. When $c_3 \geq 0$, eqn (4.12) has positive real zeros if and only if $x_1 > 0$ and $l(x_1) \leq 0$.

Proof. Recall that $l(x) = x^3 + c_1x^2 + c_2x + c_3$ and the local minimum occurs at:

$$x_1 = \frac{-c_1 + \sqrt{D}}{3}, \qquad \text{where } D = c_1^2 - 3c_2.$$

Case 1: Suppose $x_1 \leq 0$ or $x_1 > 0$ and $l(x_1) > 0$.
If $x_1 \leq 0$, then $l(x)$ is increasing for $x \geq x_1$. Since $l(0) = c_3 \geq 0$ and $l(x)$ increases thereafter, we conclude that $l(x) > 0$ for all $x > 0$. Hence, eqn (4.12) has no positive real root.

Case 2: Suppose $x_1 > 0$ and $l(x_1) \leq 0$.
Then $l(x)$ has a local minimum at x_1 and is increasing for $x > x_1$. Since $l(x_1) \leq 0$, and $\lim_{x\to\infty} l(x) = \infty$, there must exist $x_0 > x_1$ such that $l(x_0) = 0$. Thus, eqn (4.12) has at least one positive real root.

Therefore, the condition $x_1 > 0$ and $l(x_1) \leq 0$ is both necessary and sufficient for eqn (4.12) to have a positive real zero.

Lemma 1. Suppose eqn (4.14) defines x_1. Then:

- If $c_3 < 0$, eqn (4.12) has at least one positive real root.
- If $c_3 \geq 0$ and $D = c_1^2 - 3c_2 < 0$, eqn (4.12) has no positive real root.
- If $c_3 \geq 0$ and $D \geq 0$, eqn (4.12) has positive real roots if and only if $x_1 > 0$ and $l(x_1) \leq 0$.

Proof. Assume that eqn (4.12) admits positive real roots. Without loss of generality, let x_1, x_2, x_3 be positive real roots of eqn (4.12). Then eqn (4.11) has three positive roots corresponding to:

$$\omega_1 = \sqrt{x_1}, \quad \omega_2 = \sqrt{x_2}, \quad \omega_3 = \sqrt{x_3}.$$

From eqn (4.10), we have:

$$\sin \omega\tau = \frac{A_2\omega - \omega^3}{d}.$$

Solving for τ, we obtain:

$$\tau = \frac{1}{\omega}\left[\sin^{-1}\left(\frac{A_2\omega - \omega^3}{d}\right) + 2(j-1)\pi\right], \quad j = 1, 2, 3, \ldots$$

Now consider the sequence:

$$\tau_{\mathrm{k}}^{(j)} = \frac{1}{\omega_{\mathrm{k}}}\left[\sin^{-1}\left(\frac{A_2\omega_{\mathrm{k}} - \omega_{\mathrm{k}}^3}{d}\right) + 2(j-1)\pi\right], \; k = 1, 2, 3; \; j = 0, 1, 2, \ldots$$

Therefore, each pair $\tau_{\mathrm{k}}^{(j)}, i\omega_{\mathrm{k}}$ represents a purely imaginary root of eqn (4.11).

Let $\tau_{\mathrm{k}}^{(j)}$ be defined for $k = 1, 2, 3$ and $j \in \mathbb{N}_0$. Observe that:

$$\lim_{j\to\infty} \tau_{\mathrm{k}}^{(j)} = \infty.$$

Define the minimal value of $\tau_{\mathrm{k}}^{(j)}$ across all roots and j as

$$\tau_0 = \tau_{k_0}^{(j_0)} = \min_{\substack{k=1,2,3 \\ j\in\mathbb{Z}_{\geq 0}}} \left[\tau_{\mathrm{k}}^{(j)}\right], \quad \omega_0 = \omega_{k_0}, \quad x_0 = x_{k_0}. \tag{4.15}$$

Lemma 2. Suppose that $A_1 > 0$, $(A_3 + d) > 0$, and $A_1A_2 - (A_3 + d) > 0$.

- If $c_3 \geq 0$ and $D = c_1^2 - 3c_2 < 0$, then every zero of eqn (4.7) has negative real part for all $\tau \geq 0$.
- If $c_3 < 0$ or $c_3 \geq 0$, $x_1 > 0$ and $l(x_1) \leq 0$, then every zero of eqn (4.7) has negative real part for all $\tau \in [0, \tau_0)$.

Proof. If $\tau = 0$, then eqn (4.7) becomes:

$$\lambda^3 + (A_1 + B_1)\lambda^2 + (A_2 + B_2)\lambda + (A_3 + B_3) = 0. \tag{4.16}$$

By Hurwitz's criterion, condition *(H1)* ensures that all roots of eqn (4.7) are located in the left half of the complex plane (i.e., have negative real parts) if and only if:

$$(A_3 + B_3) > 0, \quad (A_1 + B_1)(A_2 + B_2) > (A_3 + B_3).$$

When $c_3 \geq 0$ and $D = c_1^2 - 3c_2 < 0$, Lemma 1(b) implies that eqn (4.7) has no root with zero real part for all $\tau \geq 0$.

If $c_3 < 0$ or $c_3 \geq 0$, $x_1 > 0$ and $l(x_1) \leq 0$ (by Lemma 1(a, c)), then for all $\tau \neq \tau_{\text{k}}^{(j)}$, $k = 1, 2, 3$ and $j \geq 1$, eqn (4.7) does not admit any root with zero real part, and the minimum value of τ for which this occurs is τ_0. Thus, all roots of eqn (4.7) have real parts that are negative for all $\tau \in [0, \tau_0)$.

Now consider the purely imaginary roots of (4.7) given by:

$$\lambda(\tau) = \psi(\tau) + i\omega(\tau). \tag{4.17}$$

Let these roots satisfy:

$$\psi(\tau_0) = 0, \quad \omega(\tau_0) = \omega_0.$$

To prove that $\lambda(\tau_0)$ is a purely imaginary root of eqn (4.7) with $\tau = \tau_0$, we differentiate:

$$l''(x_0) \neq 0.$$

This confirms that the pair $(\tau_0, \lambda(\tau_0))$ satisfies the transversality condition, completing the proof.

Lemma 3. Consider $x_0 = \omega_0^2$. If $\tau = \tau_0$, then

$$\text{Sign}[\psi'(\tau_0)] = \text{Sign}[l''(x_0)].$$

Proof. Substitute τ into eqn (4.7) and differentiate with respect to τ

$$\frac{d\lambda}{d\tau}\left[3\lambda^2 + 2A_1\lambda + A_2 - \tau(B_1\lambda^2 + B_2\lambda + B_3)\right.$$
$$\left. + (2B_1\lambda + B_2)\right] e^{-\lambda\tau} = \lambda(B_1\lambda^2 + B_2\lambda + B_3)e^{-\lambda\tau},$$

Then:

$$\left(\frac{d\lambda}{d\tau}\right)^{-1} = \frac{(3\lambda^2 + 2A_1\lambda + A_2)e^{\lambda\tau}}{\lambda(B_1\lambda^2 + B_2\lambda + B_3)} + \frac{(2B_1\lambda + B_2)}{\lambda(B_1\lambda^2 + B_2\lambda + B_3)} - \frac{\tau}{\lambda}$$

From eqn (4.9)–(4.11), we have:

$$\mu'(\tau_0) = \text{Re}\left[\frac{(3\omega_0^2 + 2A_1\omega_0 + A_2)e^{\lambda\tau}}{\lambda(B_1\lambda^2 + B_2\lambda + B_3)}\right] + \text{Re}\left[\frac{(2B_1\lambda + B_2)}{\lambda(B_1\lambda^2 + B_2\lambda + B_3)}\right] - \frac{1}{\Delta}\left[3\omega_0^6 + 2c_1\omega_0^4 + c_2\omega_0^2\right]$$

Here,

$$\Delta = \left[(B_3 - B_1\omega_0^2)^2 + (B_2\omega_0)^2\right], \quad \Delta > 0, \quad \omega_0 > 0.$$

Therefore,

$$\text{Sign}[\psi'(\tau_0)] = \text{Sign}[l''(x_0)].$$

4.5 Sensitivity Analysis of State Variables with Respect to Model Parameters

Sensitivity analysis represents the impact on the stability behaviour of the state variables when we vary the parameter values at variance with the time lag that is τ. For instance, the partial derivatives of the system (P_S, P_I, H) with respect to δ (herbivores gain by consuming infectious plants) lead to the following set of sensitivity equations:

$$\frac{dS_1}{dt} = \left[-3P_S^2 + 2(b+1)P_S + (b-\alpha)P_I - 2P_SP_I\right]S_1 + \left[-P_S^2 + (b-\alpha)P_S\right]S_2. \tag{4.18}$$

$$\frac{dS_2}{dt} = \alpha P_I S_1 + (\alpha P_S - \beta H - \gamma)S_2 - \beta P_I S_3. \tag{4.19}$$

$$\frac{dS_3}{dt} = -\mu S_3 + \delta H(t-\tau)S_2(t-\tau) + \delta P_I(t-\tau)S_3(t-\tau). \tag{4.20}$$

where,

$$S_1 = \frac{\partial P_S}{\partial \delta}, \quad S_2 = \frac{\partial P_I}{\partial \delta}, \quad S_3 = \frac{\partial H}{\partial \delta}.$$

4.6 Numerical Results and Discussion

To investigate the dynamic behavior and stability of the eco-epidemiological model, we incorporate the following parameter values:

$$\alpha = 0.5, \quad \beta = 0.1, \quad \gamma = 0.05, \quad \delta = 0.55, \quad \mu = 0.1, \quad b = -5$$

4.6.1 Stability transitions under varying time delay

Figure 4.1 illustrates the time evolution of the system variables: susceptible plants, infected plants, and herbivores when the time delay $\tau = 0$. The solution trajectory converges to the non-zero equilibrium point $E^* = (0.8013, 0.1818, 3.5127)$, indicating that the system is absolutely stable in the absence of delay. All populations reach steady states without oscillations. In Figure 4.2, we observe the dynamics when the time delay τ is less than the critical threshold (here, $\tau < 10.6$). The system converges asymptotically to the equilibrium $E^* = (0.7888, 0.1949, 3.4142)$. While transient oscillations are visible initially, they gradually die out, demonstrating asymptotic stability. This phase space diagram, as illustrated in Figure 4.3, confirms the result of Figure 4.2. The trajectory spirals inward toward the equilibrium point in the 3D state space $(P_\text{S}, P_\text{I}, H)$, further validating the asymptotic stability of the system when $\tau < 10.6$. Figure 4.4 depicts the time series evolution of the system variables when the time delay exceeds the critical value (i.e., $\tau \geq 10.6$). The solution no longer converges to a steady state but exhibits sustained oscillations around the equilibrium $E^* = (0.5014, 0.4673, 1.4156)$. This behavior is indicative of a Hopf bifurcation, where the system loses its stability and transitions into a limit cycle. The 3D phase portrait shown in Figure 4.5 reveals a closed trajectory, characteristic of a stable limit cycle.

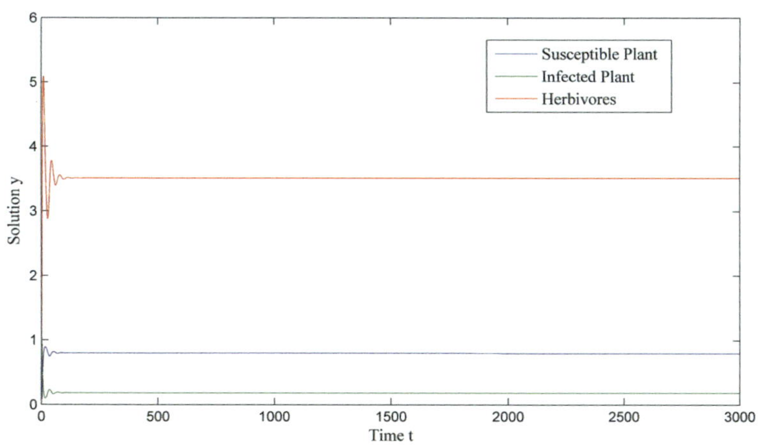

Figure 4.1 $E^*(0.8013, 0.1818, 3.5127)$ is absolutely stable when there is no time lag, that is when $\tau = 0$.

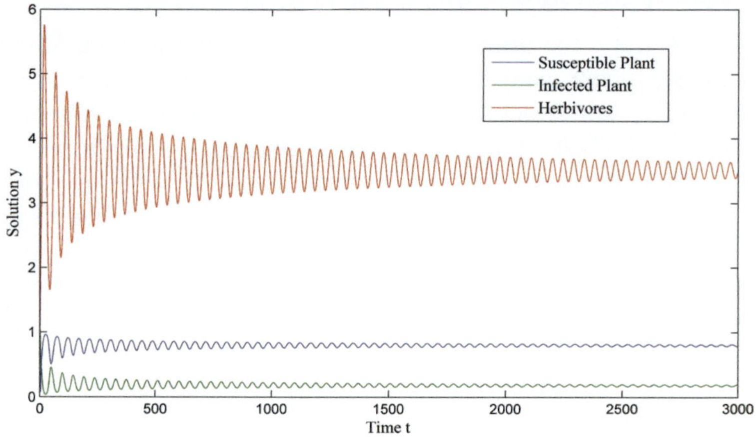

Figure 4.2 $E^*(0.7888,\ 0.1949,\ 3.4142)$ is asymptotically stable when the time lag is less than the threshold value, that is $\tau < 10.6$.

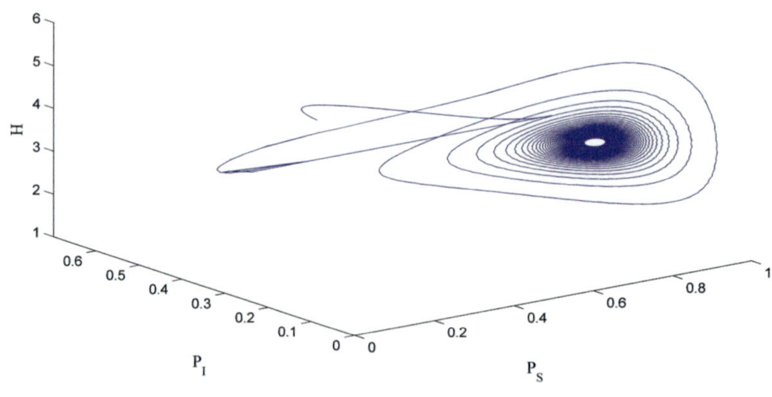

Figure 4.3 Phase space diagram showing the asymptotic stability of $E^*(0.7888,\ 0.1949,\ 3.4142)$ when the time lag is less than the threshold value, that is $\tau < 10.6$.

This further confirms that the Hopf bifurcation occurs as the time delay crosses the threshold, and the system exhibits periodic behavior instead of converging to a fixed point.

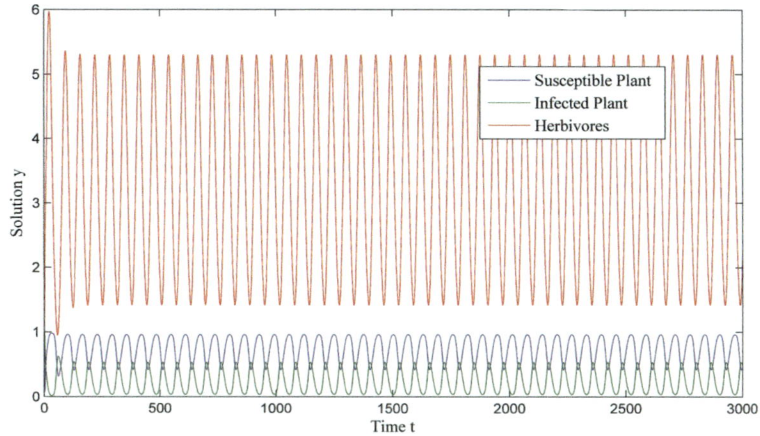

Figure 4.4 $E^*(0.5014,\ 0.4673,\ 1.4156)$ loses stability, and a Hopf bifurcation is observed when the time lag surpasses the threshold point, that is $\tau \geq 10.6$.

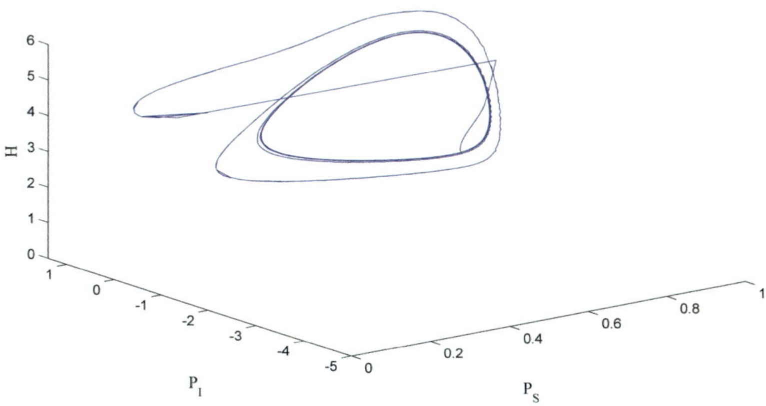

Figure 4.5 Phase space diagram showing the Hopf bifurcation of $E^*(0.5014,\ 0.4673,\ 1.4156)$ when the time lag crosses the threshold value, that is $\tau \geq 10.6$.

4.6.2 Sensitivity analysis with respect to parameter δ

Figures 4.6–4.8 illustrate the sensitivity of the system's state variables, including susceptible plants (P_{S}), infectious plants (P_{I}), and herbivores

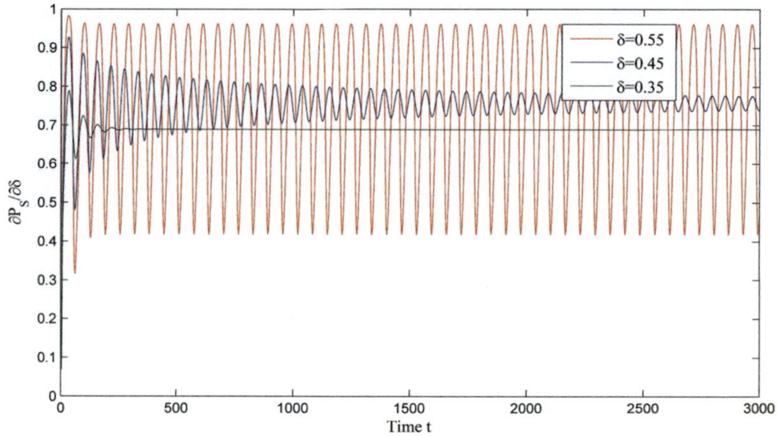

Figure 4.6 Time series graph of change in the density of susceptible plant P_S with respect to variation in the herbivore gain from consuming infectious plants (δ).

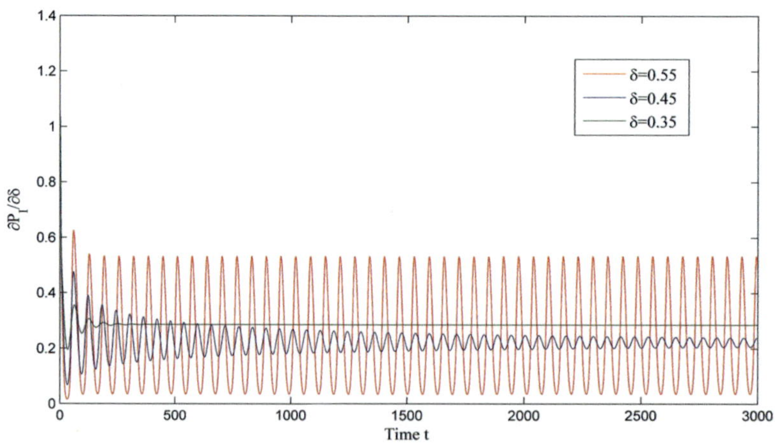

Figure 4.7 Time series graph of change in the density of infectious plant P_I with respect to variation in the herbivores' gain by consuming infectious plants (δ).

(H), with respect to the herbivores' gain from consuming infectious plants, denoted by the parameter δ.

Figure 4.6 shows the sensitivity of P_S to variations in δ. As δ decreases from 0.55 to 0.45 and further to 0.35, the oscillations in $\partial P_S/\partial\delta$ diminish

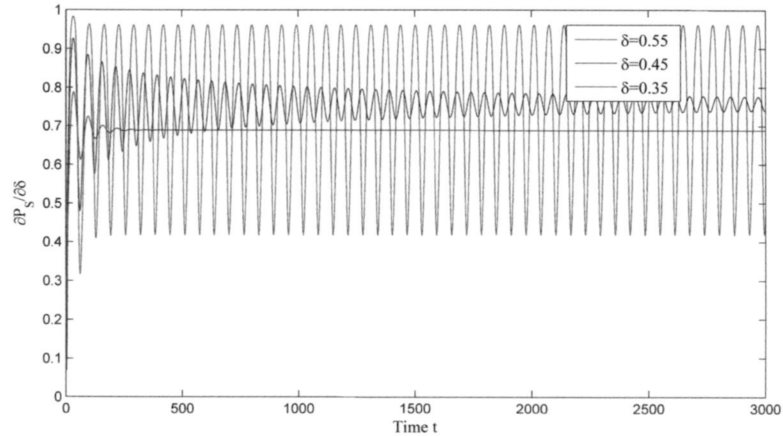

Figure 4.8 Time series graph of change in density of herbivores H with respect to change in the herbivores gain by consuming infectious plant (δ).

significantly. This behavior indicates that the system transitions from an unstable regime exhibiting oscillations (Hopf bifurcation) to a more stable regime, eventually becoming absolutely stable as δ becomes sufficiently small.

Figure 4.7 displays the sensitivity of the infected plant population P_I with respect to δ. A similar trend is observed: for $\delta = 0.55$, the system exhibits pronounced oscillations, indicative of limit cycle behavior. As δ is reduced, these oscillations decrease in amplitude and eventually fade out, reflecting a transition to asymptotic and then absolute stability.

Figure 4.8 portrays the sensitivity of the herbivore population H with respect to δ. The pattern is consistent with the other state variables. High δ values lead to large oscillations in $\partial H/\partial \delta$, which gradually subside as δ is decreased.

These results demonstrate a clear progression in system behavior:

- At higher δ (e.g., $\delta = 0.55$), the system undergoes a Hopf bifurcation, resulting in sustained oscillations.
- When δ is moderately reduced (e.g., $\delta = 0.45$), the system becomes asymptotically stable, with oscillations that gradually diminish over time.
- At lower values (e.g., $\delta = 0.35$), the system achieves absolute stability, where all perturbations decay without oscillation.

This sensitivity analysis confirms that δ plays a critical role in the dynamics of the eco-epidemiological system. As herbivores' gain from consuming infectious plants decreases, the system transitions from instability through a Hopf bifurcation to a stable equilibrium. Biologically, this implies that while high herbivore gain initially stimulates growth and reproduction, it may later lead to intense competition and plant decline, thereby destabilizing the system. Reducing δ limits this overexploitation, allowing the ecosystem to stabilize.

4.6.3 Sensitivity Analysis with Respect to Mortality Rate μ

Figures 4.9–4.11 demonstrate the sensitivity of the system's state variables, including susceptible plants (P_S), infectious plants (P_I), and herbivores (H), to changes in the mortality rate of herbivores (μ). Figure 4.9 illustrates the variation in $\partial P_S/\partial\mu$, showing how the density of susceptible plants responds to different values of μ. As μ increases from 0.1 to 0.13 and then to 0.17, the oscillations in sensitivity decrease significantly. This pattern indicates a transition from oscillatory (Hopf bifurcation) behavior to asymptotic and ultimately absolute stability. Figure 4.10 shows the sensitivity of the infected plant population P_I with respect to μ. Initially, for lower mortality ($\mu = 0.1$), the system exhibits strong oscillations. As μ increases, these oscillations diminish and the solution stabilizes, reflecting a shift to a more stable regime.

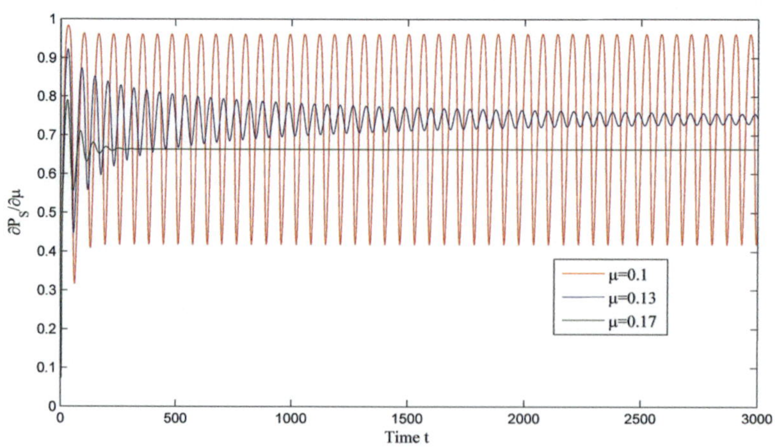

Figure 4.9 Time series graph of change in the density of susceptible plant P_S with respect to variation in the mortality rate of herbivores (μ).

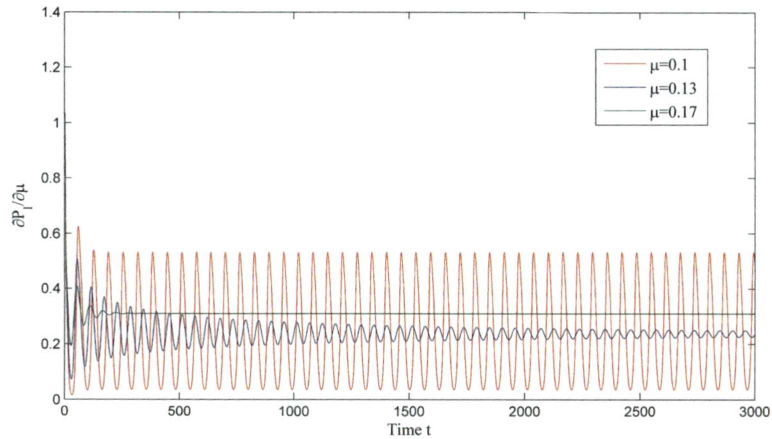

Figure 4.10 Time series graph of change in the density of infectious plant P_I with respect to variation in the mortality rate of herbivores (μ).

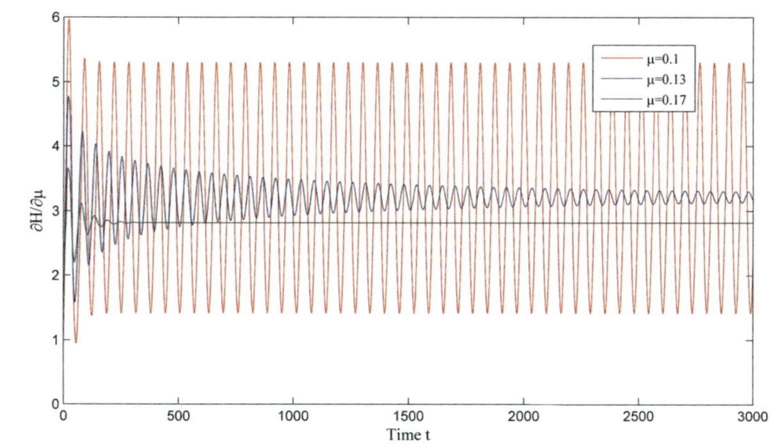

Figure 4.11 Time series graph of change in the density of infectious plant P_I with respect to variation in the mortality rate of herbivores (μ).

Figure 4.11 captures the sensitivity of the herbivore population H with respect to μ. The system's behavior again follows a similar trend: increasing μ leads to reduced oscillatory response and convergence to a steady state, confirming the stabilizing effect of increased herbivore mortality.

Overall, these results indicate that an increase in the mortality rate of herbivores (μ) leads to a change in system dynamics from Hopf bifurcation to asymptotic stability, and finally to absolute stability. Biologically, this implies that as herbivores consuming infectious plants experience higher mortality, the population of infectious plants declines, leading to a rise in susceptible plants. Consequently, the system as a whole transitions to a stable equilibrium.

4.7 Conclusion

The eco-epidemiological model presented in this study incorporates three interacting populations: susceptible plants, infectious plants, and herbivores. The analysis considers the influence of a weak Allee effect and adopts a Holling type-I functional response to describe the interaction dynamics.

A key assumption of the model is that susceptible plants can become infectious. The system is governed by delay differential equations to account for the time-dependent nature of biological processes such as gestation, which rely on historical states. The analysis focuses on the stability of the system's non-zero equilibrium points. It is observed that:

- The system is absolutely stable when there is no time delay (i.e., $\tau = 0$).
- When the delay τ is less than the critical threshold ($\tau < 10.6$), the system remains asymptotically stable.
- However, when the delay exceeds the threshold ($\tau \geq 10.6$), the system loses stability and undergoes a Hopf bifurcation.

To support these findings, phase space diagrams illustrating the asymptotic stability and Hopf bifurcation are provided. Additionally, time series plots representing the dynamic behavior of the system are included to visualize the system's evolution under different time delays.

References

[1] Yongxue Chen and Yong Jiang. An eco-epidemiological model with infectious disease in food chain. *International Journal of Bifurcation and Chaos*, 21(07):1935–1952, 2011.

[2] Alfred James Lotka. *Elements of physical biology*. Williams & Wilkins, 1925.

[3] Vito Volterra. Variations and fluctuations of the number of individuals in animal species living together. *ICES Journal of Marine Science*, 3(1):3–51, 1928.

[4] KP Hadeler and HI Freedman. Predator-prey populations with parasitic infection. *Journal of mathematical biology*, 27(6):609–631, 1989.

[5] Mainul Haque and Ezio Venturino. An ecoepidemiological model with disease in predator: the ratio-dependent case. *Mathematical methods in the Applied Sciences*, 30(14):1791–1809, 2007.

[6] Chandrali Baishya, Sindhu J Achar, P Veeresha, and DG Prakasha. Dynamics of a fractional epidemiological model with disease infection in both the populations. *Chaos: An Interdisciplinary Journal of Nonlinear Science*, 31(4), 2021.

[7] Mainul Haque. A predator–prey model with disease in the predator species only. *Nonlinear Analysis: Real World Applications*, 11(4):2224–2236, 2010.

[8] Shuvojit Mondal, Abhijit Lahiri, and Nandadulal Bairagi. Analysis of a fractional order eco-epidemiological model with prey infection and type 2 functional response. *Mathematical Methods in the Applied Sciences*, 40(18):6776–6789, 2017.

[9] Yanni Xiao and Lansun Chen. Modeling and analysis of a predator–prey model with disease in the prey. *Mathematical biosciences*, 171(1):59–82, 2001.

[10] Horst R Thieme, Thanate Dhirasakdanon, Zhun Han, and Roy Trevino. Species decline and extinction: synergy of infectious disease and allee effect? *Journal of Biological Dynamics*, 3(2-3):305–323, 2009.

[11] Abdul-Aziz Yakubu. Allee effects in a discrete-time sis epidemic model with infected newborns. *Journal of Difference Equations and Applications*, 13(4):341–356, 2007.

[12] Warder Clyde Allee. Animal aggregations. *The Quarterly Review of Biology*, 2(3):367–398, 1927.

[13] Pankaj Kumar and Rupali Verma. Impact of delay parameter on plant–herbivore dynamics under allee effect. In *Journal of Physics: Conference Series*, volume 2267, page 012018. IOP Publishing, 2022.

[14] F Courchamp, BT Grenfell, and TH Clutton-Brock. Impact of natural enemies on obligately cooperative breeders. *Oikos*, 91(2):311–322, 2000.

[15] Mikko Kuussaari, Ilik Saccheri, Mark Camara, and Ilkka Hanski. Allee effect and population dynamics in the glanville fritillary butterfly. *Oikos*, pages 384–392, 1998.

[16] Brian A Lerch, Ben C Nolting, and Karen C Abbott. Why are demographic allee effects so rarely seen in social animals? *Journal of Animal Ecology*, 87(6):1547–1559, 2018.

[17] Philip A Stephens and William J Sutherland. Consequences of the allee effect for behaviour, ecology and conservation. *Trends in ecology & evolution*, 14(10):401–405, 1999.

[18] Philip A Stephens, William J Sutherland, and Robert P Freckleton. What is the allee effect? *Oikos*, pages 185–190, 1999.

[19] Mei-Hui Wang and Mark Kot. Speeds of invasion in a model with strong or weak allee effects. *Mathematical biosciences*, 171(1):83–97, 2001.

[20] Saber N Elaydi and Robert J Sacker. Population models with allee effect: a new model. *Journal of Biological Dynamics*, 4(4):397–408, 2010.

[21] Brian Dennis. Allee effects: population growth, critical density, and the chance of extinction. *Natural Resource Modeling*, 3(4):481–538, 1989.

[22] Preety Kalra and Pankaj Kumar. Role of delay in plant growth dynamics: A two compartment mathematical model. In *AIP conference proceedings*, volume 1860. AIP Publishing, 2017.

[23] Pankaj Kumar et al. Delay differential equation model of forest biomass and competition between wood-based industries and synthetic-based industries. *Mathematical Methods in the Applied Sciences*, 46(9), 2023.

[24] Pankaj Kumar et al. Investigating the impact of toxicity on plant growth dynamics through the zero of a fifth-degree exponential polynomial: A mathematical model using dde. *Chaos, Solitons & Fractals*, 171:113457, 2023.

[25] Richard R Vance. Predation and resource partitioning in one predator–two prey model communities. *The American Naturalist*, 112(987):797–813, 1978.

5

Impacts of Pesticides and External Interventions on Crop Production: A Modeling Approach

Manisha Yadav [1], Pradeep Malik[1], Bhawana Yadav[1], and Salman Saud Alsaeed[2]

[1]Department of Mathematics, Faculty of Applied and Basic Sciences, SGT University, India
[2]Department of Mathematics, College of Science, Jouf University, Saudi Arabia
E-mail: yadavmanisha98@gmail.com; pradeepmaths@gmail.com; bhawanaysn089@gmail.com; ssalsaeed@ju.edu.sa

Abstract

This study develops a dynamic mathematical framework to analyze the impact of pest control measures and external environmental influences on agricultural output. The proposed formulation assumes that crop growth follows a logistic pattern but is hindered by pest populations that lower yield. To counteract these effects and enhance productivity, pest control strategies and external interventions are incorporated, with the latter affecting both the crop's growth rate and its maximum sustainable capacity. The investigation focuses on identifying the equilibrium states of the system and assessing their stability using the Routh–Hurwitz stability criterion, ensuring predictable behavior under varying conditions. The basic reproduction number (R_0) is derived to assess the potential for insect population persistence and the effectiveness of control measures. Numerical simulations are conducted in MATLAB to validate our analytical findings and examine the system's behavior over time under different parameter values. Results indicate that insecticides can significantly reduce pest populations and improve yields in the short term. However, excessive reliance on chemical control may

DOI: 10.1201/9788743806691-5

lead to pest resistance, destabilizing the ecosystem and reducing long-term productivity.

Keywords: Pest management, fractional-order differential equation, Hopf bifurcation and Routh–Hurwitz stability criterion

5.1 Introduction

Agriculture plays a fundamental role in ensuring food security and economic stability, but is increasingly challenged by pest infestations, environmental shifts, and the overuse of chemical treatments [1, 2]. Maintaining a balance between high productivity and ecological sustainability is essential for long-term success.

This chapter develops a nonlinear mathematical model to explore crop–pest interactions, insecticide applications, and external interventions. While insecticides are effective in pest control, their excessive use can degrade soil quality, disrupt biodiversity, and contribute to pest resistance, ultimately increasing production costs [3, 4]. Additionally, climate change alters ecological conditions, making pest management more complex [5].

As a sustainable alternative, intercropping enhances soil fertility and supports beneficial insect populations, reducing pest pressure naturally [6]. Mathematical tools such as fractional-order differential equations and stability analysis help examine pest control strategies and intervention effectiveness [7–12]. The next-generation matrix approach, widely used in epidemiological modeling, has been adapted for analyzing pest spread and control measures [10]. Recent research underscores the value of fractional-order models in optimizing sustainable agricultural practices [13–20]. Mathematical frameworks designed initially for epidemiological and social systems have also been applied to pest dynamics and resource management, providing new insights into sustainable farming strategies [21–25].

By combining advanced mathematical modeling with ecological principles, this chapter emphasizes innovative pest management techniques that enhance agricultural productivity while maintaining environmental balance.

5.2 Preliminaries

Fractional-order derivatives are a unique class of derivatives that represent the rate of change of a function, allowing the analysis of natural phenomena that integer-order derivatives cannot sufficiently capture. Certain properties of fractional-order derivatives are defined based on the domain of the power function. This part presents a few key concepts and conclusions from the fractional operators.

This part presents a few key concepts and conclusions from the fractional operators.

Definition 1 [20, 22, 25], the *Liouville–Caputo fractional derivative operator* of order α ($\alpha \geq 0$) and $n \in \mathbb{N} \cup \{0\}$ is presented as

$$\begin{aligned} D_{\mathrm{m}}^{\alpha} f(m) &= T_{\mathrm{m}}^{n-\alpha} - D_{\mathrm{m}}^{n} f(m) \\ &= \frac{1}{\Gamma(n-\alpha)} \int_0^m (m-t)^{n-\alpha-1} f^n(t)\, dt, \quad n \in \mathbb{Z}^+ \text{ and } m > 0, \end{aligned}$$

where $\Gamma(\cdot)$ is a Gamma function. For example, let $f(m) = m^l$, then we can get:

$$\begin{aligned} D_{\mathrm{m}}^{\alpha} m^l &= 0, \quad l < \alpha, \\ D_{\mathrm{m}}^{\alpha} m^l &= \frac{\Gamma(l+1)}{\Gamma(l-\alpha+1)} m^{l-\alpha}, \quad l \geq \alpha. \end{aligned}$$

Definition 2 [20, 22, 25], suppose that f is a function (may or may not be differentiable), and ω is a real number such that $\omega > 0$. Then, the *Caputo derivative* of the order ω with power law is defined as

$$D_{\mathrm{m}}^{\alpha} = \frac{1}{\Gamma(\alpha)} \int_0^m (m-t)^{\alpha-1} f(t)\, dt.$$

As the Caputo approach is more suited and effective for addressing physical issues, authors have primarily compared Caputo-type fractional derivatives to the RL derivative [20, 25].

5.3 Construction of the Model

Figure 5.1 illustrates the complex ecological relationships among pests, herbivores, predators, soil organisms, and agrochemicals (such as insecticides, herbicides, and fungicides) and their impact on crop yield and health. Insecticides are intended to regulate herbivores and predators; however, their use can lead to pest resurgence by diminishing predator populations. Herbicides primarily target weeds but may also affect algae and contribute to water contamination. Fungicides help control pathogens but can inadvertently harm

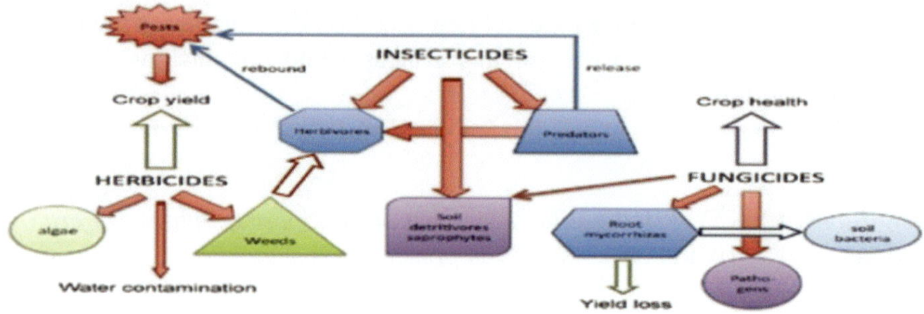

Figure 5.1 Diagram depicting the primary effects of pesticide use on soil quality, plant growth, and arthropod communities.

beneficial root mycorrhizas and soil bacteria, potentially reducing crop yield. The diagram highlights both the direct and indirect effects of these chemicals on agricultural ecosystems, emphasizing the importance of a balanced strategy for sustainable crop production.

This study employs a mathematical model to quantify these interactions through differential equations that describe population dynamics, chemical impacts, and crop yield responses. The figure represents the relationships among pests, herbivores, predators, soil organisms, and agrochemicals (insecticides, herbicides, and fungicides), illustrating their influence on crop yield, crop health, and environmental contamination. It also highlights feedback mechanisms such as pest resurgence, predator suppression, and the unintended effects of chemical applications on soil ecosystems.

The total population $N(t)$ is represented by four mutually exclusive compartments, including $A(t)$ denotes crop biomass, $S(t)$ denotes soil nutrient concentration, $P(t)$ denotes the pest population, and $F(t)$ denotes applied external efforts. Thus, it follows:

$$N(t) = A(t) + S(t) + P(t) + F(t). \tag{5.1}$$

All the variables and assumed parameters involved throughout the study are considered non-negative:

$$\begin{cases} \frac{d^\rho A}{dt^\rho} = rA\left(1 - \frac{A}{K}\right) - \alpha_1 PA^2 + \beta FA + \beta_1 FA^2, \\ \frac{d^\rho S}{dt^\rho} = \theta\alpha AS - \delta S^2 - \lambda\varphi_1 PS, \\ \frac{d^\rho P}{dt^\rho} = \varphi S - \varphi_0 P - \varphi_1 PS, \\ \frac{d^\rho F}{dt^\rho} = \mu(K - A) - \mu_0 F, \end{cases}$$

with initial conditions $A(0) = A_0 > 0$, $S(0) = S_0 > 0$, $P(0) = P_0 > 0$, and $F(0) = F_0 > 0$.

The study incorporates key parameters to model crop–insect interactions and control strategies. The intrinsic growth rate r and carrying capacity K define crop dynamics during insect consumption rate α, conversion efficiency θ, and intra-specific competition mortality δ influence pest populations.

Control measures include insecticide application and external interventions. The spraying rate influences the effectiveness of insecticides φ, natural depletion φ_0, the amount absorbed by insects φ_1, and insect mortality due to insecticides λ determine pesticide effectiveness. External efforts are governed by their application rate μ and depletion μ_0.

The model also considers the negative impact of insecticides on crops α_0 and the beneficial effects of external efforts through growth rate enhancement β and yield improvement β_1. These parameters help evaluate pest management strategies and their effects on agricultural productivity.

5.4 Model Analysis

5.4.1 Lemma (generalized mean value theorem)

Suppose $g(t) \in R[a,b]$ and the Liouville–Caputo fractional operator $D_t^{\rho} g(t) \in R(a,b]$ for $0 < \rho \leq 1$. Then, for $0 \leq t \leq \tau$, for all $t \in (a,b]$, we obtain:

$$g(t) = g(b) + \frac{1}{\Gamma(\rho)} D_t^{\rho} g(t)(t-b)^{\rho}.$$

5.4.2 Remark

If $g(t) \in R[0,b]$ and the Liouville–Caputo fractional operator $D_t^{\rho} g(t) \in R(a,b]$ for $0 < \rho \leq 1$, then by the above Lemma, if $D_t^{\rho} g(t) \leq 0$ for all $t \in (0,b]$, then $g(t)$ is a decreasing function. Conversely, if $D_t^{\rho} g(t) \geq 0$ for all $t \in (0,b]$, then $g(t)$ is an increasing function.

5.4.3 Theorem

The biologically feasible region of the system is defined in $\mathbb{R}_+^4$ as

$$Y = (A(t), S(t), P(t), F(t)) \in \mathbb{R}_+^4 : 0 \leq A + S + P + F \leq \frac{\mu}{\mu_0} K.$$

Here, the conditions for existence, uniqueness and boundedness are satisfied, ensuring that all solutions remain within the feasible domain Y.

Proof. The uniqueness and existence of solutions are examined by demonstrating the positive invariance of the set Y:

$$\left.\frac{d^\rho A}{dt^\rho}\right|_{A=0} = 0 \geq 0$$
$$\left.\frac{d^\rho S}{dt^\rho}\right|_{S=0} = 0 \geq 0$$
$$\left.\frac{d^\rho P}{dt^\rho}\right|_{P=0} = \Phi S \geq 0$$
$$\left.\frac{d^\rho F}{dt^\rho}\right|_{F=0} = \mu(K - A) \geq 0.$$

Along each hyperplane that bounds the non-negative orthant in $\mathbb{R}^4_+$, the vector field directs inward. Given the condition $N = A + S + P + F$, it follows that:

$$\frac{d^\rho N}{dt^\rho} = \frac{d^\rho A}{dt^\rho} + \frac{d^\rho S}{dt^\rho} + \frac{d^\rho P}{dt^\rho} + \frac{d^\rho F}{dt^\rho}.$$

Then, we have:

$$\frac{d^\rho N}{dt^\rho} = \mu K - \mu_0 N.$$

By using the Laplace transform in the above equation, we get:

$$N(t) = \left(-\frac{\mu}{\mu_0}k + N(0)\right)\mathcal{E}_{\mathrm{rm}\rho}(-\mu_0 t^\rho) + \frac{\mu}{\mu_0}k,$$

where $\mathcal{E}_\rho(-\mu_0 t^\rho)$ represents the *Mittag-Leffler* function. Since $0 \leq \mathcal{E}_\rho(-\mu_0 t^\rho) \leq 1$, if $N(0) \leq \frac{\mu}{\mu_0}k$, then $N(t) \leq \frac{\mu}{\mu_0}k$, ensuring that the closed set Y remains positively invariant under the system. The compartment population can be expressed in normalized form by redefining N in terms of the current state variables:

$$u = \frac{A}{N}, \quad v = \frac{S}{N}, \quad w = \frac{P}{N}, \quad z = \frac{F}{N}.$$

The population can be normalized in non-dimensional form as

$$\frac{\partial^\rho u}{\partial t^\rho} = ru\left(1 - \frac{u}{k}\right) - \alpha uv - \alpha_1 uw^2 + \beta zu + \beta_1 zu^2,$$
$$\frac{\partial^\rho v}{\partial t^\rho} = \theta\alpha uv - \delta v^2 - \lambda\varphi_1 wv,$$

$$\frac{\partial^{\rho} w}{\partial t^{\rho}} = \varphi v - \varphi_0 w - \varphi_1 w v,$$

$$\frac{\partial^{\rho} z}{\partial t^{\rho}} = \mu(k-u) - \mu_0 z^4,$$

with initial conditions $u(0) = u_0 \geq 0$, $v(0) = v_0 \geq 0$, $w(0) = w_0 \geq 0$, $z(0) = z_0$.

5.5 Equilibria Points

A point $x = x^*$ of a dynamical system $m\frac{dx}{dt} = f(x)$ is called an equilibrium point if $f(x^*) = 0$. There are two equilibria in the Caputo fractional model. The infected-free equilibrium K_0, in which insecticides are eradicated from the crop, is given as follows:

$$K_0 = \left(\frac{\mu}{\mu_0}K, 0, 0, 0\right).$$

And, there is another type of equilibrium known as an *endemic equilibrium* $K^* = (A^*, S^*, P^*, F^*)$ where,

$$A^* = \frac{\delta + \lambda\varphi_1 E^*}{\theta\alpha},$$

$$S^* = \frac{\varphi_0}{\varphi - \varphi_1},$$

$$P^* = \frac{\varphi E^* - \varphi_0}{\varphi_1},$$

$$F^* = \frac{\mu K - \mu E^*}{\mu_0}.$$

5.5.1 Calculating basic reproduction number (BRN)

The most crucial element in identifying the kind of effect is the basic reproduction number, R_0. Actually, during the whole time period, R_0 is defined as the total number of individuals infected as a result of one sick person. Thus, the chance of a disease spreading is measured by R_0, a non-dimensional variable. Consequently, the probability of the disease spreading is represented by the dimensionless number R_0. There are multiple methods for figuring out R_0 in terms of epidemic spread. We apply the next-generation matrix technique in our study work.

The system equations can be rewritten as

$$\frac{dy}{dx} = F_{\rm i}(X) - \lambda_{\rm i}(Y) \quad \text{where } i = 1, 2, 3, 4 \quad \text{and} \quad X = \begin{bmatrix} A \\ S \\ P \\ F \end{bmatrix}.$$

Here, $F_{\rm i}(X)$ indicates the rate at which insects affect the population arising in the compartment $i^{\rm th}$, $\lambda_{\rm i}(Y)$ indicates the insects spread from the compartment $i^{\rm th}$ to other affected compartments, and shedding is taken into account.

The Jacobian matrices of

$$F_{\rm i}(X) = \begin{bmatrix} -\alpha AS \\ 0 \\ 0 \\ 0 \end{bmatrix}, \quad \lambda_{\rm i}(Y) = \begin{bmatrix} 0 \\ 0 \\ 0 \\ 0 \end{bmatrix} \quad \text{at DFE Point } K_1 = (K, 0, 0, 0)$$

are given by

$$M = \left(\frac{\partial F_{\rm i}}{\partial X_{\rm k}}\right)_{K_1} = \begin{bmatrix} 0 & \alpha k & \alpha k \\ 0 & 0 & 0 \\ 0 & 0 & 0 \end{bmatrix}, \tag{5.2}$$

and

$$L = \left(\frac{\partial \lambda_{\rm i}}{\partial Y_{\rm k}}\right)_{K_1} = \begin{bmatrix} 0 & \alpha k & \alpha k \\ 0 & 0 & 0 \\ 0 & 0 & 0 \end{bmatrix}$$

provide an estimate of the number of secondary infections that will occur $i^{\rm th}$ compartment as a result of an infected object. $R_0 = r_1 + r_2 + r_3$, where r_1 represent susceptible to infected routes and r_2 and r_3 represent exposure to the environment and infect crop production.

5.5.2 Biological significance

This work's two primary concerns are the shedding parameters a, b and their indirect effects on the susceptible population γ. As a result, a threshold value of the parameters a, b, and γ is determined, resulting in a reduction in the epidemic size.

5.5.3 Shedding parameters

From the expression of R_0,

$$\frac{\partial R_0}{\partial a} = \frac{\mu\gamma}{\beta z(\beta + \kappa + \epsilon\rho)}, \tag{5.3}$$

$$\frac{\partial R_0}{\partial b} = \frac{\mu\gamma\epsilon\rho}{\beta z(\pi + \eta + \beta + \sigma)(\beta + \kappa + \epsilon\rho)}. \tag{5.4}$$

Thus, R_0 increases with the shedding parameters, a and b. The critical values of a, b corresponding to $R_0 = 1$ are given by

$$a^* = \frac{\beta z(\pi + \eta + \beta + \sigma)(\beta + \kappa + \epsilon\rho) - \mu\epsilon\rho(\delta z + \gamma b)}{\mu\gamma(\pi + \eta + \beta + \sigma)}, \tag{5.5}$$

$$b^* = \frac{\beta z(\pi + \eta + \beta + \sigma)(\beta + \kappa + \epsilon\rho) - \mu\epsilon\rho(\delta z + \gamma a(\pi + \eta + \beta + \sigma))}{\mu\gamma\epsilon\rho}. \tag{5.6}$$

If $a < a^*$, $b < b^*$ then $R_0 < 1$ and $R_0 > 1$ for $a > a^*$, $b > b^*$. Therefore, reducing the shedding parameters below their critical values can help mitigate the impact of the pandemic.

5.5.4 Rate of pathogen transmission from the environment via shedding

The basic reproduction number R_0 grows as the parameter γ increases, given by

$$\frac{\partial R_0}{\partial \gamma} = \frac{\mu(\epsilon\rho b + a(\pi + \eta + \beta + \sigma))}{\beta z(\pi + \eta + \beta + \sigma)(\beta + \kappa + \epsilon\rho)}. \tag{5.7}$$

Further, the threshold value of γ is:

$$\gamma^* = \frac{\beta z(\pi + \eta + \beta + \sigma)(\beta + \kappa + \epsilon\rho) - \mu\epsilon\rho\delta z}{\mu(\epsilon\rho b + a(\pi + \eta + \beta + \sigma))}. \tag{5.8}$$

Hence, to ensure that R_0 does not exceed 1, γ must remain below γ^*; otherwise, the disease will continue to spread.

5.6 Local Stability of DFE (Disease-free-equilibrium)

The Jacobian matrix J corresponding to system (1) is derived as follows:

$$J = \begin{bmatrix} a_{11} & -\alpha A & -\alpha_1 A^2 & \beta A + \beta_1 A^2 \\ \theta\alpha S & \theta\alpha A - 2\delta S & -\lambda\varphi_1 P - \lambda\varphi_1 S & 0 \\ 0 & \varphi - \varphi_1 P & -(\varphi_0 + \varphi_1 S) & 0 \\ -\mu & 0 & 0 & -\mu_0 \end{bmatrix},$$

where $a_{11} = r - 2r\dfrac{A}{K} - \alpha S - 2\alpha_1 PA + \beta F + 2\beta_1 FA$. At the interior equilibrium E^*, the corresponding Jacobian matrix is given by

$$J_{E^*} = \begin{bmatrix} \dfrac{rA^*}{K} - \alpha_1 P^* A^* + \beta_1 F^* A^* & -\alpha A^* & -\alpha_1 A^{*2} & \dfrac{\beta A^* + \beta_1 A^{*2}}{2} \\ \theta\alpha S^* & -\delta S^* & -\lambda\varphi_1 S^* & 0 \\ 0 & \varphi - \varphi_1 P^* & -(\varphi_0 + \varphi_1 S^*) & 0 \\ -\mu & 0 & 0 & -\mu_0 \end{bmatrix}.$$

The characteristic polynomial for J_{E^*} is written as

$$X^4 + A_1 X^3 + A_2 X^2 + A_3 X + A_4 = 0,$$

where the characteristic polynomial for J_{E^*} is obtained as

$$X^4 + A_1 X^3 + A_2 X^2 + A_3 X + A_4 = 0,$$

here:

$$\begin{aligned}
A_1 &= \mu_0 + \frac{rA^*}{K} + \alpha_1 P^* A^* - \beta_1 F^* A^* + \delta S^* + \varphi_0 + \varphi_1 S^*, \\
A_2 &= \mu A^*(\beta + \beta_1 A^*) + \mu_0 \left(\frac{rA^*}{K} + \alpha_1 P^* A^* - \beta_1 F^* A^*\right) \\
&\quad + \mu_0 \delta S^*(\varphi_0 + \varphi_1 S^*) + (\delta S^* + \varphi_0 + \varphi_1 S^*) \\
&\quad \left(\frac{rA^*}{K} + \alpha_1 P^* A^* - \beta_1 F^* A^*\right) + \delta S^*(\varphi_0 + \varphi_1 S^*) \\
&\quad + \lambda\varphi_1 S^*(\varphi - \varphi_1 P^*) + \theta\alpha^2 A^* S^*. \\
A_3 &= \mu A^*(\beta + \beta_1 A^*)\delta S^*(\varphi_0 + \varphi_1 S^*) + \lambda\varphi_1 S^*(\varphi - \varphi_1 P^*) \\
&\quad + \mu_0 \delta S^*(\varphi_0 + \varphi_1 S^*)\left(\frac{rA^*}{K} + \alpha_1 P^* A^* - \beta_1 F^* A^*\right) \\
&\quad + \mu_0 \alpha_1 \theta\alpha A^* S^*(\varphi - \varphi_1 P^*) \\
&\quad + 2\varphi_1 \mu_0 S^*(\varphi - \varphi_1 P^*)\left(\frac{rA^*}{K} + \alpha_1 P^* A^* - \beta_1 F^* A^*\right) \\
&\quad + \theta\alpha^2 A^* S^* \mu_0(\varphi_0 + \varphi_1 S^*), \\
A_4 &= \mu A^*(\beta + \beta_1 A^*)\delta S^*(\varphi_0 + \varphi_1 S^*) + \lambda\varphi_1 S^*(\varphi - \varphi_1 P^*) \\
&\quad + \mu_0 \delta S^*(\varphi_0 + \varphi_1 S^*)\left(\frac{rA^*}{K} + \alpha_1 P^* A^* - \beta_1 F^* A^*\right) \\
&\quad + \mu_0 \alpha_1 \theta\alpha A^* S^*(\varphi - \varphi_1 P^*)
\end{aligned}$$

$$+ \lambda\varphi_1\mu_0 S^*(\varphi - \varphi_1 P^*)\left(\frac{rA^*}{K} + \alpha_1 P^* A^* - \beta_1 F^* A^*\right)$$
$$+ \theta\alpha^2 A^* S^* \mu_0(\varphi_0 + \varphi_1 S^*).$$

5.7 Routh–Hurwitz (RH) Stability Criterion for Fractional Order System

According to [6, 7], the Routh and Hurwitz criterion for stability is explained. The system is stable when all of the characteristic polynomial roots lie in the complex plane's left half. Consider the given polynomial:

$$P(X) = X^n + a_{n-1}X^{n-1} + \cdots + a_1 X + a_0, \tag{5.9}$$

where the coefficients a_i ($i = 0, 1, \ldots, n-1$) are all real constants. Using these coefficients, we form Hurwitz matrices H_k for the given system of equations as

$$H_2 = \begin{pmatrix} a_{n-1} & a_{n-3} \\ a_n & a_{n-2} \end{pmatrix}, \tag{5.10}$$

$$H_3 = \begin{pmatrix} a_{n-1} & a_{n-3} & a_{n-5} \\ a_n & a_{n-2} & a_{n-4} \\ 0 & a_{n-1} & a_{n-3} \end{pmatrix}, \tag{5.11}$$

$$H_4 = \begin{pmatrix} a_{n-1} & a_{n-3} & a_{n-5} & a_{n-7} \\ a_n & a_{n-2} & a_{n-4} & a_{n-6} \\ 0 & a_{n-1} & a_{n-3} & a_{n-5} \\ 0 & a_n & a_{n-2} & a_{n-4} \end{pmatrix}, \tag{5.12}$$

$$H_5 = \begin{pmatrix} a_{n-1} & a_{n-3} & a_{n-5} & a_{n-7} & a_{n-9} \\ a_n & a_{n-2} & a_{n-4} & a_{n-6} & a_{n-8} \\ 0 & a_{n-1} & a_{n-3} & a_{n-5} & a_{n-7} \\ 0 & a_n & a_{n-2} & a_{n-4} & a_{n-6} \\ 0 & 0 & a_{n-1} & a_{n-3} & a_{n-5} \end{pmatrix}. \tag{5.13}$$

For all $k = 1, 2, \ldots, n$, all roots of the polynomial are either negative or have negative real parts if and only if the determinants of all Hurwitz matrices remain positive, that is $\det(H_k) > 0$ for all k.

The Routh–Hurwitz criterion for polynomials of degree $n = 2, 3, 4$, and 5 is outlined as follows:

- **For $n = 2$:**

$$\Delta_1 = a_{n-1} > 0, \quad \Delta_2 = \det(H_2) > 0.$$

- **For** $n = 3$**:**

$$\Delta_1 = a_{n-1} > 0, \quad \Delta_2 = \det(H_2) > 0, \quad \Delta_3 = \det(H_3) > 0.$$

- **For** $n = 4$**:**

$$\Delta_1 = a_{n-1} > 0, \Delta_2 = \det(H_2) > 0, \Delta_3 = \det(H_3) > 0,$$
$$\Delta_4 = \det(H_4) > 0.$$

- **For** $n = 5$**:**

$$\Delta_1 = a_{n-1} > 0, \Delta_2 = \det(H_2) > 0, \Delta_3 = \det(H_3) > 0,$$
$$\Delta_4 = \det(H_4) > 0, \Delta_5 = \det(H_5) > 0.$$

According to [7], the Routh–Hurwitz matrix of a stable polynomial is totally non-negative.

5.8 Existence of Hopf Bifurcation

The given characteristic polynomial is:

$$X^4 + A_1(\varphi)X^3 + A_2(\varphi)X^2 + A_3(\varphi)X + A_4(\varphi) = 0. \tag{5.14}$$

The condition at $\varphi = \varphi_c$ is:

$$\frac{A_3(\varphi_c)}{A_1(\varphi_c)A_2(\varphi_c)} - A_3(\varphi_c) - A_1^2(\varphi_c)A_4(\varphi_c) = 0. \tag{5.15}$$

Then, at $\varphi = \varphi_c$, the characteristic polynomial can be derived as

$$\left(X^2 + \frac{A_3}{A_1}\right)\left(X^2 + A_1X + \frac{A_1A_4}{A_3}\right) = 0. \tag{5.16}$$

The given equation has four roots, denoted as X_i ($i = 1, 2, 3, 4$) with a pair of purely imaginary roots given by $X_{1,2} = \pm i\omega_0$, with $\omega_0 = \left(\frac{A_3}{A_1}\right)^{1/2}$. For a Hopf bifurcation to occur, the remaining roots, X_3 and X_4, must be positioned in the left half of the complex plane, excluding $\pm i\omega_0$. To determine the nature of these two roots, the study proceeds with:

$$X_3 + X_4 = -A_1, \tag{5.17}$$
$$\omega_0^2 + X_3X_4 = A_2, \tag{5.18}$$

$$\omega_0^2(X_3 + X_4) = -A_3, \tag{5.19}$$

$$\omega_0^2 X_3 X_4 = A_4. \tag{5.20}$$

When X_3 and X_4 are complex conjugates, eqn (6.15) gives $2\text{Re}(X_3) = -A_1$, indicating that both roots have negative real parts. If X_3 and X_4 are real, analysis using eqn (5.17)–(6.16), confirms that both are negative. Thus, these roots remain in the left half of the complex plane, ensuring the occurrence of a Hopf bifurcation.

Now, consider a parameter φ in a neighborhood of φ_c, such that $\varphi \in (\varphi_c - \epsilon, \varphi_c + \epsilon)$. In this case, the roots depend on φ, and are expressed as $X_{1,2} = \eta(\varphi) \pm i\xi(\varphi)$. Substituting this value into eqn (6.13) and separating the real and imaginary components result in:

$$\eta^4 + A_1\eta^3 + A_2\eta^2 + A_3\eta + A_4 + \xi^4 - 6\eta^2\xi^2 - 3A_1\eta\xi^2 - A_2\xi^2 = 0, \tag{5.21}$$

$$4\eta\xi(\eta^2 - \xi^2) - A_1\xi^3 + 3A_1\eta\xi^2 + 2A_2\eta\xi + A_3\xi = 0. \tag{5.22}$$

Since $\xi(\varphi) \neq 0$, eqn (5.22) implies that:

$$-(4\eta + A_1)\xi^2 + 4\eta^3 + 3A_1\eta^2 + 2A_2\eta + A_3 = 0. \tag{5.23}$$

Substituting the expression for ξ^2 into eqn (6.17), yields:

$$\begin{aligned} &-64\eta^6 - 96A_1\eta^5 - \left(\frac{16}{3}A_1^2 + 2A_2\right)\eta^4 - 8\left(A_3 + 4A_1A_2\right)\eta^3 \\ &- 4\left(A_2^2 + 2A_1^2A_2 + A_1A_3 - 4A_4\right)\eta^2 - 2A_1\left(A_1A_3 + A_2^2 - 4A_4\right)\eta \\ &- \left(A_3(A_1A_2 - A_3) - A_1^2A_4\right) = 0. \end{aligned}$$

Differentiate the above with respect to φ, and recalling $\eta(\varphi_c) = 0$, we get:

$$\left.\frac{d\eta}{d\varphi}\right|_{\varphi=\varphi_c} = \left.\frac{\frac{d}{d\varphi}\left(A_3(A_1A_2 - A_3) - A_1^2A_4\right)}{-2A_1(A_1A_3 + A_2^2 - 4A_4)}\right|_{\varphi=\varphi_c}. \tag{5.24}$$

Using the value of

$$A_4(\varphi_c) = \frac{A_1(\varphi_c)A_2(\varphi_c)A_4(\varphi_c) - A_3^2(\varphi_c)}{A_1^2(\varphi_c)}$$

from eqn (6.14) in the above expression, we get:

$$\left.\frac{d\eta}{d\varphi}\right|_{\varphi=\varphi_c} = \left.\frac{\frac{d}{d\varphi}\left(A_3(A_1A_2 - A_3) - A_1^2\frac{A_1A_2A_3 - A_3^2}{A_1^2}\right)}{-2A_1(A_1A_3 + A_2^2 - 4A_4)}\right|_{\varphi=\varphi_c}. \tag{5.25}$$

It suggests that

$$\left.\frac{d\eta}{d\varphi}\right|_{\varphi=\varphi_c} = \left.\frac{\frac{d}{d\varphi}\left(A_3(A_1A_2 - A_3) - A_1^2A_4\right)}{-2A_1\left(A_1A_3 + \left(A_2 - 2\frac{A_3}{A_1}\right)^2\right)}\right|_{\varphi=\varphi_c} \neq 0, \tag{5.26}$$

provided that:

$$\left.\frac{d}{d\varphi}\left(A_3(A_1A_2 - A_3) - A_1^2A_4\right)\right|_{\varphi=\varphi_c} \neq 0. \tag{5.27}$$

Thus, the following result establishes the existence of a Hopf bifurcation.

5.9 Numerical Approx. Method

By extending the approach with Caputo's derivative, the model's solution is interpreted numerically:

$$\begin{aligned}
u_{\mathrm{n+1}} = u_0 &+ \frac{h^{\rho_1}}{\Gamma(\rho_1+2)}\Bigg[ru^p_{\mathrm{n+1}}\left(1 - \frac{u^p_{\mathrm{n+1}}}{k}\right) - \alpha u^p_{\mathrm{n+1}}v^p_{\mathrm{n+1}} \\
&- \alpha_1 w^p_{\mathrm{n+1}}\left(u^p_{\mathrm{n+1}}\right)^2 + \beta z^p_{\mathrm{n+1}}u^p_{\mathrm{n+1}} + \beta_1 z^p_{\mathrm{n+1}}\left(u^p_{\mathrm{n+1}}\right)^2\Bigg] \\
&+ \frac{h^{\rho_1}}{\Gamma(\rho_1+2)}\sum_{j=0}^{n}\rho_{1,\mathrm{j,n+1}}\Bigg[ru_{\mathrm{j}}\left(1 - \frac{u_{\mathrm{j}}}{k}\right) - \alpha u_{\mathrm{j}}v_{\mathrm{j}} - \alpha_1 w_{\mathrm{j}}(u_{\mathrm{j}})^2 \\
&+ \beta z_{\mathrm{j}}u_{\mathrm{j}} + \beta_1 z_{\mathrm{j}}(u_{\mathrm{j}})^2\Bigg], \\
v_{\mathrm{n+1}} = v_0 &+ \frac{h^{\rho_2}}{\Gamma(\rho_2+2)}\Bigg[\alpha u^p_{\mathrm{n+1}}v^p_{\mathrm{n+1}} - \delta\left(v^p_{\mathrm{n+1}}\right)^2 - \lambda\varphi_1 w^p_{\mathrm{n+1}}v^p_{\mathrm{n+1}} \\
&+ \frac{h^{\rho_2}}{\Gamma(\rho_2+2)}\sum_{j=0}^{n}\rho_{2,\mathrm{j,n+1}}\left[\alpha u_{\mathrm{j}}v_{\mathrm{j}} - \delta(v_{\mathrm{j}})^2 - \lambda\varphi_1 w_{\mathrm{j}}v_{\mathrm{j}}\right]\Bigg], \\
w_{\mathrm{n+1}} = w_0 &+ \frac{h^{\rho_3}}{\Gamma(\rho_3+2)}\Bigg[\varphi v^p_{\mathrm{n+1}} - \varphi_0 w^p_{\mathrm{n+1}} - \varphi_1 w^p_{\mathrm{n+1}}v^p_{\mathrm{n+1}} \\
&+ \frac{h^{\rho_3}}{\Gamma(\rho_3+2)}\sum_{j=0}^{n}\rho_{3,\mathrm{j,n+1}}\left[\varphi v_{\mathrm{j}} - \varphi_0 w_{\mathrm{j}} - \varphi_1 w_{\mathrm{j}}v_{\mathrm{j}}\right]\Bigg].
\end{aligned}$$

$$z_{\mathrm{n+1}} = z_0 + \frac{h^{\rho_4}}{\Gamma(\rho_4+2)}\left[\mu\left(k - u^p_{\mathrm{n+1}}\right) - \mu_0\left(z^p_{\mathrm{n+1}}\right)^4\right] + \frac{h^{\rho_4}}{\Gamma(\rho_4+2)}\sum_{j=0}^{n}\rho_{4,\mathrm{j,n+1}}\left[\mu(k-u_{\mathrm{j}}) - \mu_0(z_{\mathrm{j}})^4\right],$$

$$u^p_{\mathrm{n+1}} = u_0 + \frac{1}{\Gamma(\rho_1)}\sum_{j=0}^{n} B_{1,\mathrm{j,n+1}}\left[ru_{\mathrm{j}}\left(1-\frac{u_{\mathrm{j}}}{k}\right) - \alpha u_{\mathrm{j}}v_{\mathrm{j}} - \alpha_1 w_{\mathrm{j}}(u_{\mathrm{j}})^2 + \beta z_{\mathrm{j}}u_{\mathrm{j}} + \beta_1 z_{\mathrm{j}}(u_{\mathrm{j}})^2\right],$$

$$v^p_{\mathrm{n+1}} = v_0 + \frac{1}{\Gamma(\rho_2)}\sum_{j=0}^{n} B_{2,\mathrm{j,n+1}}\left[\theta\alpha u_{\mathrm{j}}v_{\mathrm{j}} - \delta(v_{\mathrm{j}})^2 - \lambda\varphi_1 w_{\mathrm{j}}v_{\mathrm{j}}\right],$$

$$w^p_{\mathrm{n+1}} = w_0 + \frac{1}{\Gamma(\rho_3)}\sum_{j=0}^{n} B_{3,\mathrm{j,n+1}}\left[\varphi v_{\mathrm{j}} - \varphi_0 w_{\mathrm{j}} - \varphi_1 w_{\mathrm{j}}v_{\mathrm{j}}\right],$$

$$z^p_{\mathrm{n+1}} = z + \frac{1}{\Gamma(\rho_4)}\sum_{j=0}^{n} B_{4,\mathrm{j,n+1}}\left[\mu(k-u_{\mathrm{j}}) - \mu_0(z_{\mathrm{j}})^4\right].$$

$$\rho_{\mathrm{i,j,n+1}} = \begin{cases} n^{\rho_{\mathrm{i}}+1} - (n-\rho_{\mathrm{i}})(n+1)\rho_{\mathrm{i}}, & \text{if } j=0 \\ (n-j+2)\rho_{\mathrm{i+1}} - (n-j)\rho_{\mathrm{i+1}} - 2(n-j+1)\rho_{\mathrm{i+1}}, & \text{if } 0 \le j \le n \\ 1, & \text{if } j=1 \end{cases}$$

$$B_{\mathrm{i,j,n+1}} = \frac{h^{\rho_{\mathrm{i}}}}{\rho_{\mathrm{i}}}\left((n-j+1)^{\rho_{\mathrm{i}}} - (n-j)^{\rho_{\mathrm{i}}}\right), \quad 0 \le j \le n, \quad i = 1,2,3,4,5.$$

5.10 Numerical Simulation

The numerical simulation of the proposed model is conducted using MATLAB to support the analytical findings. The parameter values are chosen based on ecological relevance to ensure a realistic representation of the system dynamics. The equilibrium points of the system are computed, and their stability is determined by evaluating the eigenvalues of the Jacobian matrix. For the numerical simulations, the parameter values are chosen as follows:

$r = 0.2$, with a carrying capacity of $K = 50$, $\alpha = 0.025$, $\theta = 0.6$, $\epsilon = 0.05$, $\varphi = 0.08$, $\varphi_0 = 0.01$, $\varphi_1 = 0.05$, $\lambda = 8$, $\mu = 0.1$, $\mu_0 = 0.01$, $\alpha_1 = 0.01$, $\beta = 0.001$, $\beta_1 = 0.01$.

For the given parameter set, the equilibrium points of system (1) are calculated as

$$E_0 = \begin{pmatrix} 0 \\ 0 \\ 0 \\ 500 \end{pmatrix}, \quad E_1 = \begin{pmatrix} 50 \\ 0 \\ 0 \\ 0 \end{pmatrix}, \quad E^* = \begin{pmatrix} 49.83 \\ 2.96 \\ 1.50 \\ 1.64 \end{pmatrix}.$$

The eigenvalues corresponding to the equilibria E_0, E_1, and E^* are calculated as

$$\begin{aligned} \Lambda_0 &= (0.7, 0, -0.01, -0.01), \\ \Lambda_1 &= (-0.01, 0.075, -0.105 \pm 1.579i), \\ \Lambda^* &= (-0.069 \pm 1.594i, -0.153 \pm 0.074i). \end{aligned}$$

The results indicate that the equilibrium E_0 is unstable due to the presence of a positive eigenvalue, while E_1 also exhibits instability as one of its eigenvalues has a positive real part. On the other hand, the interior equilibrium E^* is shown to be locally asymptotically stable since all its eigenvalues have negative real parts. Additionally, the positivity of certain key expressions, such as $A_1A_2 - A_3 - A_1^2A_4$, further reinforces the stability conclusions. The values of $A_1 = 0.44$, $A_2 = 2.62$, $A_3 = 0.78$, and $A_4 = 0.073$ are calculated, which are positive. Also, the value of $A_3(A_1A_2 - A_3) - A_1^2A_4 = 0.28$, which is also positive.

Consider the parameter values $r = \alpha_1 = 0.00001$, $\beta = 0.00001$, $\beta_1 = 0.00000001$, $\varphi_1 = 0.0005$ while keeping the remaining parameters unchanged as previously defined. It is verified that the global stability conditions outlined in Theorem 6.3 hold for these parameter values. The solution trajectories originating within the region of attraction converge towards (S^*, A^*, F^*) in the S, A, F-space, confirming that E^* is globally asymptotically stable in this space. Figure 5.2 illustrates the effect of insecticides on crop yield by plotting the time evolution of different state variables for varying levels of the insecticide parameter φ.

The curves are color-coded for different values of φ (0.03, 0.06, and 0.10), where red corresponds to the lowest insecticide level, black to the intermediate level, and blue to the highest level. In the first subplot, $A(t)$ shows an increasing trend over time, with higher insecticide levels leading to a more rapid initial rise but ultimately converging to a steady state. The second subplot depicting $S(t)$ indicates a similar pattern where the nutrient concentration stabilizes over time, but higher insecticide levels result in lower

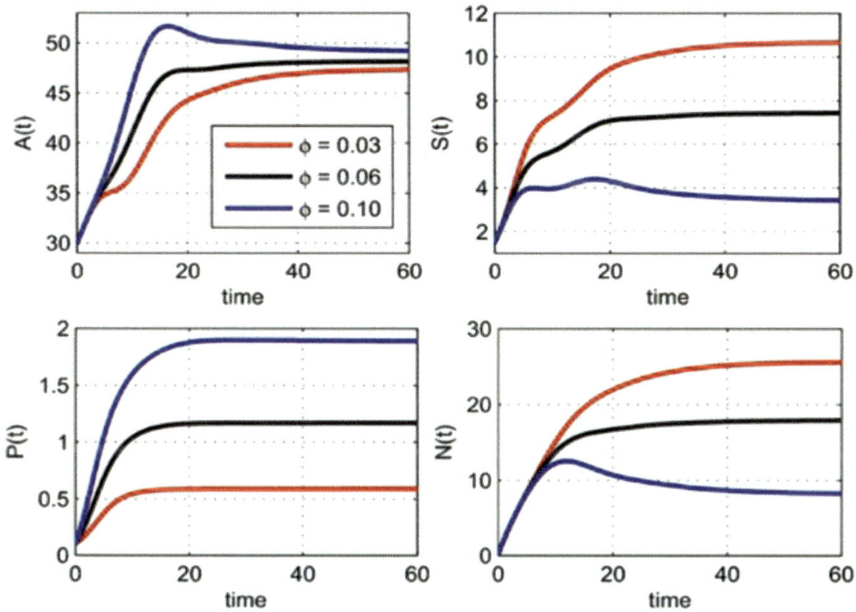

Figure 5.2 Graph showing the effect of insecticides on crop yield.

equilibrium values. The third subplot represents the pest population $P(t)$, which initially rises but then stabilizes at a lower level for higher insecticide concentrations, confirming the effectiveness of insecticide application in reducing pest density. The final subplot for $N(t)$ reveals that the natural enemy population also stabilizes, but its equilibrium value decreases with increasing insecticide levels, suggesting that excessive insecticide application may negatively impact the predator population. These findings highlight the trade-off in insecticide use, where increasing application reduces pests and enhances crop yield but may also have adverse effects on ecological balance.

5.11 Conclusion

The findings presented in this chapter underscore the multifaceted role of pesticides in agriculture. Beyond their primary function of controlling insect populations, pesticides can also influence plant metabolism, particularly protein synthesis, and alter the concentrations of essential mineral nutrients in plant tissues. However, it is crucial to acknowledge the potential residual effects of these chemicals, which may pose long-term environmental and health risks. These concerns extend beyond the scope of this discussion and warrant further investigation. Additionally, while variations exist in the impact of insecticides on plant growth, metabolism, and mineral composition, such differences are generally minimal across different insecticides and crop species. This highlights a broader pattern in pesticide-plant interactions, suggesting that while these chemicals serve critical agricultural purposes, their broader ecological and physiological consequences should not be overlooked.

References

[1] Prabhaker N. Castle S., Palumbo J. Newer insecticides for plant virus disease management. *Virus Res.*, 141(2):131–139, 2009.

[2] Adam S Davis, Jason D Hill, Craig A Chase, Ann M Johanns, and Matt Liebman. Increasing cropping system diversity balances productivity, profitability and environmental health. 2012.

[3] TK Kar, Abhijit Ghorai, and Soovoojeet Jana. Dynamics of pest and its predator model with disease in the pest and optimal use of pesticide. *Journal of theoretical biology*, 310:187–198, 2012.

[4] Kusum Lata and Arvind Kumar Kumar Misra. Modeling the effect of economic efforts to control population pressure and conserve forestry resources. *Nonlinear Analysis: Modelling and Control*, 22(4):473–488, 2017.

[5] Haiqiang Li, Hongsheng Pan, Dongmei Wang, Bing Liu, Jian Liu, Jianping Zhang, and Yanhui Lu. Intercropping with fruit trees increases population abundance and alters species composition of spider mites on cotton. *Environmental Entomology*, 47(4):781–787, 2018.

[6] Bruce Horwith. A role for intercropping in modern agriculture. *BioScience*, 35(5):286–291, 1985.

[7] E Ahmed, AMA El-Sayed, and Hala AA El-Saka. On some routh–hurwitz conditions for fractional order differential equations and their applications in lorenz, rössler, chua and chen systems. *Physics Letters A*, 358(1):1–4, 2006.

[8] Haileyesus Tessema Alemneh and Negesse Yizengaw Alemu. Mathematical modeling with optimal control analysis of social media addiction. *Infectious Disease Modelling*, 6:405–419, 2021.

[9] Hadi Delavari, Dumitru Baleanu, and Jalil Sadati. Stability analysis of caputo fractional-order nonlinear systems revisited. *Nonlinear Dynamics*, 67:2433–2439, 2012.

[10] Odo Diekmann, Johan Andre Peter Heesterbeek, and Michael G Roberts. The construction of next-generation matrices for compartmental epidemic models. *Journal of the royal society interface*, 7(47):873–885, 2010.

[11] Kai Diethelm and Neville J Ford. Analysis of fractional differential equations. *Journal of Mathematical Analysis and Applications*, 265(2):229–248, 2002.

[12] Kai Diethelm, Neville J Ford, and Alan D Freed. A predictor-corrector approach for the numerical solution of fractional differential equations. *Nonlinear Dynamics*, 29:3–22, 2002.

[13] MM El-Dessoky and Muhammad Altaf Khan. Modeling and analysis of an epidemic model with fractal-fractional atangana-baleanu derivative. *Alexandria Engineering Journal*, 61(1):729–746, 2022.

[14] MA El-Tawil and SN Huseen. The q-homotopy analysis method (q-ham). *Int. J. Appl. Math. Mech*, 8(15):51–75, 2012.

[15] Magdy A El-Tawil and Shaheed N Huseen. On convergence of the q-homotopy analysis method. *Int. J. Contemp. Math. Sci*, 8(10):481–497, 2013.

[16] VS Erturk, E Godwe, D Baleanu, P Kumar, J Asad, and A Jajarmi. Novel fractional-order lagrangian to describe motion of beam on nanowire. *Acta Physica Polonica A*, 140:265–272, 2021.

[17] Behzad Ghanbari, Sunil Kumar, and Ranbir Kumar. A study of behaviour for immune and tumor cells in immunogenetic tumour model with non-singular fractional derivative. *Chaos, Solitons & Fractals*, 133:109619, 2020.

[18] Khondoker Nazmoon Nabi, Hamadjam Abboubakar, and Pushpendra Kumar. Forecasting of covid-19 pandemic: From integer derivatives to fractional derivatives. *Chaos, Solitons & Fractals*, 141:110283, 2020.

[19] Zaid Odibat, Vedat Suat Erturk, Pushpendra Kumar, and V Govindaraj. Dynamics of generalized caputo type delay fractional differential equations using a modified predictor-corrector scheme. *Physica Scripta*, 96(12):125213, 2021.

[20] Igor Podlubny. *Fractional differential equations: an introduction to fractional derivatives, fractional differential equations, to methods of their solution and some of their applications*, volume 198. elsevier, 1998.

[21] SZ Rida, AAM Arafa, and YA Gaber. Solution of the fractional epidemic model by l-adm. *J. Fract. Calc. Appl*, 7(1):189–195, 2016.

[22] Preeti Deolia, Anuraj Singh, and Pradeep Malik. Covid-19 outbreak: A predictive mathematical study. In *Computing and Simulation for Engineers*, pages 79–88. CRC Press, 2022.

[23] Saif Ullah, Muhammad Altaf Khan, Muhammad Farooq, Zakia Hammouch, and Dumitru Baleanu. A fractional model for the dynamics of tuberculosis infection using caputo-fabrizio derivative. *Discrete and Continuous Dynamical Systems Series S*, 13:975–993, 2020.

[24] Arti Malika, Pradeep Malik, Amit Kumar Attric, and Nitendra Kumard. Fractional seir-sei epidemic model on zika virus. *Bulletin of Environmental Pharmacology and Life Sciences*, 5:482–487, 2022.

[25] Pradeep Malik and Deepika. Stability analysis of fractional order modelling of social media addiction. *Mathematical Foundations of Computing*, 6:670–690, 2023.

6

Integrating Mathematical Precision with Biological Insight in Tumor Dynamics

Satyvir Singh [1, 2] and Mukesh Kumar Awasthi[3]

[1]Applied and Computational Mathematics, RWTH Aachen University, Germany
[2]Department of Mathematics, Graphic Era Deemed to be University, India
[3]Department of Mathematics, Babasaheb Bhimrao Ambedkar University, India
E-mail: singh@acom.rwth-aachen.de; mukeshiitr.kumar@gmail.com

Abstract

This study presents a high-fidelity computational framework for modeling tumor evolution, integrating mathematical precision with biological insight. Using a mixed modal discontinuous Galerkin (DG) method, we simulate cancer cell invasion driven by proliferation, haptotaxis, extracellular matrix (ECM) degradation, and matrix-degrading enzyme (MDE) secretion. The model captures the spatiotemporal dynamics of tumor progression in both 1D and 2D domains. Through numerical experiments, we analyze the influence of key parameters such as the cancer proliferation rate and haptotactic sensitivity on tumor spread and tissue remodeling. The DG method ensures accurate resolution of invasion fronts and enables a detailed study of tumor–ECM interactions. Our findings highlight the effectiveness of computational modeling in uncovering biological mechanisms of cancer invasion. This integrative approach offers valuable insights for advancing predictive models in oncology and supports the development of targeted therapeutic strategies.

Keywords: Tumor dynamics, mathematical modeling, DG method, Haptotaxis

DOI: 10.1201/9788743806691-6

6.1 Introduction

Cancer progression is a multifaceted process governed by the dynamic interplay of biological, chemical, and mechanical factors. One of the most critical and life-threatening stages is tumor invasion, where malignant cells breach the boundaries of their tissue of origin and infiltrate surrounding healthy tissue. This invasive behavior, often a precursor to metastasis, is regulated by a complex suite of mechanisms – including cell proliferation, directed migration (haptotaxis), extracellular matrix (ECM) degradation, and matrix-degrading enzyme (MDE) secretion [1–3].

Mathematical modeling has proven to be an essential tool for dissecting and simulating these complex biological processes. Early attempts, such as those by Hill (1928) on oxygen diffusion and Thomlinson and Gray (1955) on tumor architecture, laid the groundwork for mechanistic tumor models [4, 5]. Subsequently, Araujo and McElwain (2004) provided a historical and analytical review of tumor growth models, classifying them into ODE, PDE, and hybrid discrete-continuum approaches [6]. While ODE models offer insights into population dynamics [7, 8], they fall short in addressing spatial heterogeneity – a fundamental characteristic of invasive tumors.

To address this gap, researchers have adopted reaction-diffusion PDE models that incorporate spatial and temporal dynamics of tumor cell populations, ECM, and chemical mediators. Seminal work by Gatenby and Gawlinski (1996) introduced a model accounting for acid-mediated tumor invasion [8], which was later extended by McGillen et al. [9] to include competition between normal and cancerous cells. Chaplain and Lolas [10] further advanced these models by integrating detailed descriptions of proteolytic enzyme systems, such as the urokinase plasminogen activator (uPA) cascade, essential for ECM degradation [11].

Central to many invasion models is haptotaxis, the directed movement of cancer cells along gradients of ECM density. This mechanism is pivotal in determining tumor morphology and was effectively modeled by Anderson et al. [12], Gerisch and Chaplain [13], and Painter et al. [14]. These models emphasized that the balance between cell-cell and cell-matrix adhesion governs the formation of invasive versus non-invasive tumor patterns.

In parallel, a substantial body of work has focused on numerical methods for solving the complex PDE systems that arise from these models. Traditional methods-such as finite difference, finite volume, and finite element methods-have been widely used but often struggle with accuracy and stability when dealing with sharp invasion fronts or anisotropic diffusion

[15–17]. To overcome these limitations, discontinuous Galerkin (DG) methods have emerged as a powerful alternative. These methods combine high-order accuracy with the ability to handle non-linearities and discontinuities, making them particularly suited for simulating the heterogeneous and abrupt transitions inherent in tumor invasion [18].

In this study, we propose a high-fidelity mixed modal DG framework to simulate cancer invasion dynamics in both 1D and 2D domains. Our model incorporates key biological processes – cancer cell proliferation, ECM-directed migration (haptotaxis), MDE diffusion and decay, and ECM remodeling. We perform a comprehensive numerical study, varying parameters such as proliferation rate and haptotactic sensitivity to examine their influence on tumor evolution. The DG method's modular structure and high-order precision make it ideal for capturing complex spatial patterns and invasive morphologies. This study demonstrates how the synergy between mathematical modeling and advanced numerical techniques can provide biologically relevant insights into tumor behavior. Ultimately, these high-resolution simulations lay the groundwork for predictive modeling in computational oncology, with implications for patient-specific therapy design, tumor aggressiveness assessment, and the development of novel anti-invasion treatments.

6.2 Mathematical Modeling of Tumor Invasion

Tumor invasion is a dynamic process involving interactions between cancer cells and their surrounding microenvironment. A robust mathematical model must capture key biological mechanisms such as cancer cell proliferation, migration through haptotaxis, extracellular matrix (ECM) degradation, and the secretion and diffusion of matrix-degrading enzymes (MDEs). This section presents a continuum-based reaction-diffusion model formulated through a system of partial differential equations (PDEs), offering a spatiotemporal framework to analyze tumor evolution.

6.2.1 Biological basis of the model

The model considers three interacting components:

1. *Cancer cell density* $n(\mathbf{x}, t)$: Representing the local population of malignant cells.
2. *ECM density* $f(\mathbf{x}, t)$: Indicating the fibrous structural material degraded during invasion.

3. *MDE concentration* $m(\mathbf{x}, t)$: Reflecting enzymes such as matrix metalloproteinases (MMPs) secreted by cancer cells.

These variables evolve dynamically over space and time, governed by biological processes including diffusion, directional migration (haptotaxis), cellular proliferation, ECM degradation, and enzyme-mediated remodeling. The interplay between these components drives the invasive behavior characteristic of malignant tumors.

6.2.2 Governing equations

The coupled PDE system describing tumor invasion is given as [10]

$$\begin{aligned}
\frac{\partial n}{\partial t} &= D_{\mathrm{n}}\nabla^2 n + \nabla\cdot(\chi_0 n \nabla f) + \mu n\left(1 - \frac{n}{N} - \frac{f}{F}\right) \quad \text{in}\,\Omega\times(0,T),\\
\frac{\partial f}{\partial t} &= -k_{\mathrm{f}} f + \rho_{\mathrm{f}} f\left(1 - \frac{n}{N} - \frac{f}{F}\right) \quad \text{in}\,\Omega\times(0,T),\\
\frac{\partial m}{\partial t} &= D_{\mathrm{m}}\nabla^2 m + \zeta n\left(1 - \frac{m}{M}\right) - \gamma m \quad \text{in}\,\Omega\times(0,T),
\end{aligned} \tag{6.1}$$

where time is represented by $t \in [0, T]$ and the diffusion coefficients D_{n} and D_{m} of n and m, respectively, are taken to be constants. μ is the rate at which cancer cells proliferate, and χ_0 is the haptotactic rate. The rates of ECM remodeling and degradation are represented by ρ_{f} and k_{fSP}, respectively. Moreover, ζ and γ, respectively, provide the growth and decay rates of MDE. In contrast, the maximum sustained cancer cell density, ECM density, and MDE concentration are N, F, and M, respectively. Additionally, the constants γ and $\chi_0, \mu, k, \rho, \zeta$, are positive.

6.2.3 Non-dimensionalization

The dimensionless form of a mathematical model for cancer cell invasion is crucial because it simplifies and generalizes the model, making it applicable across different scales and units. It reduces the number of parameters, highlighting the dominant processes and improving numerical stability. For this purpose, we consider $L = 0.1$ cm and $\tau = \frac{L^2}{D}$ as the characteristic length and time scales, respectively, with $D \simeq 10^{-6}$ cm^2 s^{-1} [10]. We define the dimensionless variables as

$$\tilde{n} = \frac{n}{N}, \quad \tilde{f} = \frac{f}{F}, \quad \tilde{m} = \frac{m}{M}, \quad \tilde{x} = \frac{x}{L}, \quad \tilde{t} = \frac{t}{\tau}. \tag{6.2}$$

In addition, the reference densities n and f are assumed to be 6.7×10^7 cell/cm^3 and 10^{-10} M, respectively [10]. The reference MDE concentration is left undefined because acceptable experimental values are hard to come by. When these dimensionless variables are applied to system (6.1) and the tilde is eliminated, the model equations in $\Omega \times (0, T)$ assume the dimensionless form shown below:

$$
\begin{aligned}
\frac{\partial n}{\partial t} &= \nabla \cdot (d_1 \nabla n + \chi n \nabla f) + \lambda n(1 - n - f), \\
\frac{\partial f}{\partial t} &= -\eta f m + \rho f(1 - n - f), \\
\frac{\partial m}{\partial t} &= d_2 \nabla^2 m + \alpha n(1 - m) - \beta m,
\end{aligned}
\tag{6.3}
$$

where all parameters are now dimensionless.

6.2.4 Initial and boundary conditions

The initial conditions assume a localized tumor, which are defined as

$$
\begin{aligned}
&n(\mathbf{x}, 0) = n_0(\mathbf{x}) > 0 \quad \text{in } \Omega_1 \subset \Omega, \\
&f(\mathbf{x}, 0) = f_0(\mathbf{x}), \quad m(\mathbf{x}, 0) = m_0(\mathbf{x}),
\end{aligned}
\tag{6.4}
$$

with zero elsewhere. We apply zero-flux boundary conditions:

$$
\begin{aligned}
\mathbf{n} \cdot (-d_1 \nabla n + \chi n \nabla f) &= 0, \\
\mathbf{n} \cdot \nabla f &= 0, \\
\mathbf{n} \cdot \nabla m &= 0 \quad \text{on } \partial\Omega.
\end{aligned}
\tag{6.5}
$$

These reflect a closed domain, conserving mass.

6.2.5 Model significance

This model captures self-organization of invasive fronts, ECM remodeling, and enzyme-mediated degradation. It extends classical work by Gatenby and Gawlinski [11], and Chaplain and Lolas [10], preparing the foundation for high-fidelity numerical schemes in the next section.

6.3 Numerical Framework: Mixed Modal DG Method

Simulating tumor invasion involves solving coupled, nonlinear partial differential equations (PDEs) with steep gradients, sharp moving fronts, and anisotropic behavior-features that require both stability and high spatial accuracy. Traditional finite difference and continuous finite element methods often encounter difficulties in accurately resolving such features without excessive mesh refinement. To overcome these challenges, we adopt a mixed modal miscontinuous Galerkin (DG) method, a high-order numerical technique well-suited for solving convection-diffusion-reaction systems arising in biological modeling.

Now we present the numerical framework employed to solve the cancer cell invasion model described earlier. To capture the complex behavior of the system, we utilize a mixed modal DG method, known for its robustness in handling higher-order diffusion operators and strong gradients. The formulation introduces an auxiliary variable $\mathbf{q}$, representing the spatial derivatives of the solution variables (n, f, m). This allows the original model to be rewritten in first-order form suitable for the mixed DG framework:

$$\begin{aligned} \mathbf{q} - \nabla \mathbf{W} &= 0, \\ \frac{\partial \mathbf{W}}{\partial t} - \nabla \cdot \mathbf{F}(\mathbf{W}, \mathbf{q}) &= \mathbf{S}(\mathbf{W}). \end{aligned} \tag{6.6}$$

The computational domain $\Omega \subset \mathbb{R}^2$ is partitioned into K non-overlapping square elements E_{k}, each mapped from a reference element $E_0 = [-1, 1]^2$ using bilinear transformations as

$$\begin{aligned} \vec{X}_{\mathrm{k}}(a, b) = \frac{1}{4}\big[&\vec{x}_1(1-a)(1-b) + \vec{x}_2(1-a)(1+b) \\ &+ \vec{x}_3(1+a)(1+b) + \vec{x}_4(1+a)(1-b)\big], \end{aligned} \tag{6.7}$$

where $\vec{x}_{\mathrm{i}}$ denote the coordinates of the corners of E_{k}. For uniform Cartesian grids, the Jacobian $\mathbf{J} = \Delta x \Delta y / 4$ remains constant.

We define the approximation space as

$$V_{\mathrm{h}}^{l} = \{\nu \in [L^2(D)] : \nu|_{E_{\mathrm{k}}} \in P^{l}(E_{\mathrm{k}}),\ k = 1, \ldots, K\}, \tag{6.8}$$

where $P^{l}(E_{\mathrm{k}})$ represents polynomials of degree $\leq l$ on E_{k}. Each variable is expanded using hierarchical orthogonal scaled Legendre basis functions [19]:

$$\begin{aligned} \mathbf{q}_{\mathrm{h}}|_{E_{\mathrm{k}}}(a, b, t) &\approx \sum_{i=0}^{N_l - 1} \mathbf{q}_{\mathrm{h}}^{i}(t)\, \varphi_{\mathrm{i}}(a, b), \\ \mathbf{W}_{\mathrm{h}}|_{E_{\mathrm{k}}}(a, b, t) &\approx \sum_{i=0}^{N_l - 1} \mathbf{W}_{\mathrm{h}}^{i}(t)\, \varphi_{\mathrm{i}}(a, b), \end{aligned} \tag{6.9}$$

where $\varphi_{\mathrm{i}}(a,b)$ are modal basis functions. We obtain the weak form by multiplying eqn. (6.6) by a test function φ_{h}, integrating over E_{k}, and applying integration by parts:

$$\begin{aligned}\int_{E_{\mathrm{k}}} \mathbf{q}_{\mathrm{h}}\, \varphi_{\mathrm{h}}\, dV + \int_{E_{\mathrm{k}}} \nabla\varphi_{\mathrm{h}} \cdot \mathbf{W}_{\mathrm{h}}\, dV - \int_{\partial E_{\mathrm{k}}} \varphi_{\mathrm{h}}\, \mathbf{W}_{\mathrm{h}} \cdot \mathbf{n}\, d\Gamma = 0,\\ \frac{\partial}{\partial t}\int_{E_{\mathrm{k}}} \mathbf{W}_{\mathrm{h}}\varphi_{\mathrm{h}}\, dV + \int_{E_{\mathrm{k}}} \nabla\varphi_{\mathrm{h}} \cdot \mathbf{F}(\mathbf{W}_{\mathrm{h}}, \mathbf{q}_{\mathrm{h}})\, dV - \\ \int_{\partial E_{\mathrm{k}}} \varphi_{\mathrm{h}}\, \mathbf{F}(\mathbf{W}_{\mathrm{h}}, \mathbf{q}_{\mathrm{h}}) \cdot \mathbf{n}\, d\Gamma = \int_{E_{\mathrm{k}}} \varphi_{\mathrm{h}}\, \mathbf{S}(\mathbf{W}_{\mathrm{h}})\, dV.\end{aligned} \tag{6.10}$$

Since $\mathbf{W}_{\mathrm{h}}$ and $\mathbf{q}_{\mathrm{h}}$ are discontinuous across elements, we replace boundary fluxes with numerical fluxes $\mathbf{H}^{\mathrm{aux}}$ and $\mathbf{H}^{\mathrm{alt}}$:

$$\begin{aligned}\mathbf{H}^{\mathrm{aux}} &= \frac{1}{2}(\mathbf{W}_{\mathrm{h}}^{\mathrm{int}} + \mathbf{W}_{\mathrm{h}}^{\mathrm{ext}}) + C_{11}(\mathbf{W}_{\mathrm{h}}^{\mathrm{int}} - \mathbf{W}_{\mathrm{h}}^{\mathrm{ext}}),\\ \mathbf{H}^{\mathrm{alt}} &= \frac{1}{2}\left[\mathbf{F}(\mathbf{W}_{\mathrm{h}}^{\mathrm{int}}, \mathbf{q}_{\mathrm{h}}^{\mathrm{int}}) + \mathbf{F}(\mathbf{W}_{\mathrm{h}}^{\mathrm{ext}}, \mathbf{q}_{\mathrm{h}}^{\mathrm{ext}})\right]\\ &\quad - C_{11}(\mathbf{W}_{\mathrm{h}}^{\mathrm{int}} - \mathbf{W}_{\mathrm{h}}^{\mathrm{ext}}) + C_{12}(\mathbf{q}_{\mathrm{h}}^{\mathrm{int}} - \mathbf{q}_{\mathrm{h}}^{\mathrm{ext}}).\end{aligned} \tag{6.11}$$

Substituting these into the weak form yields:

$$\begin{aligned}\int_{E_{\mathrm{k}}} \mathbf{q}_{\mathrm{h}}\, \varphi_{\mathrm{h}}\, dV + \int_{E_{\mathrm{k}}} \nabla\varphi_{\mathrm{h}} \cdot \mathbf{W}_{\mathrm{h}}\, dV - \int_{\partial E_{\mathrm{k}}} \varphi_{\mathrm{h}}\, \mathbf{H}^{\mathrm{aux}}\, d\Gamma = 0,\\ \frac{\partial}{\partial t}\int_{E_{\mathrm{k}}} \mathbf{W}_{\mathrm{h}}\varphi_{\mathrm{h}}\, dV + \int_{E_{\mathrm{k}}} \nabla\varphi_{\mathrm{h}} \cdot \mathbf{F}(\mathbf{W}_{\mathrm{h}}, \mathbf{q}_{\mathrm{h}})\, dV - \\ \int_{\partial E_{\mathrm{k}}} \varphi_{\mathrm{h}}\, \mathbf{H}^{\mathrm{alt}}\, d\Gamma = \int_{E_{\mathrm{k}}} \varphi_{\mathrm{h}}\, \mathbf{S}(\mathbf{W}_{\mathrm{h}})\, dV.\end{aligned} \tag{6.12}$$

We apply Gauss–Legendre quadrature for integration over each element and face. For polynomials of degree l, integration rules must be exact for degree $2l$ (volume) and $2l+1$ (surface). For a reference element:

$$\int_{E_{\mathrm{k}}} f(x,y)\, dxdy = \mathbf{J}\sum_{i=1}^{N_{\mathrm{q}}}\sum_{j=1}^{N_{\mathrm{q}}} \omega_{\mathrm{i}}\omega_j f(a_{\mathrm{i}}^*, b_j^*), \tag{6.13}$$

where $\omega_{\mathrm{i}}, \omega_j$ are weights and a_{i}^*, b_j^* are quadrature points. Combining the elemental contributions, we derive a semi-discrete system:

$$\mathbf{M}\frac{d\mathbf{W}_{\mathrm{h}}}{dt} = \mathbf{L}(\mathbf{W}_{\mathrm{h}}, \mathbf{q}_{\mathrm{h}}), \tag{6.14}$$

where $\mathbf{M}$ is the block-diagonal mass matrix. Since $\mathbf{M}$ is diagonal, it is inverted element-wise. The system is evolved using the third-order SSP-RK method:

$$\begin{aligned}
\mathbf{W}_{\mathrm{h}}^{(1)} &= \mathbf{W}_{\mathrm{h}}^{n} + \Delta t\, \mathbf{M}^{-1}\mathbf{L}(\mathbf{W}_{\mathrm{h}}^{n}, \mathbf{q}_{\mathrm{h}}^{n}),\\
\mathbf{W}_{\mathrm{h}}^{(2)} &= \frac{3}{4}\mathbf{W}_{\mathrm{h}}^{n} + \frac{1}{4}\left(\mathbf{W}_{\mathrm{h}}^{(1)} + \Delta t\, \mathbf{M}^{-1}\mathbf{L}(\mathbf{W}_{\mathrm{h}}^{(1)}, \mathbf{q}_{\mathrm{h}}^{(1)})\right),\\
\mathbf{W}_{\mathrm{h}}^{n+1} &= \frac{1}{3}\mathbf{W}_{\mathrm{h}}^{n} + \frac{2}{3}\left(\mathbf{W}_{\mathrm{h}}^{(2)} + \Delta t\, \mathbf{M}^{-1}\mathbf{L}(\mathbf{W}_{\mathrm{h}}^{(2)}, \mathbf{q}_{\mathrm{h}}^{(2)})\right).
\end{aligned} \tag{6.15}$$

6.4 Validation Study

To establish the accuracy and reliability of the proposed solver, we perform a validation study against a benchmark problem in the literature. Although the mixed modal DG framework has been previously tested on standard reaction-diffusion systems [19], this section focuses specifically on validating the cancer cell invasion model.

We compare our numerical results with those reported by Enderling et al. [20], who proposed a pioneering study on cancer invasion dynamics. In

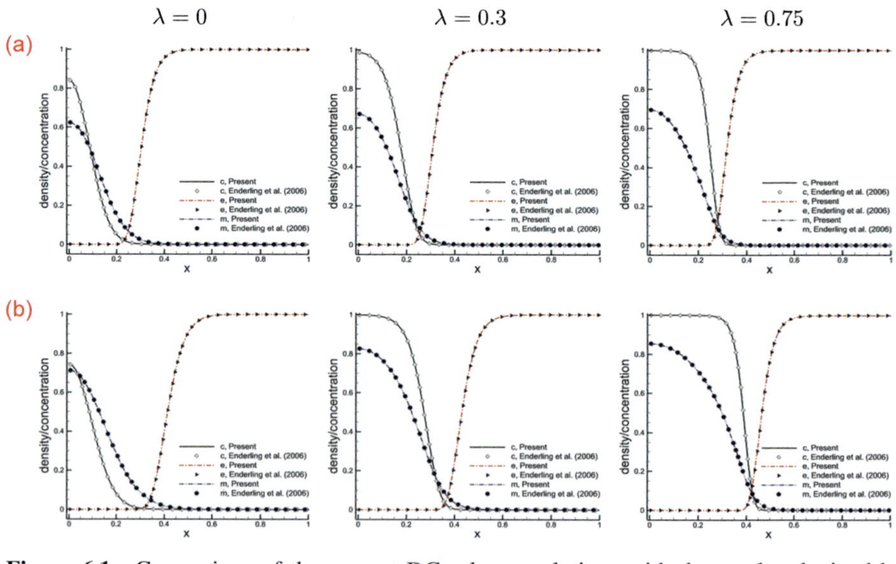

Figure 6.1 Comparison of the present DG scheme solutions with the results obtained by Enderling et al. [20] with three different cancer proliferation rates $\lambda = 0, 0.3, 0.75$ at $t = 10$ and 20.

our simulation, we adopt the same dimensionless parameter values used in their work:

$$\begin{aligned} &d_1 = 0.0001, \quad d_2 = 0.0005, \quad \chi = 0.00005, \quad \eta = 10, \\ &\rho = 0.0, \quad \alpha = 0.1, \quad \beta = 0, \end{aligned}$$

with three cancer proliferation rates $\lambda = 0, 0.3, 0.75$.

The domain is defined as $\Omega = [0, 1]$, and the model equations are solved using the mixed modal DG scheme described earlier. Figure 6.1 presents a side-by-side comparison of the cancer cell density, ECM density, and MDE concentration profiles obtained from our DG method with those reported by Enderling et al [20]. In the figure, the solid profile lines denote the computed solutions using our DG solver, while the symbols correspond to Enderling et al.'s published data. The results exhibit excellent agreement across all three variables, confirming the correctness and fidelity of the current implementation. This validation not only reinforces the robustness of the mixed modal DG approach but also supports its suitability for simulating biologically realistic tumor invasion scenarios.

6.5 Simulation Studies and Biological Insight

To assess the performance of the mixed modal DG method and to extract biologically relevant insights, we conducted a series of numerical simulations for 1D and 2D cases of the cancer cell invasion model. These simulations investigate the effects of critical parameters such as cancer cell proliferation rate and haptotactic sensitivity on the invasive behavior of tumor cells and the associated changes in ECM and MDE distributions.

6.5.1 1D simulations

To investigate the tissue dynamics of cancer cell invasion model (6.3) in 1D space, we choose the following initial conditions in the computational domain $\Omega = [0, 1]$ with $\epsilon = 0.01$:

$$\begin{aligned} n(x,0) &= \begin{cases} \exp\left(\frac{-x^2}{\epsilon}\right), & x \in [0, 0.25] \\ 0, & x \in (0.25, 1] \end{cases}, \\ f(x,0) &= 1 - 0.5\, n(x,0), \\ m(x,0) &= 0.5\, n(x,0). \end{aligned} \tag{6.16}$$

In the 1D case, we consider the spatial domain $\Omega = [0, 1]$, initialized with a localized distribution of cancer cells and homogeneous ECM and MDE profiles. The objective is to understand the individual and combined roles of cancer proliferation and haptotactic migration in shaping tumor dynamics.

6.5.1.1 Effect of proliferation rate (λ)

Here, we show the impact of cancer proliferation rate (λ) on the dynamics of cancer cell invasion model. For this purpose, we take into account the following parameters' values: $d_1 = 0.0001$, $d_2 = 0.0005$, $\chi = 0.01$, $\eta = 10$, $\rho = 0, \alpha = 0.5$, $\beta = 0.4$ with four different λ values ($\lambda = 0, 0.5, 1, 1.5$). Figure 6.2 illustrates the λ impacts on the tissue dynamics of cancer cell invasion by showing the evolution of cancer cell density, ECM density, and MDE

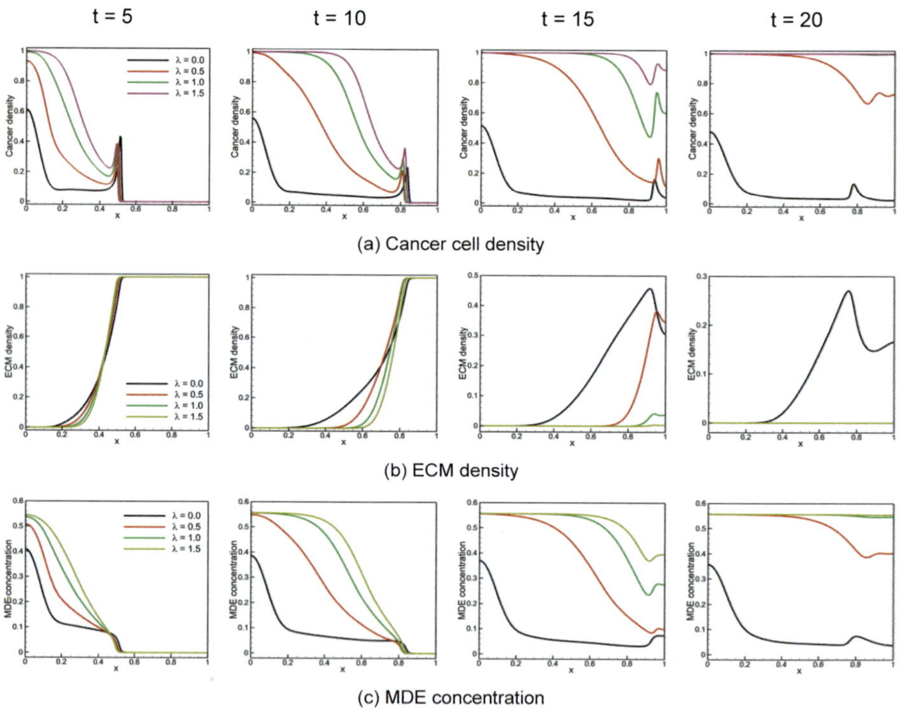

Figure 6.2 Impact of cancer proliferation rate ($\lambda = 0, 0.5, 1.0, 1.5$) on the tissue dynamics of 1D cancer cell invasion model: Profiles of (a) cancer cell density, (b) ECM density, and (c) MDE concentration with $d_1 = 0.0001$, $d_2 = 0.0005$, $\chi = 0.01$, $\eta = 10$, $\rho = 0, \alpha = 0.5$, $\beta = 0.4$ at $t = 5, 10, 15, 20$.

concentration over time ($t = 5, 10, 15,$ and 20). According to the cancer cell density profiles in Figure 6.2(a), a more aggressive invasion is produced by $\lambda = 1.5$ than by lower values, indicating that higher proliferation rates cause tumors to expand more quickly. Initially, the tumor stays contained at $t = 5$, but as time goes on, increased rates of proliferation speed up the spread and produce a more distinct invasion front. As can be seen from the ECM density profiles in Figure 6.2(b), the proliferation rate has a major impact on ECM degradation. ECM is more severely degraded at higher λ, especially near the tumor front, while it is mostly intact at lower λ. By $t = 20$, for $\lambda = 1.5$, the tumor-invaded region's extracellular matrix is almost completely gone, which promotes the spread of the disease. The MDE concentration patterns, which are displayed in Figure 6.2(c), demonstrate that enzyme synthesis is first concentrated close to the tumor area and then expands as the invasion advances. Higher rates of proliferation result in longer-lasting MDE release, which promotes tumor movement and ECM disintegration. These findings validate that the pace of proliferation is a critical factor in determining the rate of cancer invasion, ECM remodeling, and enzyme activity, all of which impact the overall course of cancer.

6.5.1.2 Effect of haptotactic coefficient

Here, we analyze the impact of the haptotactic coefficient (χ) on the dynamics of 1D cancer cell invasion model. For this purpose, we consider the following parameters values: $d_1 = 0.0001$, $d_2 = 0.0005$, $\lambda = 0.01$, $\eta = 10$, $\rho = 0, \alpha = 0.5$, $\beta = 0.4$ with four different haptotaxis coefficient values $\chi = 0.00001, 0.001, 0.01,$ and 0.1. Figure 6.3 illustrates the χ impact on the cancer cell density, ECM density, and MDE concentration at different time instants ($t = 5, 10, 15,$ and 20). Haptotaxis describes the directed movement of cancer cells in response to gradients in ECM density, making χ a crucial parameter in tumor invasion dynamics. The profiles of cancer cell density are shown in Figure 6.3(a). The tumor stays mostly contained at early times ($t = 5$), but as time goes on, a substantially faster invasion front is caused by larger values of χ ($\chi = 0.1$), while slow or limited migration is caused by lower values ($\chi = 0.00001$). Higher haptotactic sensitivity cancer cells ($\chi = 0.1$) have propagated aggressively over the domain by $t = 20$, while lower χ values exhibit restricted mobility. This suggests that cancer cells can invade more quickly and precisely when their haptotactic responses are higher, which makes it easier for them to go toward areas with a lot of extracellular matrix. ECM density profiles are displayed in Figure 6.3(b),

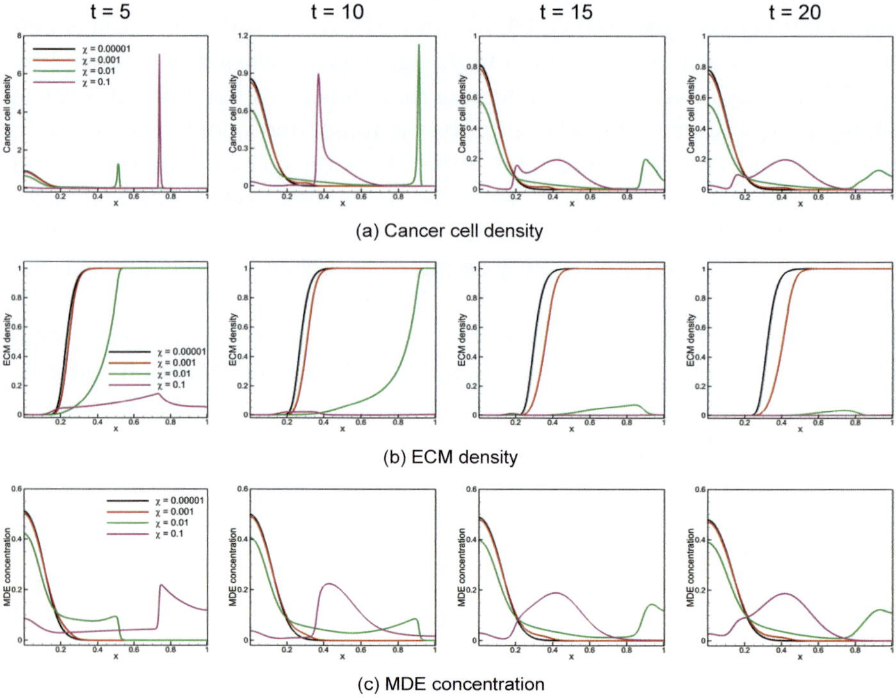

Figure 6.3 Impact of haptotactic coefficient ($\chi = 0.00001, 0.001, 0.01, 0.1$) on the dynamics of 1D cancer cell invasion model: Profiles of (a) cancer cell density , (b) ECM density, and (c) MDE concentration with $d_1 = 0.0001$, $d_2 = 0.0005$, $\lambda = 0.01$, $\eta = 10$, $\rho = 0, \alpha = 0.5$, $\beta = 0.4$ at $t = 5, 10, 15, 20$.

emphasizing the deterioration process brought on by cancer cell invasion. Since cancer cells do not move very far from their original location, the extracellular matrix (ECM) is mostly intact for smaller χ values. ECM deterioration is more noticeable at greater χ, though, especially along the invasion front, where cancer cells are aggressively dissolving the matrix to make room for additional movement. ECM is nearly completely destroyed in areas where cancer cells have infiltrated by $t = 20$, for $\chi = 0.1$, indicating a substantial association between haptotaxis and ECM remodeling. MDE concentration profiles are shown in Figure 6.3(c), which demonstrates the strong relationship between MDE secretion and the location of cancer cells. MDE spreads quickly at greater χ, which promotes broad ECM disintegration. MDE stays close to the tumor area in the early stages ($t = 5$), but as cancer cells spread farther ($t = 10, 15,$ and 20), the concentration of the enzyme increases in tandem, speeding up the breakdown of extracellular matrix and promoting tumor invasion.

6.5.2 2D simulations

To explore the dynamics of cancer cell invasion model (6.3) in 2D space, the following initial conditions are considered with $\epsilon = 0.0025$:

$$\begin{aligned} n(x,y,0) &= \exp\left(-\frac{(x-0.5)^2+(y-0.5)^2}{\epsilon}\right), \\ f(x,y,0) &= 1-0.5\,n(x,y,0), \\ m(x,y,0) &= 0.5\,n(x,y,0). \end{aligned} \tag{6.17}$$

For this problem, a square domain $\Omega = [0,1] \times [0,1]$ is considered. The simulations are performed with grid points 150×150 and time-step $\Delta = 0.001$.

6.5.2.1 Overall tissue dynamics

We illustrate here the overall tissue dynamics in 2D cancer cell invasion model. For this numerical experiment, the following parameters are considered: $d_1 = 0.0001$, $d_2 = 0.0005$, $\chi = 0.00001, \lambda = 1.0$, $\eta = 10$, $\rho = 0, \alpha = 0.5$, $\beta = 0.4$. Figure 6.4 presents the time evolution of the 2D cancer cell invasion model, showing the dynamic interactions between cancer cell density, ECM density, and MDE concentration at different time instants ($t = 5, 10, 15,$ and 20). The 3D surface plots visualize the spatial distribution and progression of cancer invasion over time, demonstrating how the tumor expands, degrades the ECM, and secretes MDEs. At $t = 0$, the cancer cell density first produces a confined peak in Figure 6.4(a), which represents a tiny tumor. The cancer cells multiply and spread outward over time ($t = 5, 10,$ and 15), enlarging the surrounding area. An increase in density at the center and the development of a wider invasion front are characteristics of tumor growth that suggest active spreading brought on by proliferation and haptotactic migration toward areas rich in extracellular matrix. The ECM density progression is depicted in Figure 6.4(b), which demonstrates the increasing deterioration brought on by cancer cell invasion. The ECM, which gives the tissue structural support, is largely intact at $t = 0$. ECM degradation increases surrounding the tumor site, especially at the growing invasion front, when cancer cells invade ($t = 5, 10,$ and 15). ECM remodeling is a critical element in tumor growth, as seen by the considerable

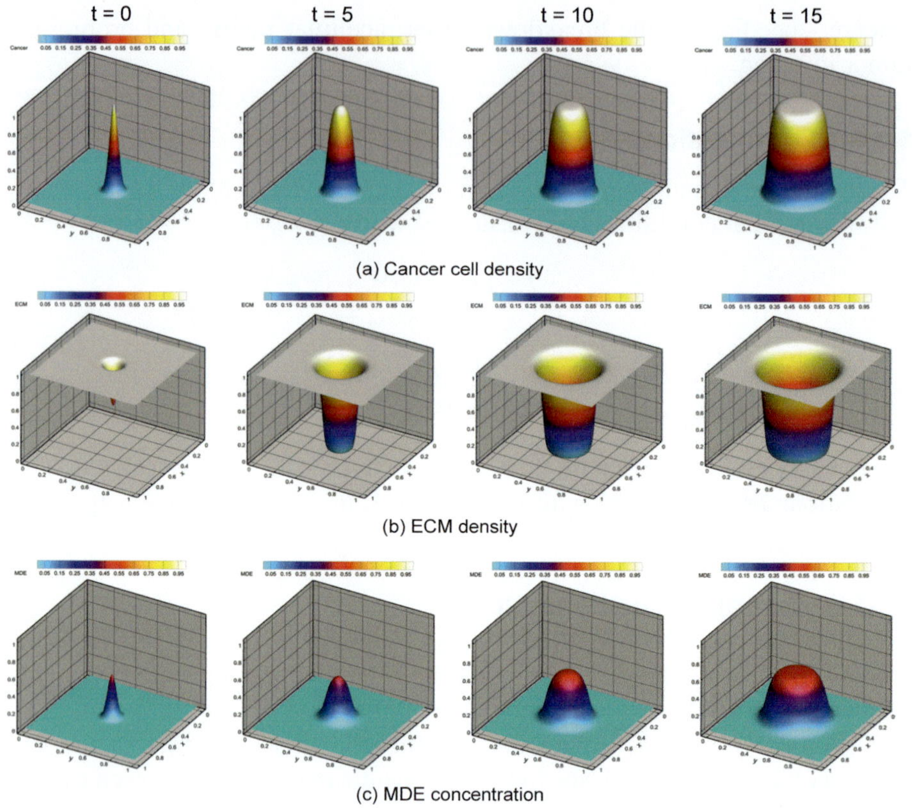

Figure 6.4 2D cancer cell invasion model: time evolution of (a) cancer cell density, (b) ECM density, and (c) MDE concentration contours with $d_1 = 0.0001$, $d_2 = 0.0005$, $\chi = 0.00001, \lambda = 1.0$, $\eta = 10$, $\rho = 0, \alpha = 0.5$, $\beta = 0.4$.

ECM depletion that occurs by $t = 15$ in areas where cancer cells have moved. The MDE concentration over time is shown in Figure 6.4(c), emphasizing the part matrix-degrading enzymes play in promoting ECM degradation. MDE production follows the migration pattern of cancer cells as they spread, but initially it is concentrated close to the tumor ($t = 0$). The MDE distribution at $t = 15$ indicates that the tumor front, where ECM degradation is most severe, is where enzymatic activity is highest. This strengthens the positive feedback loop that links MDE secretion, ECM degradation, and cancer growth.

Figure 6.5 presents the 1D profiles of cancer cell density, ECM density, and MDE concentration along the centerline of the computational domain ($y = 0.5$) at different time instants ($t = 0, 5, 10,$ and 15). These profiles,

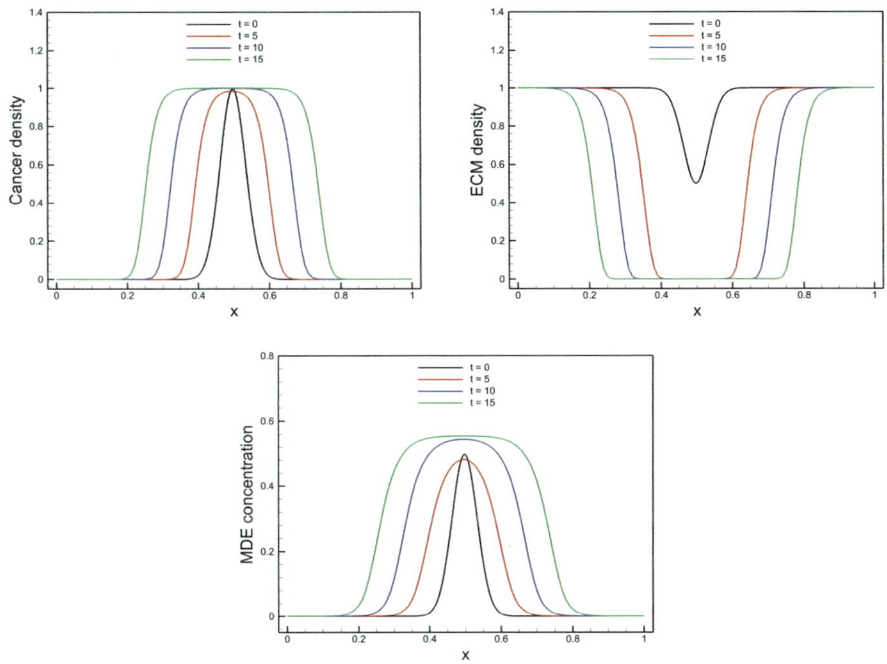

Figure 6.5 2D cancer cell invasion model: profiles of cancer cell density, ECM density, and MDE concentration taken along the centerline of computational domain ($y = 0.5$) at different time instants.

which offer a cross-sectional view of how tumor progression changes over time, are taken from the 2D cancer invasion model displayed in Figure 6.4. The density profile of cancer cells is displayed in the first plot. At $t = 0$, cancer cells are found close to the domain's center, peaking at $x = 0.5$. While keeping a high center density, the tumor widens and spreads outward with time ($t = 5, 10,$ and 15). As time passes, the profile widens, signifying ongoing tumor invasion and proliferation into the surrounding tissue. Later time steps demonstrate the directed migration of cancer cells by making the leading edges of the tumor front more noticeable. The ECM density profile is depicted in the second plot. ECM is evenly dispersed throughout the domain at $t = 0$. However, ECM breakdown is visible in the core region as cancer cells invade ($t = 5, 10,$ and 15). Where cancer cells are present, the ECM density dramatically drops, creating a V-shaped depletion zone. ECM degradation spreads outward over time, demonstrating how the tumor gradually reorganizes and disassembles the extracellular environment to allow for further invasion. The MDE concentration profile is seen in the third plot. MDE levels are initially low and centered close to the tumor site. MDE secretion rises and spreads throughout time, resulting in a wider distribution at subsequent time steps. The feedback process in which cancer cells generate enzymes to break down ECM and promote migration is strengthened by the peak concentration of MDE staying in line with the tumor front.

6.5.2.2 Effect of cancer proliferation rate

Figure 6.6 illustrates the effect of different cancer proliferation rates ($\lambda = 0, 0.5, 1.0, \text{and} 1.5$) on the tumor progression in a 2D cancer cell invasion

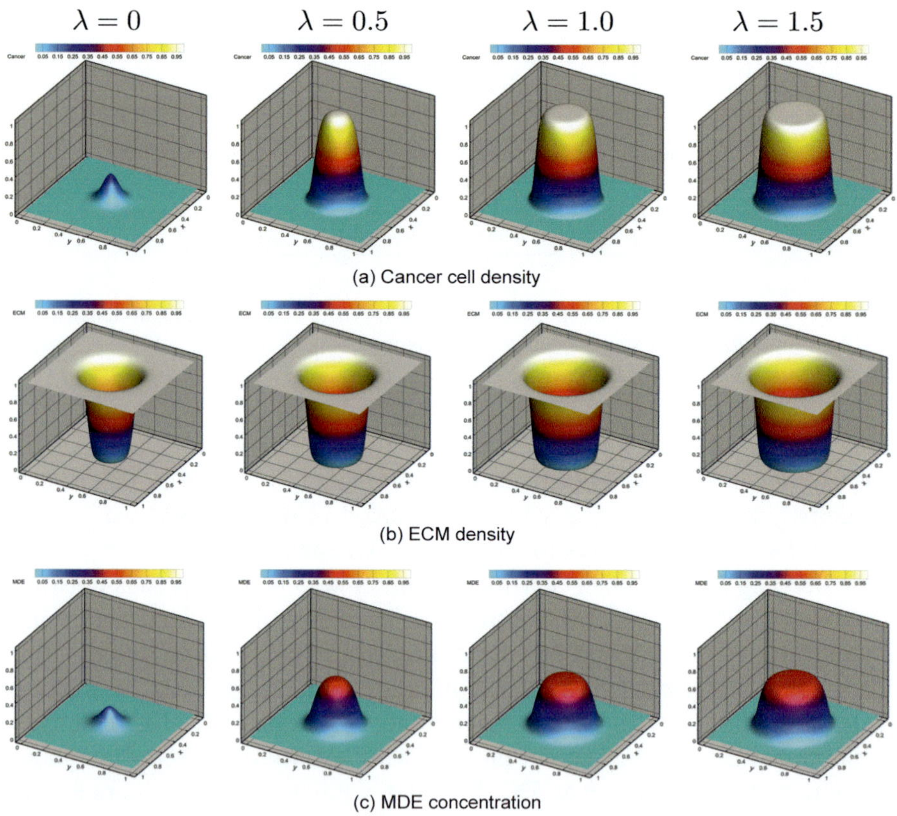

Figure 6.6 2D cancer cell invasion model: Contours of (a) cancer cell density, (b) ECM density, and (c) MDE concentration with $d_1 = 0.0001$, $d_2 = 0.0005$, $\chi = 0.01$, $\eta = 10$, $\rho = 0, \alpha = 0.5$, $\beta = 0.4$ and different cancer proliferation rate ($\lambda = 0, 0.5, 1.0, 1.5$) at $t = 15$.

model at $t = 15$. The other simulation parameters are considered as follows: $d_1 = 0.0001$, $d_2 = 0.0005$, $\chi = 0.01$, $\eta = 10$, $\rho = 0, \alpha = 0.5$, $\beta = 0.4$. The figure presents the spatial distributions of cancer cell density, ECM density, and MDE concentration to analyze how proliferation influences tumor expansion, ECM degradation, and enzymatic activity. Figure 6.6(a) displays the cancer cell density, highlighting the effect of proliferation on tumor growth. For $\lambda = 0$, the tumor remains localized with minimal expansion, while for increasing values of λ, the cancer cell population spreads outward, forming a well-defined invasion front. Higher proliferation rates lead to faster and more extensive tumor expansion, with $\lambda = 1.5$ exhibiting the largest tumor region. The results confirm that cancer proliferation plays a key role in determining the speed and extent of invasion. The ECM density is depicted in Figure 6.6(b), which demonstrates how tumor cells break down the extracellular matrix to make invasion easier. Since the tumor is not actively growing at $\lambda = 0$, the extracellular matrix is mostly intact. ECM breakdown, however, intensifies with increasing λ, especially near the tumor front where cancer cells migrate and invade. ECM is nearly completely depleted in the invaded region for $\lambda = 1.5$, indicating that more aggressive ECM remodeling results from quicker proliferation. The MDE concentration, which is essential for ECM degradation, is shown in Figure 6.6(c). At first, MDE secretion is small and confined when $\lambda = 0$. MDE synthesis escalates when λ rises, extending beyond the tumor boundary and accelerating the breakdown of the extracellular matrix. MDEs are substantially enriched close to the invasion front at higher proliferation rates ($\lambda = 1.5$), which speeds up ECM degradation and permits more widespread tumor growth.

6.6 Conclusion and Outlook

In this study, we presented a high-fidelity computational framework for modeling cancer cell invasion by integrating a biologically grounded reaction-diffusion model with a robust mixed modal DG numerical scheme. The model encapsulates key mechanisms of tumor progression, including proliferation, haptotaxis, ECM degradation, and MDE secretion, and provides a spatiotemporal platform for exploring tumor dynamics in 1D and 2D domains.

Through extensive numerical simulations, we examined the influence of critical parameters such as the cancer cell proliferation rate and haptotactic sensitivity on the overall invasion pattern and matrix remodeling. The mixed modal DG method effectively captured steep gradients, nonlinearity, and sharp invasion fronts, offering accurate and stable numerical solutions.

Validation against benchmark models confirmed the accuracy and reliability of our approach. The results demonstrate that proliferation significantly enhances tumor spread and ECM breakdown, while higher haptotactic sensitivity leads to faster, more directed cell migration. 2D simulations illustrated how complex tumor morphologies emerge from the interaction of these mechanisms, with the spatial distribution of MDE closely following tumor progression and reinforcing ECM degradation.

Future work could extend this framework to include additional biological phenomena such as angiogenesis and immune response. Integrating patient-specific data and machine learning techniques may enhance its predictive capability and support clinical applications in therapy planning and cancer prognosis.

References

[1] D. Hanahan and R.A. Weinberg. Hallmarks of cancer: the next generation. *Cell*, 144(5):646–674, 2011.

[2] P. Friedl and K. Wolf. Tumour-cell invasion and migration: diversity and escape mechanisms. *Nature Reviews Cancer*, 3(5):362–374, 2003.

[3] E. Sahai. Mechanisms of cancer cell invasion. *Current opinion in genetics & development*, 15(1):87–96, 2005.

[4] A. V. Hill. The diffusion of oxygen and lactic acid through tissues. *Proc. R. Soc. London, Ser. B*, 104:39–96, 1928.

[5] R.H. Thomlinson and L.H. Gray. The histological structure of some human lung cancers and the possible implications for radiotherapy. *Br. J. Cancer*, 9:539–549, 1955.

[6] R.P. Araujo and D.L.S. McElwain. A history of the study of solid tumour growth: the contribution of mathematical modelling. *Bulletin of mathematical biology*, 66(5):1039–1091, 2004.

[7] H. McAneney and S.F.C. O'Rourke. Investigation of various growth mechanisms of solid tumour growth within the linear-quadratic model for radiotherapy. *Physics in Medicine & Biology*, 52(4):1039–1054, 2007.

[8] H. McAneney and S.F.C. O'Rourke. Simple ode models of tumor growth and anti-angiogenic or radiation treatment. *Mathematical and Computer Modelling*, 33(12–13):1297–1305, 2001.

[9] J.B. McGillen, E.A. Gaffney, N.K. Martin, and P.K. Maini. A general reaction–diffusion model of acidity in cancer invasion. *Journal of mathematical biology*, 68:1199–1224, 2014.

[10] M.A.J. Chaplain and G Lolas. Mathematical modelling of cancer cell invasion of tissue: the role of the urokinase plasminogen activation system. *Math. Models Methods Appl. Sci.*, 15:1685–1734, 2005.

[11] R.A. Gatenby and E.T. Gawlinski. A reaction-diffusion model of cancer invasion. *Cancer Res.*, 56:5745–5753, 1996.

[12] A.R.A. Anderson, M.A.J. Chaplain, E.L. Newman, R.J.C. Steele, and A.M. Thompson. Mathematical modelling of tumor invasion and metastasis. *J. Theor. Med.*, 2:129–154, 2000.

[13] A. Gerisch and M.A.J. Chaplain. Mathematical modelling of cancer cell invasion of tissue:local and non-local models and the effect of adhesion. *J. Theoret. Biol.*, 250:684–704, 2008.

[14] K.J. Painter, N.J. Armstrong, and J.A. Sherratt. The impact of adhesion on cellular invasion processes in cancer and development. *J. Theoret. Biol.*, 264:1057–1067, 2010.

[15] S. Ganesan and S. Lingeshwaran. Galerkin finite element method for cancer invasion mathematical model. *Comput. Math. Appl.*, 73:2603–2617, 2017.

[16] M. Dehghan and N. Narimani. An element-free galerkin meshless method for simulating the behavior of cancer cell invasion of surrounding tissue. *Appl. Math. Model.*, 59:500–513, 2018.

[17] F. Asadi-Mehregan, P. Assari, and M. Dehghan. Simulation of the cancer cell growth and their invasion into healthy tissues using local radial basis function method. *Eng. Anal. Bound. Elem.*, 163:56–68, 2024.

[18] Y. Epshteyn. Discontinuous galerkin methods for the chemotaxis and haptotaxis models. *J. Computat. Appl. Mathemat*, 224:168–181, 2009.

[19] S. Singh, R.C. Mittal, S.R. Thottoli, and M. Tamsir. High-fidelity simulations for turing pattern formation in multi-dimensional gray–scott reaction-diffusion system. *Appl. Mathemat. Comput.*, 452:128079, 2023.

[20] H. Enderling. Cancer stem cells and tumor dormancy. *Adv. Exp. Med. Biol.*, 734:55–71, 2013.

7

Computational Modeling of Shock-induced Instabilities in Chevron-shaped Light Gas Interface

Satyvir Singh[1,2]

[1] Institute for Applied and Computational Mathematics, RWTH Aachen University, Germany
[2] Department of Mathematics, Graphic Era Deemed to be University, India
E-mail: singh@acom.rwth-aachen.de

Abstract

This chapter presents a detailed computational study of shock-induced instabilities arising in a chevron-shaped light gas interface subjected to planar shock waves of varying strength. The interaction between the incident shock and the geometrically complex interface is modeled using a high-order discontinuous Galerkin method to solve the 2D compressible Euler equations for two-component flows. The simulations capture the nonlinear evolution of Richtmyer–Meshkov instability, emphasizing the effects of shock strength on interface deformation, vorticity generation, and enstrophy growth. A comprehensive flow analysis reveals that higher Mach numbers lead to enhanced baroclinic torque, stronger compressibility effects, and increased asymmetry in vortex structures. The chevron-shaped interface is particularly prone to instability due to localized curvature-induced misalignment of pressure and density gradients, which amplifies baroclinic vorticity deposition. Diagnostic measures such as integrated vorticity components and enstrophy provide quantitative insights into the instability mechanisms and their dependence on shock intensity.

Keywords: Richtmyer–Meshkov instability, Chevron-shaped interface, shock wave, discontinuous Galerkin simulations

DOI: 10.1201/9788743806691-7

7.1 Introduction

The shock-induced instabilities, particularly Richtmyer–Meshkov instability (RMI), arise when a shock wave interacts with a perturbed interface between two fluids of different densities, leading to the generation of complex flow features including vortex formation, jetting, and turbulent mixing. Initially described by Richtmyer [1] and later extended by Meshkov [2], RMI is driven by the baroclinic vorticity deposition due to the misalignment of pressure and density gradients ($\nabla p \times \nabla \rho \neq 0$). These instabilities have profound implications in various high-speed compressible flow environments such as inertial confinement fusion [3, 4], astrophysical explosions [5, 6], supersonic combustion [7], and advanced propulsion systems [8].

Most classical studies of RMI have focused on geometrically simple configurations such as planar [9], cylindrical [10, 11], spherical [12], and elliptical [13] interfaces. These have provided important insight into shock refraction, vortex pair formation, and interface deformation. However, growing interest in more realistic and complex flow configurations has led to the investigation of polygonal geometries [14, 15], which introduce stronger curvature effects, geometric asymmetries, and localized pressure-density misalignments, thereby amplifying baroclinic vorticity production. Among these, the chevron-shaped interfaces represent a particularly interesting case due to their sharp tips and multiple angles, promoting focused shock interactions and non-uniform instability growth.

Recent numerical studies by Singh et al. [16–19] and Luo et al. [14, 20] have highlighted the impact of Mach number, Atwood number, and thermal non-equilibrium effects on the evolution of RMI in square, rectangular, and triangular geometries. These works demonstrated that both shock strength and interface topology play decisive roles in the development of secondary instabilities such as Kelvin-Helmholtz roll-up, jet formation, and wave diffraction. Yet, the dynamics of shock interaction with chevron-shaped light gas interfaces remain underexplored.

Due to the limitations of experimental methods in capturing fine-scale interface dynamics and obtaining repeatable initial conditions, high-resolution numerical simulations have become a powerful alternative to study RMI [21, 22]. In particular, discontinuous Galerkin (DG) methods [23, 24] have emerged as a robust framework for solving hyperbolic systems of conservation laws. They combine the geometric flexibility of finite element methods with the shock-capturing capabilities of finite volume methods, making them particularly well-suited for compressible multi-component flows involving sharp gradients and complex interface dynamics.

Unlike convex or smoothly curved geometries, chevron-shaped interfaces induce directional asymmetries in wave propagation and vorticity deposition, often resulting in early formation of localized vortex structures and non-uniform shock refraction patterns. These features make chevron configurations an ideal testbed for examining the sensitivity of RMI growth to both geometry-induced curvature and shock strength. Furthermore, the inherent complexity of the chevron topology enables the study of secondary instability mechanisms, such as interface folding, vortex pairing, and shock-focusing, which are typically less pronounced in canonical geometries.

In this study, we employ a high-order modal DG scheme to simulate RMI in a 2D chevron-shaped light gas interface subjected to planar shock waves of varying strength. The compressible Euler equations are solved for two-component inviscid flow with accurate spatial and temporal discretization. The study focuses on capturing key flow features such as baroclinic and dilatational vorticity generation, interface deformation, enstrophy amplification, and secondary vortex dynamics. Spatially integrated metrics such as average vorticity, enstrophy, and kinetic energy are used to quantify the evolution and strength of instability. The influence of Mach number on the growth of instabilities is examined in detail. By presenting a systematic investigation into the shock-chevron interface interaction, this chapter provides valuable insights into the role of interface geometry in modulating RMI dynamics. The findings are relevant for improving predictive models in high-speed flow systems where control of shock-induced mixing is essential.

7.2 Problem Setup

This study investigates the interaction between a planar shock wave and a chevron-shaped heavy/light gas interface in a 2D rectangular domain of size $[0, 100] \times [0, 200]$ mm^2, as illustrated in Figure 7.1. The chevron interface is initialized downstream of an incident shock (IS) wave, with its leftmost tip located 5 mm from the IS front. The IS wave propagates from left to right along the x-axis and is embedded in an air medium. The IS wave begins 25 mm from the left boundary of the computational domain. The incoming Mach number M_s characterizes the strength of the IS wave. The chevron-shaped interface exhibits a characteristic zig-zag geometry. Its total height, twice the standard single-mode amplitude, is denoted by a_0, while the wavelength is fixed as $\lambda = L = 100$ mm. The geometric configuration is fully described by the vertex angle θ, which determines the slope of the chevron arms. The spatial location of the interface tip along the z-axis is given by the

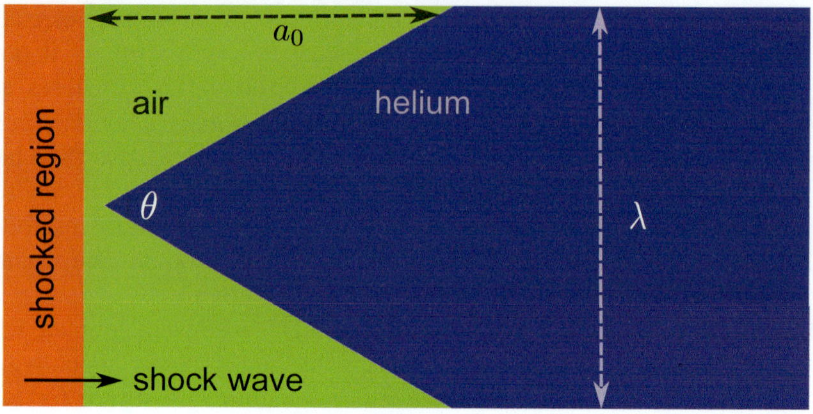

Figure 7.1 Schematic of the initial configuration for shock interaction with a chevron-shaped heavy/light gas interface. A planar shock wave propagates from left to right through a two-fluid system comprising air and helium. The Chevron-shaped interface separating the fluids is characterized by an initial wavelength λ, amplitude a_0, and vortex angle θ.

relation:

$$z = \frac{L}{8} + \frac{L}{2\tan(\theta/2)}, \tag{7.1}$$

where $\theta = 60^\circ$ is used to explore the influence of interface sharpness on instability development. To examine the influence of shock strength, three distinct Mach numbers are considered: $M_s = 1.15, 1.25,$ and 1.50. In this configuration, the left boundary is treated as the inflow, while the top, bottom, and right boundaries are designated as outlets. The initial thermodynamic conditions at the interface are set as $P_0 = 101.325$ Pa and $T_0 = 273$ K.

7.3 Governing Equations

To analyze the instability dynamics triggered by shock interaction with chevron-shaped heavy/light gas interfaces, we solve the 2D compressible Euler equations for a two-component gas system. These equations, representing conservation of mass, momentum, and energy, are formulated in a conservative form as

$$\frac{\partial U}{\partial t} + \frac{\partial F}{\partial x} + \frac{\partial G}{\partial y} = 0, \tag{7.2}$$

where $\mathbf{U}$ is the vector of conserved variables, and $\mathbf{F}$ and $\mathbf{G}$ are the inviscid fluxes in the x- and y-directions, respectively. These vectors are expressed as

$$U = \begin{bmatrix} \rho \\ \rho u \\ \rho v \\ \rho E \\ \rho\phi \end{bmatrix}, \quad F = \begin{bmatrix} \rho u \\ \rho u^2 + p \\ \rho uv \\ (\rho E + p)u \\ \rho\phi u \end{bmatrix}, \quad G = \begin{bmatrix} \rho v \\ \rho uv \\ \rho v^2 + p \\ (\rho E + p)v \\ \rho\phi v \end{bmatrix}. \tag{7.3}$$

In this formulation, ρ denotes the gas density, u and v are the flow velocities along the horizontal and vertical directions, p is the pressure, E is the specific total energy, and ϕ represents the mass fraction of the light gas species. The specific total energy is linked to pressure and kinetic energy via:

$$\rho E = \frac{p}{\gamma_{\text{mix}} - 1} + \frac{1}{2}(u^2 + v^2), \tag{7.4}$$

with γ_{mix} being the effective specific heat ratio of the gas mixture. The fluids are assumed to be ideal and calorically perfect, and the gas mixture properties are derived from the individual components. The effective heat capacity ratio γ_{mix} is computed based on the mass-weighted contributions of the two gases as follows:

$$\gamma_{\text{mix}} = \frac{C_{\text{p1}}\phi_1 + C_{\text{p2}}\phi_2}{C_{\text{v1}}\phi_1 + C_{\text{v2}}\phi_2}, \tag{7.5}$$

where ϕ_1 and $\phi_2 = 1 - \phi_1$ denote the mass fractions of the light and heavy gases, respectively. The constants $C_{\text{p1}}, C_{\text{v1}}$ and $C_{\text{p2}}, C_{\text{v2}}$ represent the specific heats at constant pressure and volume for each component. Initially, the domain is initialized with a sharp interface: $\phi_1 = 1$ inside the light gas region (helium) and $\phi_1 = 0$ in the surrounding air. This setup allows accurate resolution of interfacial instability and baroclinic torque effects post-shock impact.

7.4 Numerical Methodology

Discontinuous Galerkin (DG) methods have become increasingly popular for solving complex PDEs in fluid dynamics and related fields due to their high-order accuracy, local conservation, and ability to handle irregular meshes [25]. In this study, a high-order modal DG method based on third-order Legendre polynomials is used to solve the two-component Euler equations. This formulation efficiently captures sharp interfaces and shock-driven instabilities in Chevron-shaped gas configurations.

To discretize the governing equation, the physical domain $D \subset \mathbb{R}^2$ is partitioned into K non-overlapping square elements, denoted by $\{E_{\text{k}}\}_{k=1}^{K}$.

Each physical element E_k is mapped to a reference element $E_0 = [-1,1]^2$ using a bilinear transformation $\overrightarrow{X}_\mathrm{k} : E_0 \to E_\mathrm{k}$, defined as

$$\begin{aligned}\overrightarrow{X_\mathrm{k}}(\xi,\eta) = \frac{1}{4}[&\overrightarrow{x_1}(1-\xi)(1-\eta) + \overrightarrow{x_2}(1-\xi)(1+\eta)\\ &+\overrightarrow{x_3}(1+\xi)(1+\eta) + \overrightarrow{x_4}(1+\xi)(1-\eta)],\end{aligned} \tag{7.6}$$

where $\{\overrightarrow{x}_1, \overrightarrow{x}_2, \overrightarrow{x}_3, \overrightarrow{x}_4\}$ are the vertices of the physical element E_k. Under the assumption of a uniform Cartesian mesh, the Jacobian of this transformation is constant and given by $J = \frac{1}{4}\Delta x \Delta y$.

Now we define the local discontinuous finite element space as

$$V_\mathrm{h}^\mathrm{l} = \left\{\nu \in [L^2(D)]^5 \mid \nu|_{E_\mathrm{k}} \in [P^\mathrm{l}(E_\mathrm{k})]^5, \quad k = 1,\ldots,K\right\}, \tag{7.7}$$

where $P^\mathrm{l}(E_\mathrm{k})$ denotes the space of polynomials of degree at most l on E_k. Each conservative variable vector $\mathbf{U} \in \mathbb{R}^5$ is approximated locally in modal form by polynomials of degree l on the reference element E_0:

$$U_\mathrm{h}(x,y,t)|_{E_\mathrm{k}} = U_\mathrm{h}(\xi,\eta,t) \approx \sum_{i,j=0}^{\mathrm{l}} U_{\mathrm{i}j}\, b_{\mathrm{i}j}(\xi,\eta), \tag{7.8}$$

where $\mathbf{U}_{\mathrm{i}j}$ are the time-dependent modal coefficients associated with element E_k, and $N_l = (l+1)^2$ is the number of basis functions for an l-exact approximation. The basis functions $b_{\mathrm{i}j}(\xi,\eta)$ are chosen as scaled Legendre polynomials [13], defined by

$$b_{\mathrm{i}j}(\xi,\eta) = \frac{2^i(i!)^2}{(2i)!}P_i(\xi) \times \frac{2^j(j!)^2}{(2j)!}P_j(\eta), \quad i,j = 0,\ldots,l, \tag{7.9}$$

where $P_i(\cdot)$ is the i-th Legendre polynomial on $[-1,1]$.

The discretization proceeds by formulating the weak form of the governing equations. This is accomplished by multiplying eqn (7.2) by a test function $b_{\mathrm{i}j} \in V_\mathrm{h}^\mathrm{l}$, integrating over each element E_k, and applying integration by parts, yielding:

$$\begin{aligned}&\frac{\partial}{\partial t}\int_{E_\mathrm{k}} U_\mathrm{h} b_{\mathrm{i}j}\, dV - \int_{E_\mathrm{k}} F(U_\mathrm{h})\frac{\partial b_{\mathrm{i}j}}{\partial x}\, dV + \int_{\partial E_\mathrm{k}} b_{\mathrm{i}j} F(U_\mathrm{h})\cdot n_\mathrm{x}\, d\Gamma\\ &- \int_{E_\mathrm{k}} G(U_\mathrm{h})\frac{\partial b_{\mathrm{i}j}}{\partial y}\, dV + \int_{\partial E_\mathrm{k}} b_{\mathrm{i}j} G(U_\mathrm{h})\cdot n_\mathrm{y}\, d\Gamma = 0,\end{aligned} \tag{7.10}$$

where n_x and n_y are the components of the outward normal vector on ∂E_k in the x and y directions, respectively. Due to the discontinuous nature of the solutions at element interfaces, the physical fluxes in the surface integrals are ambiguous. To address this, the flux terms are replaced by numerical Riemann fluxes, resulting in the formulation:

$$\begin{aligned}\frac{\partial}{\partial t}\int_{E_\mathrm{k}} U_\mathrm{h} b_{\mathrm{i}j} h\, dV - \int_{E_\mathrm{k}} F(U_\mathrm{h})\frac{\partial b_{\mathrm{i}j}}{\partial x}\, dV + \int_{\partial E_\mathrm{k}} b_{\mathrm{i}j} F_\mathrm{NF}(U_\mathrm{h})\, d\Gamma \\ - \int_{E_\mathrm{k}} G(U_\mathrm{h})\frac{\partial b_{\mathrm{i}j}}{\partial y}\, dV + \int_{\partial E_\mathrm{k}} b_{\mathrm{i}j} G_\mathrm{NF}(U_\mathrm{h})\, d\Gamma = 0,\end{aligned} \tag{7.11}$$

where $\mathbf{F}_\mathrm{NF}$ and $\mathbf{G}_\mathrm{NF}$ denote the numerical fluxes in the x and y directions, respectively, computed using a suitable Riemann solver. In this work, the numerical fluxes at the elemental interfaces are computed using the HLLC Riemann solver for two-component flows [13]. Gauss–Legendre quadrature is used to assess all of the integrals in eqn (7.18) with adequate precision. Accurate quadrature rules up to degree $2l$ for volume terms and $2l + 1$ for surface terms are necessary for exact integration for P^l approximations [13]. This work employs a high-order moment limiter for the preset modal DG scheme in order to remove the erroneous numerical fluctuations from the solutions. A high-order approximation might be limited in this way without being lowered to the first order [13].

Finally, assembling all local contributions yields a semi-discrete system of ordinary differential equations in time:

$$M\frac{dU}{dt} = L(U), \tag{7.12}$$

where $\mathbf{M}$ is the block-diagonal mass matrix and $\mathbf{L}(\mathbf{U})$ is the residual vector. Owing to the diagonal structure of $\mathbf{M}$, its inversion can be performed element-wise. Time integration is carried out using an explicit third-order TVD Runge–Kutta scheme [25]:

$$\begin{aligned}U^{(1)} &= U^n + \Delta t L(U^n),\\ U^{(2)} &= \frac{3}{4}U^n + \frac{1}{4}U^{(1)} + \frac{1}{4}\Delta t L(U^{(1)}),\\ U^{n+1} &= \frac{1}{3}U^n + \frac{2}{3}U^{(2)} + \frac{2}{3}\Delta t L(U^{(2)}).\end{aligned} \tag{7.13}$$

The time step Δt is chosen according to the CFL stability condition:

$$\Delta t = \frac{\mathrm{CFL}}{2l+1}\cdot\frac{h}{|\lambda_\mathrm{max}|}, \tag{7.14}$$

where $h = \min(\Delta x, \Delta y)$, CFL is the Courant number, and λ_{max} denotes the maximum characteristic speed of the inviscid flux.

The accuracy and robustness of the present DG framework have been thoroughly validated in our earlier studies involving shock-interface interactions and compressible multicomponent flows. Benchmark test cases, including 1D Riemann problems, 2D shock refraction at inclined interfaces, and RMI in planar configurations, have demonstrated excellent agreement with available analytical solutions and reference numerical results. These validations confirm the reliability of the implemented numerical scheme in capturing sharp discontinuities, interface evolution, and complex flow features such as vortex roll-up and mixing layer growth. For detailed verification, the reader is referred to our prior works [17, 19].

7.5 Results and Discussion

7.5.1 Mesh convergence analysis

Figure 7.2 presents a quantitative assessment of the numerical scheme through a mesh convergence study. The density profiles are extracted along the horizontal centerline in the post-shock region of a chevron-shaped light gas interface subjected to a planar incident shock. These profiles correspond to the line extracted from the horizontal chevron-shaped computational domain. To evaluate spatial accuracy, simulations were performed at three different grid resolutions: 400×200, 800×400, and 1200×600. As the grid is refined, the numerical solution shows enhanced resolution of the sharp gradients associated with the transmitted shock and rarefaction waves, particularly in the interaction zone. The inset plot zooms into this critical region, illustrating that higher resolutions reduce numerical dissipation and capture finer-scale flow features. The results demonstrate consistent convergence of the DG scheme. The scheme accurately resolves shock-interface interactions and complex post-shock density structures, confirming the reliability and robustness of the proposed methodology for modeling shock-induced instabilities in compressible multicomponent flows with chevron-shaped geometries.

7.5.2 Flow field evolution

Figure 7.3 illustrates the temporal evolution of the density field for a chevron-shaped light gas interface interacting with planar shock waves of different Mach numbers: $M_s = 1.15$, 1.25, and 1.50. The snapshots correspond

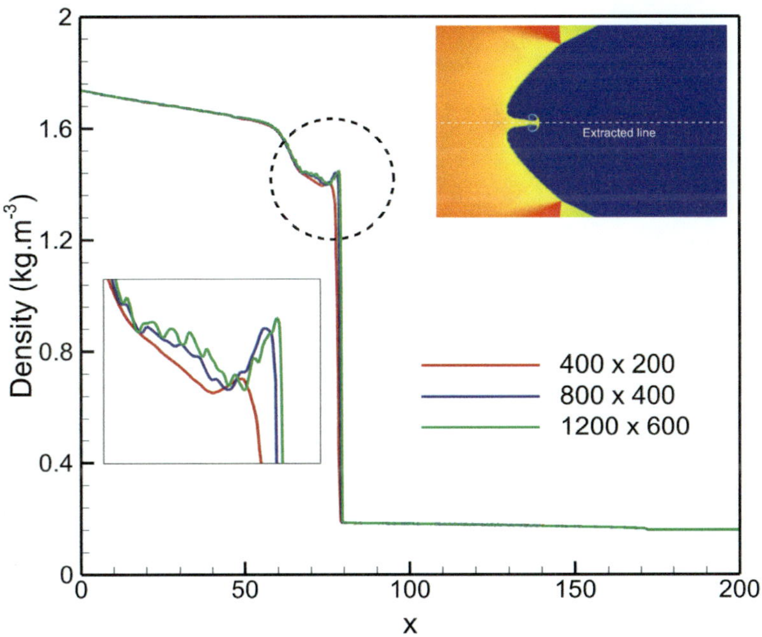

Figure 7.2 Density profile along the horizontal centerline extracted from the post-shock region of a chevron-shaped light gas interface configuration. The results are shown for three mesh resolutions: 400×200, 800×400, and 1200×600, demonstrating mesh convergence of the DG scheme.

to increasing non-dimensional times τ, capturing the transition from initial shock-interface interaction to fully developed instability. For the lowest Mach number case ($M_s = 1.15$), the transmitted shock induces mild deformation at the interface, resulting in the formation of small-amplitude vortical structures. As time progresses, the interface rolls up slightly but remains relatively stable, exhibiting weak nonlinear interactions. In the intermediate case ($M_s = 1.25$), stronger baroclinic vorticity deposition occurs due to the increased misalignment of pressure and density gradients. This leads to pronounced interface deformation and the formation of mushroom-like structures characteristic of RMI. The vortical growth intensifies over time, indicating stronger nonlinear coupling. At the highest Mach number ($M_s = 1.50$), the interaction becomes highly nonlinear. The interface rapidly destabilizes and evolves into a complex pattern with large-scale roll-ups, strong vortex pairing, and extended mixing zones. The intensity and scale of instability are significantly enhanced, highlighting the sensitivity of RMI growth to shock strength.

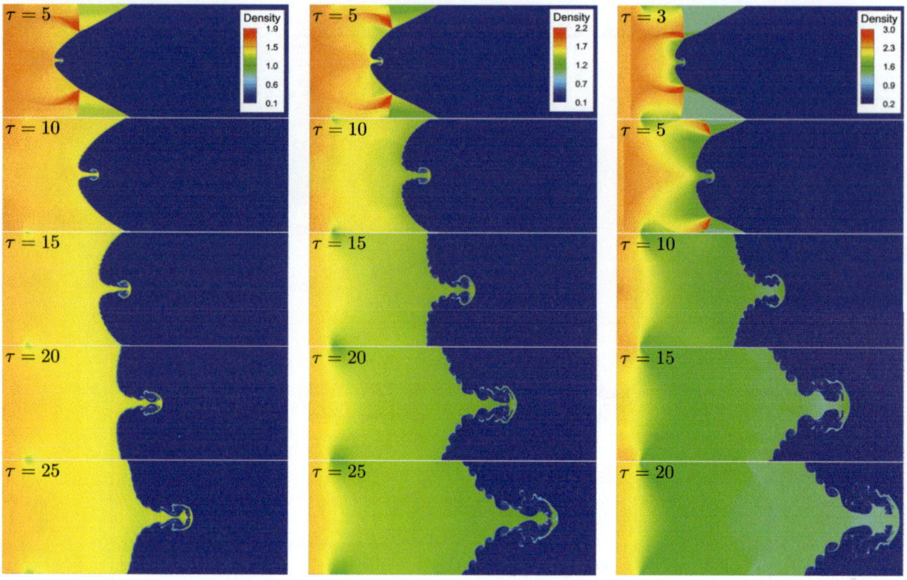

(a) $M_s = 1.15$ (b) $M_s = 1.25$ (c) $M_s = 1.50$

Figure 7.3 Time evolution of density field in the chevron-shaped light gas interface subjected to planar shock wave of varying strengths: (a) $M_s = 1.15$, (b) $M_s = 1.25$, and (c) $M_s = 1.50$.

7.5.3 Vorticity generation mechanism

The evolution of vorticity plays a pivotal role in the development of RMI, particularly in configurations involving shock interaction with structured interfaces such as chevron-shaped light gas layers. To analyze the underlying mechanisms responsible for vorticity generation in such flows, we derive the vorticity transport equation (VTE) within the framework of the inviscid compressible Euler equations employed in this study. Neglecting viscous and thermal diffusion effects, the VTE for an inviscid flow is expressed as

$$\frac{D\omega}{Dt} = \text{VSG} + \text{VSC} + \text{BAR}, \tag{7.15}$$

where the contributing terms are defined as

$$\begin{aligned} \text{VSG} &= (\omega \cdot \nabla)\mathbf{u}, \\ \text{VSC} &= -\omega(\nabla \cdot \mathbf{u}), \\ \text{BAR} &= \frac{1}{\rho^2}(\nabla\rho \times \nabla p). \end{aligned} \tag{7.16}$$

Here, $\omega = \nabla \times \mathbf{u}$ denotes the vorticity vector, $\mathbf{u}$ is the velocity field, ρ is the fluid density, and p is the pressure. The material derivative D/Dt accounts for both temporal changes and advection by the flow. Physically, the VSG term corresponds to vortex stretching and tilting due to velocity gradients along the vorticity direction. In strictly 2D flows, this contribution vanishes. The VSC term represents vorticity amplification or attenuation due to compressibility effects, becoming relevant in regions with significant flow divergence. The final and most crucial term, BAR, captures baroclinic vorticity generation arising from the misalignment between pressure and density gradients. This mechanism is especially dominant in shock-accelerated interfaces, such as those in chevron geometries, where sharp interface curvature and density contrast lead to intense baroclinic torque deposition, ultimately driving the instability and promoting small-scale mixing.

Figure 7.4 depicts the time evolution of the vorticity field for a chevron-shaped light gas interface subjected to incident shock waves of varying Mach numbers $M_s = 1.15, 1.25$, and 1.50. The vorticity contours highlight the rotational flow structures that emerge due to baroclinic torque generated by the misalignment of pressure and density gradients during shock-interface interaction. At lower shock strength ($M_s = 1.15$), baroclinic generation is relatively weak, resulting in the formation of low-intensity vorticity layers that develop into modest roll-up structures. These features remain localized

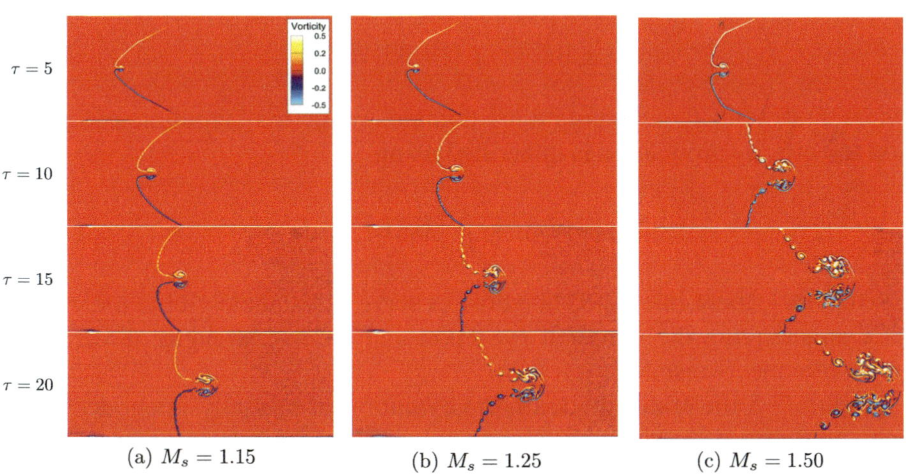

Figure 7.4 Time evolution of vorticity field in the chevron-shaped light gas interface subjected to planar shock wave of varying strengths: (a) $M_s = 1.15$, (b) $M_s = 1.25$, and (c) $M_s = 1.50$.

near the original interface and grow gradually with time. As the shock strength increases to $M_s = 1.25$, the deposited baroclinic vorticity becomes significantly stronger, leading to the formation of counter-rotating vortex pairs. These vortices undergo stretching and roll-up, marking the onset of nonlinear instability dynamics. The vorticity field shows enhanced asymmetry and large-scale coherent structures that evolve with time. In the high Mach number case ($M_s = 1.50$), intense baroclinic torque leads to rapid generation of high-magnitude vorticity, which amplifies the instability and drives complex vortex interactions, including pairing, merging, and filamentation. The resulting turbulent mixing zone becomes more extended and vigorous compared to lower Mach cases.

To systematically assess the influence of vorticity dynamics during the evolution of shock-induced instabilities, we introduce two spatially integrated metrics derived from the VTE. These global measures offer quantitative insight into the relative contributions of baroclinic torque, compressibility, and vortex stretching throughout the instability's progression in chevron-shaped interfaces. Specifically, two key quantities are considered: the integrated dilatation induced vorticity, and the integrated baroclinic vorticity. These metrics provide a comprehensive evaluation of the dominant mechanisms driving vorticity generation as observed in the vorticity field evolution shown in Figure 7.4. To evaluate the effect of compressibility, the spatially integrated magnitude of the dilatation-induced vorticity term is expressed as

$$|\omega_{\mathrm{DIL}}|(t) = \int_D |\mathrm{VSC}| \; dx \, dy, \tag{7.17}$$

where D denotes the computational domain. While the baroclinic contribution to vorticity production is quantified using

$$|\omega_{\mathrm{BARO}}|(t) = \int_D |\mathrm{BAR}| \; dx \, dy. \tag{7.18}$$

Together, these integrated diagnostics serve as valuable global indicators for characterizing the temporal evolution and relative strength of vorticity generation mechanisms.

Figure 7.5 illustrates the temporal evolution of spatially integrated vorticity components induced by dilatational and baroclinic mechanisms under varying shock wave strengths. Figure 7.5(a) shows the dilatation-induced vorticity, which arises due to compressibility effects following the shock interaction. It is evident that higher Mach numbers, particularly $M_s = 1.50$, lead to an earlier and more pronounced growth in dilatational vorticity, indicating that stronger shocks enhance compressibility-driven velocity divergence in the flow. In contrast, lower Mach numbers exhibit delayed and weaker dilatational effects. Figure 7.5(b) captures the baroclinic-induced vorticity resulting from the misalignment of pressure and density gradients across the chevron-shaped interface. Interestingly, the maximum baroclinic contribution is observed for an intermediate Mach number $M_s = 1.25$, suggesting a nonlinear interaction between the shock-induced pressure field and the interface geometry. This highlights that moderate shock strengths may provide optimal conditions for sustained baroclinic torque, especially in geometrically complex interfaces.

Figure 7.6 illustrates the temporal evolution of the extremal vorticity values – namely, the maximum and minimum vorticity – in a chevron-shaped

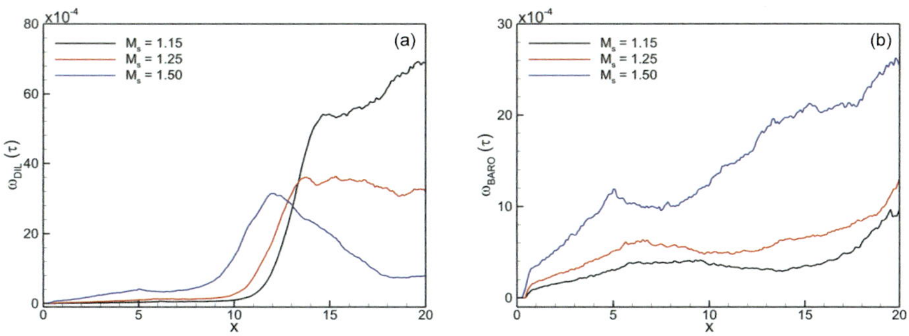

Figure 7.5 Effect of shock wave strength on the spatially integrated fields of (a) dilatation-induced vorticity, and (b) baroclinic-induced vorticity, in the chevron-shaped light gas interface.

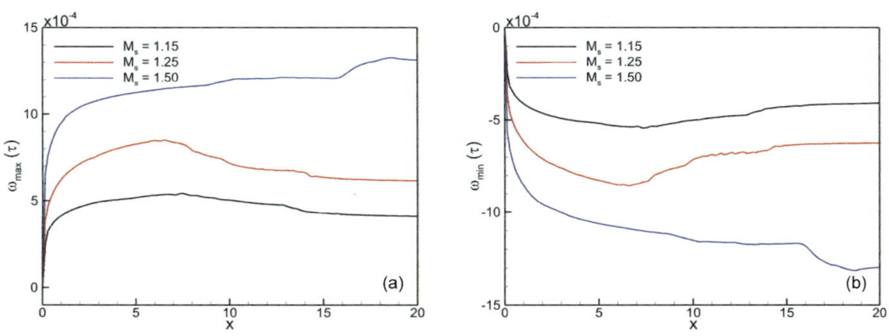

Figure 7.6 Effect of shock wave strength on the variation of (a) maximum and (b) minimum vorticity with time in the chevron-shaped light gas interface.

light gas interface subjected to shock waves of different strengths (M_s = 1.15, 1.25, and 1.50). Figure 7.6(a) shows that the maximum vorticity ω_{max} increases rapidly after shock impact, with stronger shocks generating higher peak vorticity due to intensified shear layers and localized rotational structures along the sharp curvatures of the interface. In contrast, Figure 7.6(b) reveals that the minimum vorticity ω_{min}, representing counter-rotating vortical structures, also grows in magnitude with increasing Mach number, with $M_s = 1.50$ producing the deepest negative values. These asymmetries between ω_{max} and ω_{min} suggest a growing skewness in vorticity distribution at higher Mach numbers, driven by nonlinear effects and enhanced baroclinic generation in regions of misaligned pressure and density gradients. The chevron geometry inherently facilitates such misalignments, promoting localized vorticity intensification and roll-up.

7.5.4 Evolution of enstrophy

Enstrophy is a fundamental quantity in vortex dynamics that quantifies the intensity of rotational motion within a flow field. It provides a measure of the squared vorticity and is commonly used to analyze the development of small-scale structures and turbulence. Mathematically, the enstrophy at time τ is defined as:

$$\Omega(\tau) = \frac{1}{2} \int_D \omega^2 \, dx \, dy, \tag{7.19}$$

where ω denotes the vorticity field, and D represents the spatial domain of interest.

Figure 7.7 illustrates the time evolution of the enstrophy field in a chevron-shaped light gas interface subjected to planar shock waves of varying strengths, providing a visual representation of how shock intensity influences vorticity dynamics and instability development. At the lowest Mach number ($M_s = 1.15$), the enstrophy remains localized and weak, with only modest interface deformation and limited vortex formation near the tip of the chevron. This indicates that vorticity production is minimal, and the flow stays relatively laminar. As the shock strength increases to $M_s = 1.25$, the enstrophy field intensifies significantly, with stronger vortex roll-up and broader spatial distribution of vortical structures. The interaction between the shock and the chevron-shaped interface leads to enhanced baroclinic torque, promoting the onset of asymmetric instabilities and early stages of interface breakup. For the highest shock strength ($M_s = 1.50$), the enstrophy field becomes highly developed, with densely packed, fine-scale vortical structures signifying vigorous rotational activity and strong nonlinear effects. The interface undergoes pronounced deformation and fragmentation, leading to complex turbulent mixing.

Figure 7.8 presents the temporal evolution of the spatially integrated enstrophy (Ω) in the chevron-shaped light gas interface under the influence of planar shock waves with varying strengths ($M_s = 1.15, 1.25, 1.50$). Enstrophy, defined as the integral of squared vorticity over the domain, is a key indicator of rotational activity and the development of small-scale structures in compressible, vortical flows. The figure clearly shows that enstrophy increases monotonically with time for all cases, indicating the continuous generation and amplification of vortical structures as the shock propagates and interacts with the interface. However, the rate and magnitude of enstrophy growth are strongly dependent on the shock strength. At the lowest Mach number ($M_s = 1.15$), the growth of enstrophy is relatively slow and subdued, reflecting limited baroclinic vorticity production and weak interface deformation. For the intermediate case ($M_s = 1.25$), enstrophy rises more rapidly, indicating stronger vorticity generation and enhanced

$M_s = 1.15$

$M_s = 1.25$

$M_s = 1.50$

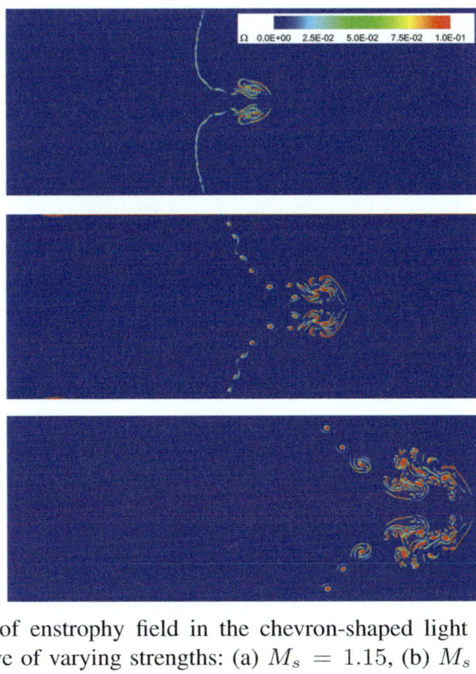

Figure 7.7 Time evolution of enstrophy field in the chevron-shaped light gas interface subjected to planar shock wave of varying strengths: (a) $M_s = 1.15$, (b) $M_s = 1.25$, and (c) $M_s = 1.50$.

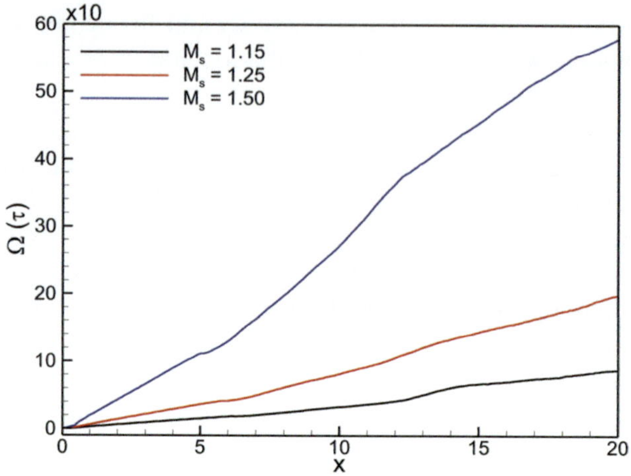

Figure 7.8 Effect of shock wave strength on the spatial integrated fields of entrophy with time in the chevron-shaped light gas interface.

interface dynamics. The highest enstrophy growth is observed at $M_s = 1.50$, where a steep and nearly linear increase suggests intense, sustained vortex generation, rapid destabilization of the interface, and a possible transition toward turbulent mixing.

7.6 Conclusions

This chapter has presented a high-fidelity numerical investigation of shock-induced instabilities in a chevron-shaped light gas interface using a high-order DG method. By solving the compressible Euler equations for two-component gas systems, the study elucidated how shock strength and interface geometry influence the evolution of Richtmyer–Meshkov-type instabilities. A detailed analysis of density, vorticity, and enstrophy fields revealed that increasing the incident Mach number enhances baroclinic vorticity deposition, intensifies interface roll-up, and promotes early transition to nonlinear instability regimes. The chevron geometry, due to its inherent curvature and angular interface features, was shown to amplify misalignments between pressure and density gradients, making it particularly susceptible to baroclinic torque and asymmetric vortex generation. Spatially integrated diagnostics of vorticity and enstrophy provided further insight into the relative contributions of compressibility and baroclinicity in driving the instability dynamics.

References

[1] R.D. Richtmyer. Taylor instability in shock acceleration of compressible fluids. *Communications on pure and applied mathematics*, 13(2):297–319, 1960.

[2] E.E. Meshkov. Instability of the interface of two gases accelerated by a shock wave. *Fluid Dynamics*, 4(5):101–104, 1969.

[3] John D Lindl, Robert L McCrory, and E Michael Campbell. Progress toward ignition and burn propagation in inertial confinement fusion. *Physics Today*, 45(9):32–40, 1992.

[4] John Lindl, Otto Landen, John Edwards, Ed Moses, NIC team, et al. Review of the national ignition campaign 2009-2012. *Physics of Plasmas*, 21(2), 2014.

[5] W David Arnett, John N Bahcall, Robert P Kirshner, and Stanford E Woosley. Supernova 1987a. *IN: Annual review of astronomy and astrophysics. Volume 27 (A90-29983 12-90). Palo Alto, CA, Annual Reviews, Inc., 1989, p. 629-700. Research supported by the University of Arizona.*, 27:629–700, 1989.

[6] Carolyn C Kuranz, H-S Park, Channing M Huntington, Aaron R Miles, Bruce A Remington, T Plewa, MR Trantham, HF Robey, Dov Shvarts, A Shimony, et al. How high energy fluxes may affect rayleigh–taylor instability growth in young supernova remnants. *Nature communications*, 9(1):1564, 2018.

[7] Joseph Yang, Toshi Kubota, and Edward E Zukoski. Applications of shock-induced mixing to supersonic combustion. *AIAA journal*, 31(5):854–862, 1993.

[8] Qingchun Yang, Juntao Chang, and Wen Bao. Richtmyer–meshkov instability induced mixing enhancement in the scramjet combustor with a central strut. *Advances in Mechanical Engineering*, 6:614189, 2014.

[9] Jeffrey W Jacobs. Shock-induced mixing of a light-gas cylinder. *Journal of Fluid Mechanics*, 234:629–649, 1992.

[10] J-F Haas and Bradford Sturtevant. Interaction of weak shock waves with cylindrical and spherical gas inhomogeneities. *Journal of Fluid Mechanics*, 181:41–76, 1987.

[11] James J Quirk and Smadar Karni. On the dynamics of a shock–bubble interaction. *Journal of Fluid Mechanics*, 318:129–163, 1996.

[12] Guillaume Layes, Georges Jourdan, and Lazhar Houas. Distortion of a spherical gaseous interface accelerated by a plane shock wave. *Physical review letters*, 91(17):174502, 2003.

[13] Satyvir Singh, Ahmed Hussein Msmali, and Mark Ian Nelson. Unfolding of shocked hydrodynamic instability at SF_6 elliptical interface: Physical insights from numerical simulations. *Computers & Fluids*, 277:106304, 2024.

[14] Zhigang Zhai, Minghu Wang, Ting Si, and Xisheng Luo. On the interaction of a planar shock with a light polygonal interface. *Journal of Fluid Mechanics*, 757:800–816, 2014.

[15] D Igra and O Igra. Shock wave interaction with a polygonal bubble containing two different gases, a numerical investigation. *Journal of Fluid Mechanics*, 889:A26, 2020.

[16] Satyvir Singh. Role of atwood number on flow morphology of a planar shock-accelerated square bubble: A numerical study. *Physics of Fluids*, 32(12), 2020.

[17] Satyvir Singh and Marco Battiato. Numerical investigation of shock mach number effects on richtmyer–meshkov instability in a heavy square bubble. *Physica D: Nonlinear Phenomena*, 453:133844, 2023.

[18] Satyvir Singh and Marco Battiato. Numerical simulations of richtmyer–meshkov instability of sf6 square bubble in diatomic and polyatomic gases. *Computers & Fluids*, 242:105502, 2022.

[19] Satyvir Singh and Manuel Torrilhon. On the shock-driven hydrodynamic instability in square and rectangular light gas bubbles: A comparative study from numerical simulations. *Physics of Fluids*, 35(1), 2023.

[20] Xisheng Luo, Minghu Wang, Ting Si, and Zhigang Zhai. On the interaction of a planar shock with an polygon. *Journal of Fluid Mechanics*, 773:366–394, 2015.

[21] VK Tritschler, XY Hu, S Hickel, and NA Adams. Numerical simulation of a richtmyer–meshkov instability with an adaptive central-upwind sixth-order weno scheme. *Physica Scripta*, 2013(T155):014016, 2013.

[22] Pooya Movahed and Eric Johnsen. A solution-adaptive method for efficient compressible multifluid simulations, with application to the richtmyer–meshkov instability. *Journal of Computational Physics*, 239:166–186, 2013.

[23] Bernardo Cockburn and Chi-Wang Shu. Runge–kutta discontinuous galerkin methods for convection-dominated problems. *Journal of scientific computing*, 16:173–261, 2001.

[24] Bernardo Cockburn, George E Karniadakis, and Chi-Wang Shu. *Discontinuous Galerkin methods: theory, computation and applications*, volume 11. Springer Science & Business Media, 2012.

[25] Satyvir Singh, Abolfazl Karchani, Tushar Chourushi, and Rho Shin Myong. A three-dimensional modal discontinuous galerkin method for the second-order boltzmann-curtiss-based constitutive model of rarefied and microscale gas flows. *Journal of Computational Physics*, 457:111052, 2022.

8

Extended Hydrodynamic Models for Rayleigh–Brillouin Scattering in Polyatomic Rarefied Gases: A Comparative Study

Anil Kumar[1], Rahul Bhattacharjee[1], Satyvir Singh[2,3], and Anirudh Singh Rana[1]

[1]Department of Mathematics, Birla Institute of Technology and Science Pilani, Pilani Campus, India
[2]Institute for Applied and Computational Mathematics, RWTH Aachen University, Germany
[3]Department of Mathematics, Graphic Era Deemed to be University, India
E-mail: p20200044@pilani.bits-pilani.ac.in;
rahulbhattacharjee926@gmail.com;
singh@acom.rwth-aachen.de; anirudh.rana@pilani.bits-pilani.ac.in

Abstract

This chapter explores modifications and extensions of the classical Navier–Stokes–Fourier (NSF) equations for polyatomic rarefied gases, focusing on their application to Rayleigh–Brillouin scattering (RBS). The analysis includes the two-temperature (2T) model, the reduced 2T model, and the polyatomic coupled constitutive relations (CCR) model, alongside the refined NSF (RNSF) framework. These models, derived from extended thermodynamics, incorporate internal energy exchange mechanisms essential for capturing non-equilibrium effects in rarefied polyatomic gas flows. A comparative study is conducted to evaluate their predictive capabilities against experimental RBS data, molecular dynamics (MD) simulations, and kinetic theory. The results highlight the strengths and limitations of each model in describing light scattering phenomena and non-equilibrium transport processes. This work provides valuable insights into the theoretical and computational modeling of polyatomic rarefied gases, supporting their

DOI: 10.1201/9788743806691-8

integration into computational fluid dynamics (CFD) frameworks for engineering applications.

Keywords: Two-temperature model, Rayleigh–Brillouin scattering, Rarefied polyatomic gases, non-equilibrium thermodynamics

8.1 Introduction

Fluid dynamics remains a cornerstone of scientific and engineering research due to its vast range of applications across various industries [1]. Its principles influence numerous disciplines, from aerodynamics and astrophysics to industrial process optimization [2–4]. The classical NSF equations, derived under the assumptions of local thermodynamic equilibrium, have been widely used for modeling fluid motion [5]. However, these equations face limitations in non-equilibrium conditions, particularly in rarefied and highly compressible flows, where higher-order effects become significant [6, 7]. While the Boltzmann equation provides a complete description of gas flow across all flow regimes, it remains computationally expensive, particularly in the transition regime, making it impractical for many engineering applications [8, 9].

To address these shortcomings, researchers have developed extended fluid dynamic models that incorporate non-equilibrium effects while maintaining computational feasibility [10–13]. These modifications enhance the accuracy of classical hydrodynamic equations, especially in scenarios where the assumptions of NSF break down. Various approaches, including higher-order moment equations, regularized models, and two-temperature formulations, have been proposed to refine our understanding of rarefied gas dynamics.

One critical area where these models find applications is the study of Rayleigh–Brillouin scattering, a fundamental optical phenomenon used to probe the thermodynamic and transport properties of gases. RBS is particularly valuable for characterizing polyatomic gases, which exhibit complex internal energy exchanges and non-equilibrium behavior [14–16]. These gases play a key role in aerospace propulsion, plasma physics, and atmospheric sensing, where accurate modeling of their transport properties is essential. The ADM-Aeolus satellite, for instance, employs RBS techniques to monitor global wind patterns, improving weather forecasting [17, 18].

In this study, we focus on modifications and extensions of the classical NSF equations that retain five or six field variables [19–22]. These models are selected due to their reduced mathematical complexity compared to other extended fluid dynamics models, making them more practical for computational applications. Their relative simplicity enables seamless integration into open-source software such as OpenFOAM [23] and commercial CFD platforms like COMSOL and ANSYS, enhancing their accessibility and usability in engineering applications.

In this chapter, we conduct a comparative analysis of these modified and extended NSF-based models in the context of RBS. Specifically, we examine the predictive capabilities of the two-temperature (2T) model [20], the reduced 2T model [22], and the polyatomic CCR model [21], alongside existing models such as RNSF [19]. The comparison is performed against experimental data and theoretical benchmarks, providing valuable insights into their effectiveness in rarefied gas dynamics and their potential integration into CFD frameworks.

8.2 Macroscopic Equations for Polyatomic Gases

This section presents a concise summary of the macroscopic transport equations for polyatomic gases without delving into their derivation. Detailed discussions on their formulation and properties can be found in the literature [19–22]. Throughout this chapter, we employ tensor index notation with the Einstein summation convention for clarity and consistency.

The fundamental conservation laws governing mass, momentum, and energy in polyatomic gases are essential for describing their macroscopic behavior. The mass conservation equation ensures the continuity of mass and is given by

$$\frac{\partial \rho}{\partial t} + \frac{\partial \rho v_\mathrm{i}}{\partial x_\mathrm{i}} = 0, \tag{8.1}$$

where ρ represents the mass density, and v_i denotes the macroscopic velocity. The momentum conservation equation describes the balance of forces within the gas, accounting for both convective and stress contributions:

$$\frac{\partial \rho v_\mathrm{i}}{\partial t} + \frac{\partial \rho v_\mathrm{i} v_\mathrm{j}}{\partial x_\mathrm{j}} + \frac{\partial p_\mathrm{ij}}{\partial x_\mathrm{j}} = 0. \tag{8.2}$$

Here, p_ij is the pressure tensor, which consists of three key components: the isotropic pressure $p = \rho\theta$, the excess normal stress (or dynamic pressure) $\Pi = \rho\vartheta$, and the shear stress σ_ij. The pressure tensor can be expressed as

$$p_\mathrm{ij} = (p + \Pi)\delta_\mathrm{ij} + \sigma_{\langle \mathrm{ij} \rangle}, \tag{8.3}$$

where δ_{ij} is the Kronecker delta function, and the angular brackets $\langle ij \rangle$ denote the symmetric and traceless part of the tensor.

The energy conservation equation accounts for the transport of internal energy, contributions from the pressure tensor, and the heat flux:

$$\frac{3+\delta}{2}\left(\frac{\partial \rho\theta}{\partial t}+\frac{\partial \rho\theta v_{\text{i}}}{\partial x_{\text{i}}}\right)+p_{\text{ij}}\frac{\partial v_{\text{j}}}{\partial x_{\text{i}}}+\frac{\partial q_{\text{i}}}{\partial x_{\text{i}}}=0. \tag{8.4}$$

Here, δ denotes the number of internal degrees of freedom, θ represents the temperature, while q_{i} is the heat flux, which plays a significant role in energy transport, particularly in rarefied and non-equilibrium gas dynamics. These governing equations serve as the foundation for analyzing macroscopic properties of polyatomic gases and their behavior under different thermodynamic conditions.

The conservation laws are the fundamental equations of continuum mechanics. However, they cannot be solved directly in their current form, as they contain the stress tensor σ_{ik}, dynamic pressure Π, and heat flux q_{i} as unknowns. To close the system, constitutive relations are required, which express these unknowns in terms of the primary variables ρ, v_{i}, and θ. The classical hydrodynamic equations describing viscous flow are derived from the NSF equations. In this framework, the shear stress, dynamic pressure, and heat flux are described by Newton's and Fourier's laws, respectively, as follows:

$$\sigma_{\text{ij}}=-2\mu\frac{\partial v_{\langle i}}{\partial x_{\text{j}\rangle}},\quad \Pi=-\mu_{\text{b}}\frac{\partial v_{\text{i}}}{\partial x_{\text{i}}},\quad q_{\text{i}}=-\kappa\frac{\partial \theta}{\partial x_{\text{i}}}. \tag{8.5}$$

Here, μ, μ_{b}, and κ denote the shear viscosity, bulk viscosity, and thermal conductivity of the gas, respectively. These transport coefficients may depend on the local density ρ and temperature θ, affecting the macroscopic behavior of the gas [24].

8.2.1 Two-temperature model [20]

The two-temperature model incorporates two separate non-equilibrium temperatures: the translational temperature (θ^{tr}) and the internal temperature (θ^{in}), which correspond to the molecular degrees of freedom associated with translational motion and internal energy storage, respectively. These temperatures play a crucial role in describing energy distribution in rarefied gas flows. The translational temperature is linked to the kinetic energy of molecular movement, while the internal temperature accounts for the energy contained in vibrational, rotational, or electronic modes. Under non-equilibrium conditions, these temperatures can deviate significantly from each other; however, in thermal equilibrium, they coincide with the thermodynamic temperature ($\theta = RT$), where R represents the gas constant. Furthermore, the temperature θ is related to the translational and internal temperatures through the expression $\frac{3+\delta}{2}\theta = \frac{3}{2}\theta^{\mathrm{tr}} + \frac{\delta}{2}\theta^{\mathrm{in}}$ and the dynamic temperature is defined as $\theta^{\mathrm{tr}} - \theta$ [25].

The governing equation for dynamic temperature is given by [20]

$$\frac{3}{2}\left\{\frac{\partial \rho\vartheta}{\partial t} + \frac{\partial \rho\vartheta v_{\mathrm{k}}}{\partial x_{\mathrm{k}}}\right\} + \frac{\delta}{3+\delta} p_{\mathrm{ik}} \frac{\partial v_{\mathrm{i}}}{\partial x_{\mathrm{k}}} - \frac{3}{3+\delta}\frac{\partial q_{\mathrm{k}}^{\mathrm{in}}}{\partial x_{\mathrm{k}}} + \frac{\delta}{3+\delta}\frac{\partial q_{\mathrm{k}}^{\mathrm{tr}}}{\partial x_{\mathrm{k}}} = \mathcal{P}_{\mathrm{prod}}, \tag{8.6}$$

where $\mathcal{P}_{\mathrm{prod}}$ represents the production term.

It should be noted that the system of eqn (8.1), (8.2), (8.4), and (8.6) involves the pressure tensor and heat flux as unknowns and other variables $(\rho, v_{\mathrm{i}}, \theta, \vartheta)$ as additional unknowns. Consequently, constitutive equations are required for establishing relations between $\sigma_{\mathrm{ij}}, q_{\mathrm{i}}^{\mathrm{tr}}, q_{\mathrm{i}}^{\mathrm{in}}$ quantities and the variables $\rho, v_{\mathrm{i}}, \theta, \vartheta$. Such constitutive relations can be derived by enforcing the second law inequality [20, 10, 26], as outlined below:

$$\sigma_{\mathrm{ik}} = -2\mu \frac{\partial v_{\langle i}}{\partial x_{\mathrm{k}\rangle}}, \tag{8.7a}$$

$$q_{\mathrm{k}}^{\mathrm{tr}} = -\left(\frac{\zeta_{11}}{(\theta+\vartheta)^2} + \frac{\delta^2\zeta_{12}}{(\delta\theta - 3\vartheta)^2}\right)\frac{\partial \theta}{\partial x_{\mathrm{k}}} - \left(\frac{\zeta_{11}}{(\theta+\vartheta)^2} - \frac{3}{\delta}\frac{\delta^2\zeta_{12}}{(\delta\theta - 3\vartheta)^2}\right)\frac{\partial \vartheta}{\partial x_{\mathrm{k}}}, \tag{8.7b}$$

$$q_{\mathrm{k}}^{\mathrm{in}} = -\left(\frac{\zeta_{12}}{(\theta+\vartheta)^2} + \frac{\delta^2\zeta_{22}}{(\delta\theta - 3\vartheta)^2}\right)\frac{\partial \theta}{\partial x_{\mathrm{k}}} - \left(\frac{\zeta_{12}}{(\theta+\vartheta)^2} - \frac{3}{\delta}\frac{\delta^2\zeta_{22}}{(\delta\theta - 3\vartheta)^2}\right)\frac{\partial \vartheta}{\partial x_{\mathrm{k}}}. \tag{8.7c}$$

The coefficients μ, ζ_{11}, ζ_{22}, and ζ_{12} are arbitrary non-negative phenomenological coefficients that enforce the relations between heat fluxes and temperature in the system. Specifically, ζ_{11}, ζ_{22}, and ζ_{12} form the Onsager coefficient matrix ζ_{ij}, defined as:

$$\zeta_{\mathrm{ij}} = \begin{bmatrix} \zeta_{11} & \zeta_{12} \\ \zeta_{12} & \zeta_{22} \end{bmatrix}. \tag{8.8}$$

This symmetric positive definite matrix governs thermal diffusion with respect to translational and internal temperatures and is closely related to the thermal conductivity of gases [27]. The relations in eqn (8.7) along with eqn (8.1), (8.2), (8.4), and (8.6) constitute the two-temperature model equations for polyatomic gases.

The values of the arbitrary non-negative coefficients μ, ζ_{11}, ζ_{22}, and ζ_{12}, as well as the production terms $\mathcal{P}_{\text{prod}}$, are determined by comparing them with various models in the literature [28, 25, 20, 27]. These production terms and phenomenological coefficients vary depending on the specific collision models used. In this context, our consideration is based on a comparison with the models proposed by Rahimi and Struchtrup [19, 25]:

$$\zeta_{11} = \frac{5\kappa}{5+\delta}(\theta^{\text{tr}})^2,\ \zeta_{12} = 0,\ \zeta_{22} = \frac{\delta\kappa}{5+\delta}(\theta^{\text{in}})^2,\ \mathcal{P}_{\text{prod}} = -\frac{\delta}{3+\delta}\frac{p\rho\vartheta}{\mu_{\text{b}}}, \tag{8.9}$$

which are detailed in [20].

8.2.2 Reduced 2T model [22]

In this simplified model, we assume a linear relationship between internal and translational heat fluxes, expressed as $q_{\text{i}}^{\text{tr}} = \frac{5}{\delta} q_{\text{i}}^{\text{in}}$. The concept of choosing a specific linear relation between heat fluxes q_{i}^{tr} and q_{i}^{in} was inspired by the order-of-magnitude method proposed by Rahimi and Struchtrup [30]. This method revealed that the magnitude of this relation can be approximated as $O(Kn^{1+\epsilon})$, where ϵ is a magnification parameter that ranges from 0 to 1.

By incorporating this assumption, the dynamic temperature eqn (8.6) reduces to:

$$\frac{3}{2}\left\{\frac{\partial \rho\vartheta}{\partial t} + \frac{\partial \rho\vartheta v_{\text{k}}}{\partial x_{\text{k}}}\right\} + \frac{\delta}{3+\delta} p_{\text{ik}} \frac{\partial v_{\text{i}}}{\partial x_{\text{k}}} + \frac{2\delta}{(3+\delta)(5+\delta)}\frac{\partial q_{\text{k}}}{\partial x_{\text{k}}} = \mathcal{P}_{\text{prod}}. \tag{8.10}$$

The system of eqn (8.1), (8.2), (8.4), and (8.10) is closed using the following constitutive relations:

$$\sigma_{\text{ij}} = -2\mu \frac{\partial v_{\langle i}}{\partial x_{\text{j}\rangle}}, \tag{8.11}$$

$$q_{\text{j}} = -\kappa\left(\frac{\partial \theta}{\partial x_{\text{j}}} + \frac{5-3\xi^2}{5+\delta\xi^2}\frac{\partial \vartheta}{\partial x_{\text{j}}}\right), \tag{8.12}$$

where $\xi = (\delta\theta + \delta\vartheta)\,/\,(\delta\theta - 3\vartheta)$.

8.2.3 Polyatomic-CCR [21]

The polyatomic-CCR framework extends classical hydrodynamic models by incorporating higher-order corrections into the stress and heat flux equations while ensuring compliance with the second law of thermodynamics. By enforcing entropy balance, the CCR model derives coupled constitutive relations for the stress tensor, dynamic pressure, and heat flux, enhancing the predictive capability of the NSF equations in non-equilibrium regimes. The constitutive relations are given as

$$\sigma_{ij} = -2\mu\left[\frac{\partial v_{\langle i}}{\partial x_{j\rangle}} + \frac{\alpha}{p}\left(\frac{\partial q_{\langle i}}{\partial x_{j\rangle}} - \frac{q_{\langle i}}{2p\theta}\frac{\partial (p\theta)}{\partial x_{j\rangle}}\right)\right], \tag{8.13a}$$

$$\Pi = -\mu_b\left[\frac{\partial v_i}{\partial x_i} + \frac{\beta}{p}\left(\frac{\partial q_i}{\partial x_i} - \frac{q_i}{2p\theta}\frac{\partial (p\theta)}{\partial x_i}\right)\right], \tag{8.13b}$$

$$q_j = -\kappa\left[\frac{\partial \theta}{\partial x_j} + \frac{\alpha}{\rho}\left(\frac{\partial \sigma_{ij}}{\partial x_i} - \frac{\sigma_{ij}}{2p\theta}\frac{\partial (p\theta)}{\partial x_i}\right)\right.$$
$$\left. + \frac{\beta}{\rho}\left(\frac{\partial \Pi}{\partial x_j} - \frac{\Pi}{2p\theta}\frac{\partial (p\theta)}{\partial x_j}\right)\right]. \tag{8.13c}$$

Here, $\alpha = \beta = \frac{2}{5+\delta}$. The constitutive relations in eqn (8.13), together with the conservation laws, form the basis of the CCR framework for polyatomic gases.

8.2.4 Refined Navier–Stokes–Fourier equations [19]

This model extends the classical NSF framework by incorporating first-order corrections $O(Kn^1)$ into the conservation laws, specifically eqn (8.1), (8.2), and (8.4).

To maintain consistency at first order, the balance equation for ϑ, which is of order $O(Kn^\epsilon)$, must include terms up to $O(Kn^{1-\epsilon})$, leading to the following equation:

$$\frac{3}{2}\rho\frac{\partial \vartheta}{\partial t} + \frac{3}{2}\rho v_i\frac{\partial \vartheta}{\partial x_i} + \frac{\delta}{3+\delta}\rho(\theta+\vartheta)\frac{\partial v_i}{\partial x_i} = -\frac{\delta}{(3+\delta)}\frac{p\rho\vartheta}{\mu_b}. \tag{8.14}$$

For closure at first order, the classical NSF equations for the stress deviator σ_{ij} and heat flux q_i, given by eqn (8.5)(1) and (8.5)(3), are retained. These equations relate the stress and heat flux to velocity and temperature gradients.

While the classical NSF framework models five field variables $\{\rho, v_i, \theta\}$, the RNSF equations introduce an additional independent variable, ϑ, leading to a six-variable system $\{\rho, v_i, \theta, \vartheta\}$. In this formulation, the parameter δ is assumed to be constant.

8.3 Linear-dimensionless Equations

To analyze the macroscopic equations, we consider the problem of RBS, where light interacts with density fluctuations in a gas. This phenomenon serves as an ideal test case because it eliminates gas-wall interactions, allowing us to focus solely on the behavior of the macroscopic equations. Additionally, in this 1D and linearizable framework, analytical solutions are always attainable.

To proceed, we introduce dimensionless variables by considering small perturbations around a uniform equilibrium state, defined by the reference mass density ρ_0 and reference temperature θ_0. This formulation facilitates the linearization of the governing equations, simplifying their analysis. The dimensionless variables are defined as follows:

$$\begin{aligned} \rho &= \rho_0 (1+\hat{\rho}), \theta = \theta_0(1+\hat{\theta}), v_i = \sqrt{\theta_0}\hat{v}_i, \\ \sigma_{ij} &= \rho_0\theta_0\hat{\sigma}_{ij}, q_i = \rho_0 \left(\sqrt{\theta_0}\right)^3 \hat{q}_i \text{ and } x_i = L\hat{x}_i, \end{aligned} \tag{8.15}$$

where L is a characteristic length scale, and the hat notation indicates dimensionless quantities. For clarity, hats are omitted in the linearized equations.

8.3.1 Linearized conservation laws

By linearizing the conservation laws around the equilibrium state, the following set of linearized equations is obtained:

$$\frac{\partial \rho}{\partial t} + \frac{\partial v}{\partial x} = 0, \tag{8.16a}$$

$$\frac{\partial v}{\partial t} + \frac{\partial \rho}{\partial x} + \frac{\partial \theta}{\partial x} + \frac{\partial \vartheta}{\partial x} + \frac{\partial \sigma}{\partial x} = 0, \tag{8.16b}$$

$$\frac{3+\delta}{2}\frac{\partial \theta}{\partial t} + \frac{\partial v}{\partial x} + \frac{\partial q}{\partial x} = 0. \tag{8.16c}$$

8.3.2 Linearized NSF equations

The classical NSF equations in linearized and dimensionless form are given as

$$\sigma = -\frac{4}{3} Kn \frac{\partial v}{\partial x}, \quad \Pi = -\nu kn \frac{\partial v}{\partial x}, \quad q = -\frac{C_p}{Pr} Kn \frac{\partial \theta}{\partial x}. \tag{8.17}$$

where the Knudsen number appears as a reference viscosity term $Kn = \mu_0 \sqrt{\theta_0}/(p_0 L)$, ν is ratio of bulk to shear viscosity, $Pr = C_{\mathrm{p}}\mu/\kappa$ is the Prandtl number, and C_{p} is the specific heat at constant pressure.

8.3.3 2T Model equations

The 2T model introduces an additional dynamic temperature equation. The balance equation for dynamic temperature is given by:

$$\frac{3}{2}\frac{\partial \vartheta}{\partial t} + \frac{\delta}{3+\delta}\frac{\partial v}{\partial x} - \frac{3}{3+\delta}\frac{\partial q^{\mathrm{in}}}{\partial x} + \frac{\delta}{3+\delta}\frac{\partial q^{\mathrm{tr}}}{\partial x} = \mathcal{P}_{\mathrm{prod}}, \tag{8.18}$$

and constitutive relations are:

$$\sigma = -\frac{4}{3} Kn \frac{\partial v}{\partial x}, \tag{8.19a}$$

$$q^{\mathrm{tr}} = -\left(\zeta_{11} + \zeta_{12}\right)\frac{\partial \theta}{\partial x} - \left(\zeta_{11} - \frac{3}{\delta}\zeta_{12}\right)\frac{\partial \vartheta}{\partial x}, \tag{8.19b}$$

$$q^{\mathrm{in}} = -\left(\zeta_{12} + \zeta_{22}\right)\frac{\partial \theta}{\partial x} - \left(\zeta_{12} - \frac{3}{\delta}\zeta_{22}\right)\frac{\partial \vartheta}{\partial x}. \tag{8.19c}$$

8.3.4 Linearized reduced 2T model equations

In the reduced 2T model, the dynamic temperature equation simplifies to:

$$\frac{3}{2}\frac{\partial \vartheta}{\partial t} + \frac{\delta}{3+\delta}\frac{\partial v}{\partial x} + \frac{2\delta}{(3+\delta)(5+\delta)}\frac{\partial q}{\partial x} = \mathcal{P}_{\mathrm{prod}}, \tag{8.20}$$

and constitutive relations take the form:

$$\sigma = -\frac{4}{3} Kn \frac{\partial v}{\partial x}, \quad q = -\frac{C_p}{Pr} Kn \left(\frac{\partial \theta}{\partial x} + \frac{2}{5+\delta}\frac{\partial \vartheta}{\partial x}\right). \tag{8.21}$$

8.3.5 Linearized polyatomic-CCR

The polyatomic-CCR model for polyatomic gases in 1D, linearized, and dimensionless form is given by:

$$\sigma = -\frac{4}{3} Kn \left(\frac{\partial v}{\partial x} + \alpha \frac{\partial q}{\partial x} \right), \tag{8.22a}$$

$$\Pi = -\nu kn \left(\frac{\partial v}{\partial x} + \beta \frac{\partial q}{\partial x} \right), \tag{8.22b}$$

$$q = -\frac{C_p}{Pr} Kn \left(\frac{\partial \theta}{\partial x} + \alpha \frac{\partial \sigma}{\partial x} + \beta \frac{\partial \Pi}{\partial x} \right). \tag{8.22c}$$

8.3.6 Linearized refined NSF model

In the RNSF equations, the dynamic temperature equation in 1D, linearized, and dimensionless form is given by:

$$\frac{3}{2} \frac{\partial \vartheta}{\partial t} + \frac{\delta}{3+\delta} \frac{\partial v}{\partial x} = -\frac{\delta}{(3+\delta)} \frac{\vartheta}{\nu Kn}. \tag{8.23}$$

This formulation provides a framework for understanding the behavior of rarefied polyatomic gases, capturing the influence of dynamic temperature effects, non-equilibrium heat transfer, and bulk viscosity on the system's evolution.

8.4 Rayleigh–Brillouin Light Scattering

When light passes through a gas, a portion of it is scattered due to interactions with the gas molecules [31]. The spectral distribution of this scattered light depends on the underlying scattering mechanisms. RBS arises from density fluctuations in the gas and combines two primary components: Rayleigh scattering, caused by the thermal motion of gas molecules leading to a Doppler shift, and Brillouin scattering, associated with the exchange of energy between light and acoustic modes in the medium.

8.4.1 Spontaneous Rayleigh–Brillouin scattering

In spontaneous RBS, incident light with wave vector $\boldsymbol{k}_{\mathrm{i}}$ interacts with spontaneous density fluctuations in gas molecules, resulting in scattered light with wave vector $\boldsymbol{k}_s$. The scattering wave number K is given by

$K = |\boldsymbol{k}_{\rm i} - \boldsymbol{k}_{\rm s}| = 2|\boldsymbol{k}_{\rm i}|sin(\theta/2)$. where θ is the scattering angle, and the corresponding scattering wavelength is $L = 2\pi/K$.

To obtain the spectra of spontaneous and coherent RBS, we apply the Laplace transform in time and the Fourier transform in space for spontaneous RBS, or the Fourier transform in both directions for coherent RBS. This transforms eqn (8.20) and (8.21) into the following matrix form [32, 27]:

$$\begin{pmatrix} -\mathrm{i}\Omega & 2\pi\mathrm{i} & 0 & 0 \\ 2\pi\mathrm{i} & \frac{16Kn\pi^2}{3} - \mathrm{i}\Omega & 2\pi\mathrm{i} & 2\pi\mathrm{i} \\ 0 & 2\pi\mathrm{i} & \frac{2Kn\pi^2(5+\delta)}{Pr} - \frac{(3+\delta)}{2}\mathrm{i}\Omega & \frac{4Kn\pi^2}{Pr} \\ 0 & \frac{2\mathrm{i}\pi\delta}{3+\delta} & \frac{4Kn\pi^2\delta}{Pr(3+\delta)} & M_{44} \end{pmatrix} \begin{pmatrix} \tilde{\rho} \\ \tilde{v} \\ \tilde{\theta} \\ \tilde{\vartheta} \end{pmatrix} = \begin{pmatrix} P_1 \\ P_2 \\ 0 \\ 0 \end{pmatrix}, \tag{8.24}$$

where $M_{44} = \frac{2Kn\pi^2(9+5\delta)}{Pr(3+\delta)} - \frac{3\mathrm{i}\Omega}{2} + \frac{\delta}{3Kn\nu + Kn\delta\nu}$ is a element of the coefficient matrix. The variables $\tilde{\rho}, \tilde{v}, \tilde{\theta}$, and $\tilde{\vartheta}$ represent the spectra of the perturbation quantities. For spontaneous RBS, the initial density perturbation gives $P_1 = 1$ and $P_2 = 0$, while in coherent RBS, an external induced acceleration leads to $P_1 = 0$ and $P_2 = 1$ [32].

This non-homogeneous system (eqn (8.24)) is formulated for the 2T model, and similar derivations apply to the reduced 2T, polyatomic-CCR, NSF, and RNSF equations. By solving this system, we obtain the density fluctuation spectra, which characterize both spontaneous and coherent RBS. In particular, the spontaneous RBS spectrum is given by $\mathcal{S}_{\rm s} = \Re(\tilde{\rho})$, representing the real part of the density fluctuation spectrum.

8.4.2 Coherent Rayleigh–Brillouin scattering

Coherent RBS occurs when light scatters due to density fluctuations induced by the interference of two pump laser beams [15]. Unlike spontaneous RBS, where scattering occurs in all directions, coherent RBS benefits from phase matching, resulting in a significantly higher signal-to-noise ratio. This technique enables high-speed gas dynamic measurements on nanosecond timescales using rapidly chirped, pulsed optical lattices [16]. In coherent RBS, the spectrum is proportional to the square of density variations, given by $\mathcal{S}_c = |(\tilde{\rho})|^2$.

8.4.3 Results and interpretation

In this section, we perform a comparative analysis of the spontaneous and coherent Rayleigh–Brillouin spectra for various polyatomic gases, including

Table 8.1 This table presents the gas dynamic properties at room temperature

Gases	δ	Pr	ν
O_2	2.13	0.717	0.37
N_2	2.01	0.7368	0.41
CH_4	3	0.7706	1.33

nitrogen (N_2), oxygen (O_2), and methane (CH_4). The analysis is conducted using existing macroscopic models, with gas-dynamic parameters set at room temperature.

To obtain specific numerical results, we utilize the transport properties of these gases, as listed in Table 8.1. The ratio of shear and bulk viscosity values was interpolated from prior studies [33–35], while the internal degrees of freedom and Prandtl number data were taken from [36, 20] for nitrogen, [37, 20] for oxygen, and [36] for methane.

The Rayleigh–Brillouin spectra obtained from macroscopic models are compared with experimental data, molecular dynamics (MD) simulations, and other theoretical approaches, including the reduced two-temperature model, polyatomic coupled constitutive relations (CCR) model, NSF equations, and regularized NSF (RNSF) equations. This comparative study provides insights into the accuracy and applicability of these models in predicting RBS spectra and assessing their consistency with experimental observations.

We now present the scattered light spectrum of these gases at room temperature while considering different values of the non-dimensional parameter y. This parameter is defined as

$$y = \sqrt{\frac{RT_0}{2}} \frac{\rho_0}{\mu_0 K}. \tag{8.25}$$

Physically, y represents the ratio of the fluctuation length scale to the mean free path of gas molecules. As $y \to 0$, the system approaches the collisionless (Knudsen) regime, where intermolecular collisions become negligible, and the spectrum exhibits only a Rayleigh peak. Conversely, for $y \geq 1$, collective molecular interactions become significant, leading to the emergence of both the Rayleigh peak and the Brillouin doublets in the spectrum.

Additionally, we introduce the dimensionless frequency x, defined as:

$$\text{reduced frequency} \quad x = \sqrt{\frac{2}{3}} \frac{\Omega}{v_0 K}, \tag{8.26}$$

where v_0 denotes the equilibrium thermal velocity. This normalization facilitates a clearer interpretation of frequency-dependent scattering characteristics across different gases. In Figures 8.1–8.3, the spontaneous RBS spectra obtained from the 2T model, reduced 2T model, and polyatomic CCR model [21] are compared with those from the NSF and RNSF models [19]. These results are further contrasted with the extended hydrodynamic theory proposed by Hammond and Wiggins [38] and the translational hydrodynamic theory from Desai and Kapral [39]. The comparisons are conducted for three different values of the non-dimensional parameter y: 18.27, 4.46, and 2.70.

The red circle symbols represent the spectral predictions of the extended hydrodynamic theory, which, as per Hammond and Wiggins, exhibit strong agreement with experimental data (with curves normalized at zero frequency shift). The downward triangle symbols correspond to the predictions of the translational hydrodynamic theory [39].

Figure 8.1 demonstrates a complete agreement between all models and both hydrodynamic theories for $y = 18.27$. In Figure 8.2, corresponding to $y = 4.46$, the results of the 2T model and reduced 2T model closely align with the predictions of both hydrodynamic theories, whereas the polyatomic CCR, NSF, and RNSF models show noticeable deviations. Finally, Figure 8.3 illustrates the spontaneous RBS spectra for $y = 2.70$. As seen in the figure, the 2T model maintains good quantitative agreement with both hydrodynamic theories, whereas the predictions of other models deviate significantly from the hydrodynamic results [38, 39].

In Figure 8.4, the spontaneous scattered light spectrum from (N_2) for $y = 5$ is compared across all models against the Tenti six-moment model (Tenti s6) [14], which closely matches experimental data [40, 15]. While all models align well with Tenti s6, the polyatomic CCR model slightly overestimates the Brillouin peak. Figure 8.5 presents the (N_2) spectrum for $y = 1.725$, where all models are compared with MD data, which fully agrees with direct simulation Monte Carlo [41]. The 2T model exhibits superior agreement with MD when compared to the other theories, whereas the polyatomic CCR model overpredicts the Brillouin peak.

Figure 8.6 depicts the (O_2) spectrum, comparing all models with MD data [41]. Again, the 2T model exhibits the closest match for $y = 1.633$, outperforming other models. Overall, Figures 8.1–8.6, confirm that the 2T model accurately describes spontaneous RBS in polyatomic gases and effectively captures the transition between hydrodynamic and kinetic regimes.

We now compare the coherent RBS spectra in (N_2) obtained from all models with the results of the seven-moment molecular gas model (Pan s7)

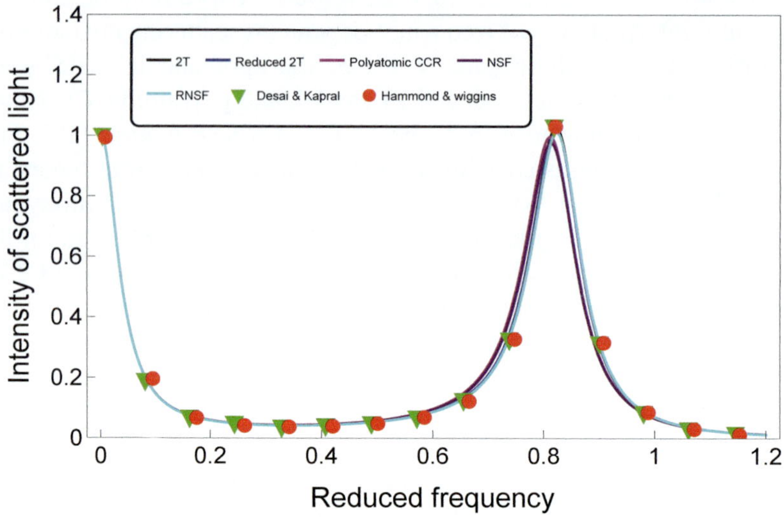

Figure 8.1 This figure illustrates the normalized spectra of spontaneous RBS in (CH_4) for $y = 18.27$, with the Rayleigh peak as the reference, demonstrating complete agreement between all models and both hydrodynamic theories.

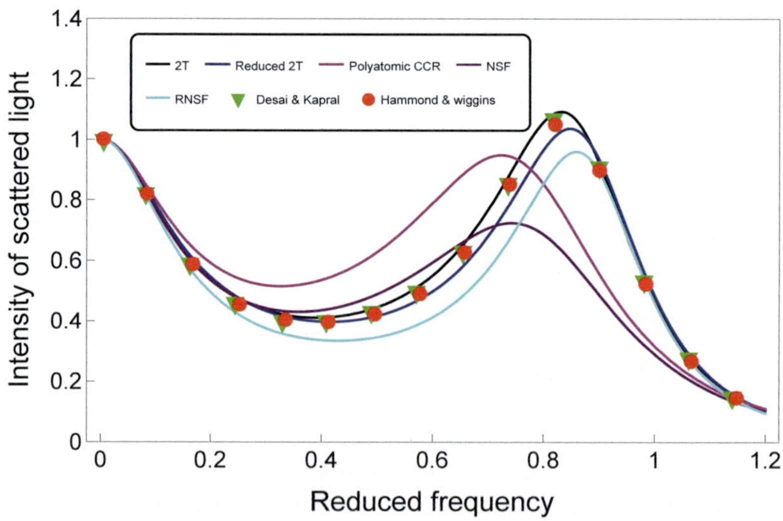

Figure 8.2 This figure illustrates the normalized spectra of spontaneous RBS in (CH_4) for $y = 4.46$, with the Rayleigh peak as the reference, showing that the 2T and reduced 2T models closely match hydrodynamic predictions, while the polyatomic CCR, NSF, and RNSF models exhibit noticeable deviations.

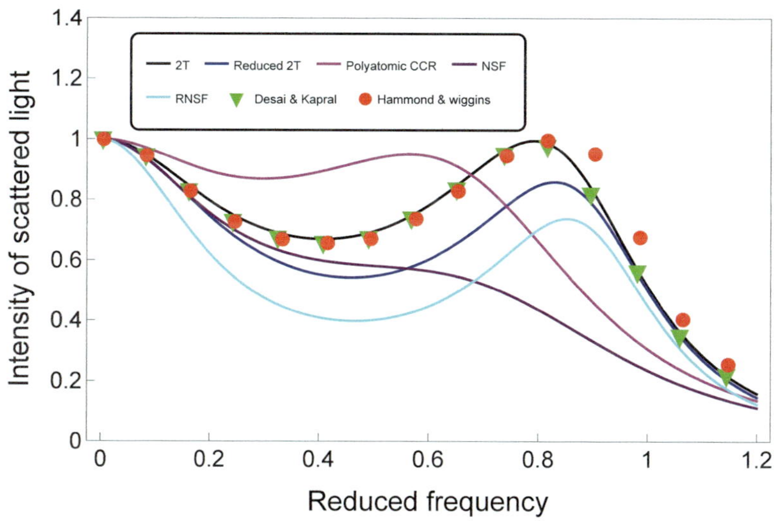

Figure 8.3 This figure shows the normalized spontaneous RBS spectra in (CH_4) for $y = 2.70$, using the Rayleigh peak as a reference. The 2T model closely aligns with hydrodynamic theories, while other models deviate significantly.

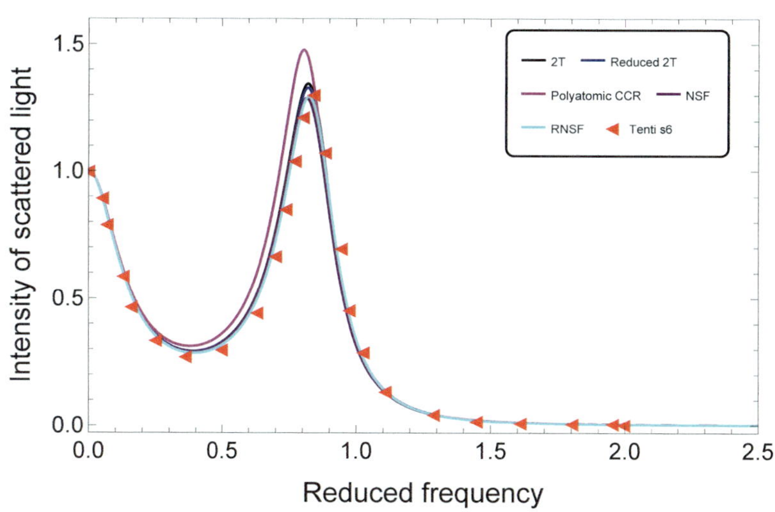

Figure 8.4 This figure presents the spontaneous scattered light spectrum from N_2 at $y = 5$, referenced to the Rayleigh peak, and compares all models with the Tenti s6 [14].

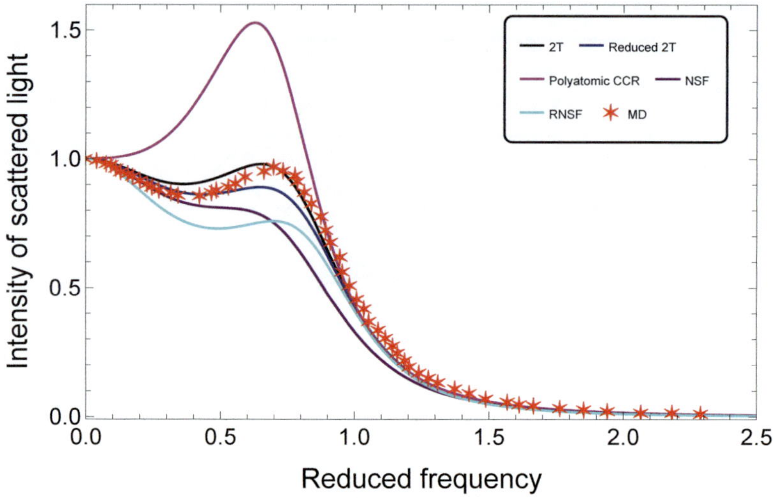

Figure 8.5 This figure presents the (N_2) spectrum for $y = 1.725$, referenced to the Rayleigh peak. All models are compared with MD data, which closely aligns with direct simulation Monte Carlo [41]. The 2T model agrees best with MD, while the polyatomic CCR model overestimates the Brillouin peak.

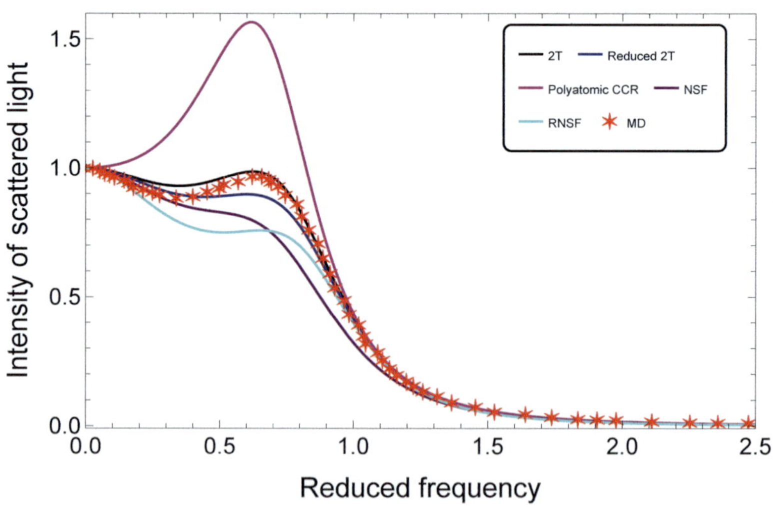

Figure 8.6 This figure presents the (O_2) spectrum for $y = 1.633$, referenced to the Rayleigh peak. All models are compared with MD data, which closely aligns with direct simulation Monte Carlo [41]. The 2T model agrees best with MD, while the polyatomic CCR model overestimates the Brillouin peak.

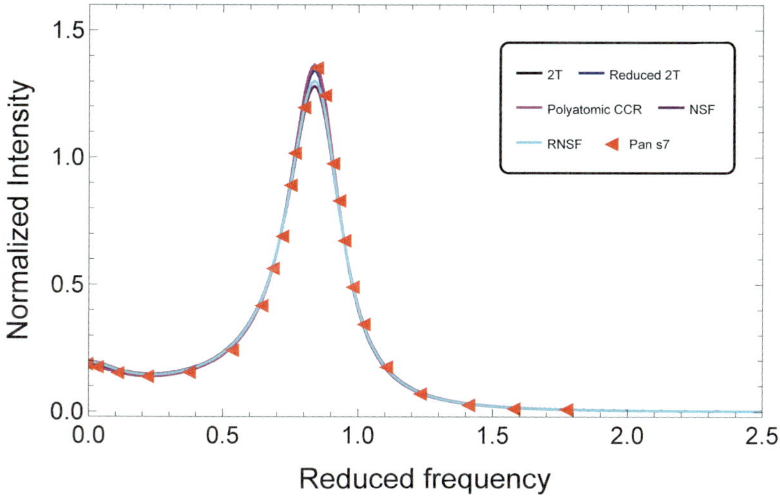

Figure 8.7 This figure shows the coherent RBS spectrum for (N_2) at $y = 5$, with each curve normalized to an area of $1/2$. All models yield nearly identical results, closely matching the Pan s7 model [42] for $y = 5$.

[42] in Figures 8.7–8.9. The Pan s7 model shows reasonable agreement with experimental data [42]. The comparisons are conducted for three different values of the non-dimensional parameter y: 5, 3, and 1.75.

Figure 8.7 demonstrates that all models produce nearly identical results, making them indistinguishable from the Pan s7 model for $y = 5$. In Figure 8.8, for $y = 3$, the 2T and polyatomic CCR models show better agreement with Pan s7 compared to the other models, which underpredict the Brillouin peak.

Figure 8.9 shows noticeable differences among all models for the small value of $y = 1.75$. The 2T, reduced 2T, NSF, and RNSF models exhibit similar behavior at the Rayleigh and Brillouin peaks, whereas the polyatomic CCR model deviates. Specifically, at the Rayleigh peak, the polyatomic CCR model predicts a lower intensity than all other models, while at the Brillouin peaks, it predicts higher values. The 2T model shows better agreement with Pan s7 compared to other models but still underpredicts the Brillouin peak.

For $y = 3$ and $y = 1.75$, the polyatomic CCR model generally predicts slightly sharper Brillouin peaks, suggesting slower dissipation processes. Overall, these results indicate the need for further modifications to the NSF equations to improve predictions of coherent RBS for small y.

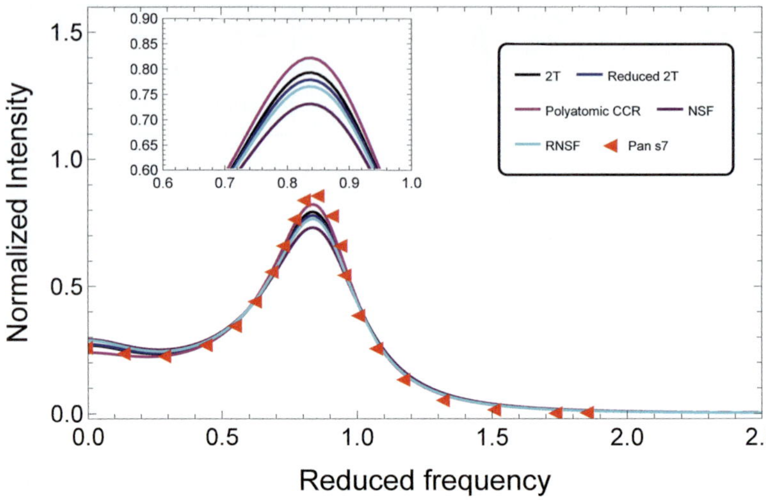

Figure 8.8 This figure illustrates the coherent RBS spectrum for (N_2) at $y = 3$, with each curve normalized to an area of $1/2$. The 2T and polyatomic CCR models show better agreement with Pan s7 [42], while the other models underpredict the Brillouin peak.

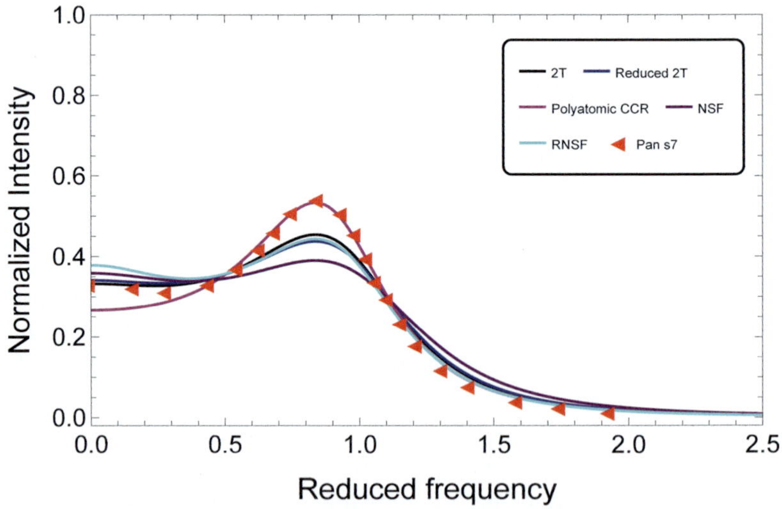

Figure 8.9 This figure illustrates the coherent RBS spectrum for (N_2) at $y = 1.75$, with each curve normalized to $1/2$. Noticeable differences are observed among all models for this small value of $y = 1.75$.

8.5 Conclusion

This chapter explored various extended hydrodynamic models for RBS in polyatomic rarefied gases, focusing on the 2T model, the reduced 2T model, the polyatomic CCR model, and the RNSF equations. By comparing these models with experimental data, molecular dynamics simulations, and theoretical benchmarks, their ability to capture non-equilibrium effects in gas flows was assessed.

The results highlight that the 2T model provides a more accurate representation of spontaneous and coherent RBS spectra across different regimes, particularly in the transition from hydrodynamic to kinetic behavior. The reduced 2T model offers a simplified yet effective alternative, while the CCR and RNSF models exhibit certain limitations in predicting Brillouin peaks at lower values of the non-dimensional parameter y. The study further reinforces the necessity of incorporating higher-order moment equations and energy exchange mechanisms to improve the fidelity of macroscopic gas models.

Overall, this comparative study advances our understanding of polyatomic rarefied gases and their interaction with light scattering phenomena. The insights gained here contribute to the development of more accurate CFD models, facilitating their application in engineering and atmospheric sciences. Future work could explore further refinements to these models, particularly in the context of high-fidelity kinetic simulations and experimental validation.

References

[1] F. Durst and I. Arnold, *Fluid mechanics: an introduction to the theory of fluid flows*, vol. 675. Springer, 2008.

[2] E. S. Cole, *Design and Validation of an Aerodynamic System for a Formula Sae® Vehicle Using Vehicle Dynamic Simulation and Experimentation.* West Virginia University, 2023.

[3] T. Gerkema, J. Zimmerman, L. Maas, and H. Van Haren, "Geophysical and astrophysical fluid dynamics beyond the traditional approximation," *Reviews of Geophysics*, vol. 46, no. 2, 2008.

[4] J. Wutz, B. Waterkotte, K. Heitmann, and T. Wucherpfennig, "Computational fluid dynamics (cfd) as a tool for industrial uf/df tank optimization," *Biochemical Engineering Journal*, vol. 160, p. 107617, 2020.

[5] O. Darrigol, *Worlds of flow: A history of hydrodynamics from the Bernoullis to Prandtl.* Oxford University Press, 2005.

[6] D. A. Lockerby and J. M. Reese, "On the modelling of isothermal gas flows at the microscale," *Journal of Fluid Mechanics*, vol. 604, pp. 235–261, 2008.

[7] C. J. GREENSHIELDS and J. M. REESE, "The structure of shock waves as a test of brenner's modifications to the navier–stokes equations," *Journal of Fluid Mechanics*, vol. 580, p. 407–429, 2007.

[8] S. Chapman and T. Cowling, "The mathematical theory of non-uniform gases, cambridge univ," *Press, Cambridge, England*, vol. 12, no. 22, p. 119, 1970.

[9] C. Cercignani, "Theory and application of the boltzmann equation," *(No Title)*, 1975.

[10] A. S. Rana, V. K. Gupta, and H. Struchtrup, "Coupled constitutive relations: a second law based higher-order closure for hydrodynamics," *Proceedings of the Royal Society A: Mathematical, Physical and Engineering Sciences*, vol. 474, no. 2218, p. 20180323, 2018.

[11] M. L. Reddy, S. K. Dadzie, R. Ocone, M. K. Borg, and J. M. Reese, "Recasting navier–stokes equations," *Journal of Physics Communications*, vol. 3, no. 10, p. 105009, 2019.

[12] H. Brenner, "Navier–stokes revisited," *Physica A: Statistical Mechanics and its Applications*, vol. 349, no. 1-2, pp. 60–132, 2005.

[13] W. Reitz, "A review of:"beyond equilibrium thermodynamics" hc öttinger, john wiley & sons inc., 111 river st., hoboken, nj 07030. 2005, 625 pages, isbn: 0471666580.," 2006.

[14] G. Tenti, C. Boley, and R. C. Desai, "On the kinetic model description of rayleigh–brillouin scattering from molecular gases," *Canadian Journal of Physics*, vol. 52, no. 4, pp. 285–290, 1974.

[15] X. Pan, M. N. Shneider, and R. B. Miles, "Coherent rayleigh-brillouin scattering in molecular gases," *Physical Review A*, vol. 69, no. 3, p. 033814, 2004.

[16] A. Gerakis, M. Shneider, and P. Barker, "Single-shot coherent rayleigh–brillouin scattering using a chirped optical lattice," *Optics letters*, vol. 38, no. 21, pp. 4449–4452, 2013.

[17] O. Reitebuch, C. Lemmerz, E. Nagel, U. Paffrath, Y. Durand, M. Endemann, F. Fabre, and M. Chaloupy, "The airborne demonstrator for the direct-detection doppler wind lidar aladin on adm-aeolus. part i: Instrument design and comparison to satellite instrument," *Journal of Atmospheric and Oceanic Technology*, vol. 26, no. 12, pp. 2501–2515, 2009.

[18] X. Zhai, U. Marksteiner, F. Weiler, C. Lemmerz, O. Lux, B. Witschas, and O. Reitebuch, "Rayleigh wind retrieval for the aladin airborne demonstrator of the aeolus mission using simulated response calibration," *Atmospheric Measurement Techniques*, vol. 13, no. 2, pp. 445–465, 2020.

[19] B. Rahimi and H. Struchtrup, "Refined navier–stokes–fourier equations for rarefied polyatomic gases," in *International Conference on Nanochannels, Microchannels, and Minichannels*, vol. 46278, p. V001T01A001, American Society of Mechanical Engineers, 2014.

[20] A. Kumar and A. S. Rana, "H-theorem and boundary conditions for two-temperature model: Application to wave propagation and heat transfer in polyatomic gases," *Physical Review E*, vol. 108, no. 6, p. 065103, 2023.

[21] A. S. Rana and S. Barve, "A second-order constitutive theory for polyatomic gases: theory and applications," *Journal of Fluid Mechanics*, vol. 958, p. A23, 2023.

[22] A. Kumar and A. S. Rana, "Capturing non-equilibrium in hypersonic flows: Insights from a two-temperature model in polyatomic rarefied gases," *Physics of Fluids*, vol. 36, no. 10, 2024.

[23] O. Borm, A. Jemcov, and H.-P. Kau, "Density based navier–stokes solver for transonic flows," in *Proceedings of 6th OpenFOAM workshop, penn state university, USA*, 2011.

[24] T. Ruggeri, M. Sugiyama, *et al.*, *Rational extended thermodynamics beyond the monatomic gas*. Springer, 2015.

[25] B. Rahimi and H. Struchtrup, "Macroscopic and kinetic modelling of rarefied polyatomic gases," *Journal of Fluid Mechanics*, vol. 806, pp. 437–505, 2016.

[26] S. R. De Groot and P. Mazur, *Non-equilibrium thermodynamics*. Courier Corporation, 2013.

[27] W. Marques Jr and G. Kremer, "Spectral distribution of scattered light in polyatomic gases," *Physica A: Statistical Mechanics and its Applications*, vol. 197, no. 3, pp. 352–363, 1993.

[28] K. Aoki, M. Bisi, M. Groppi, and S. Kosuge, "Two-temperature navier–stokes equations for a polyatomic gas derived from kinetic theory," *Physical Review E*, vol. 102, no. 2, p. 023104, 2020.

[29] V. Djordjić, G. Oblapenko, M. Pavić-Čolić, and M. Torrilhon, "Boltzmann collision operator for polyatomic gases in agreement with

experimental data and dsmc method," *Continuum Mechanics and Thermodynamics*, vol. 35, no. 1, pp. 103–119, 2023.
[30] B. Rahimi and H. Struchtrup, "Capturing non-equilibrium phenomena in rarefied polyatomic gases: A high-order macroscopic model," *Physics of Fluids*, vol. 26, no. 5, 2014.
[31] R. D. Mountain, "Spectral distribution of scattered light in a simple fluid," *Reviews of Modern Physics*, vol. 38, no. 1, p. 205, 1966.
[32] L. Wu, *Rarefied gas dynamics: kinetic modeling and multi-scale simulation*. Springer Nature, 2022.
[33] M. S. Cramer, "Numerical estimates for the bulk viscosity of ideal gases," *Physics of fluids*, vol. 24, no. 6, 2012.
[34] X.-D. Li, Z.-M. Hu, and Z.-L. Jiang, "Continuum perspective of bulk viscosity in compressible fluids," *Journal of Fluid Mechanics*, vol. 812, pp. 966–990, 2017.
[35] B. Sharma and R. Kumar, "Estimation of bulk viscosity of dilute gases using a nonequilibrium molecular dynamics approach," *Physical Review E*, vol. 100, no. 1, p. 013309, 2019.
[36] S. Singh, A. Karchani, K. Sharma, and R. S. Myong, "Topology of the second-order constitutive model based on the boltzmann–curtiss kinetic equation for diatomic and polyatomic gases," *Physics of Fluids*, vol. 32, no. 2, 2020.
[37] B. Raton, *CRC Handbook of Chemistry and Physics*. CRC Press, 1999.
[38] C. Hammond Jr and T. Wiggins, "Rayleigh–brillouin scattering from methane," *The Journal of Chemical Physics*, vol. 65, no. 7, pp. 2788–2793, 1976.
[39] R. C. Desai and R. Kapral, "Translational hydrodynamics and light scattering from molecular fluids," *Physical Review A*, vol. 6, no. 6, p. 2377, 1972.
[40] R. Xia, X. He, T. Wu, X. Zhang, and H. Yu, "Theoretical and experimental analysis of spontaneous gaseous rayleigh-brillouin scattering spectra," *Optik*, vol. 127, no. 18, pp. 7262–7267, 2016.
[41] Q. Ma, C. Yang, D. Bruno, and J. Zhang, "Molecular simulation of rayleigh-brillouin scattering in binary gas mixtures and extraction of the rotational relaxation numbers," *Physical Review E*, vol. 104, no. 3, p. 035109, 2021.
[42] X. Pan, *Coherent Rayleigh–Brillouin scattering*. PhD thesis, Princeton University, 2003.

9

Nodal Discontinuous Galerkin Framework for Solving Grad-17 Moment Equations in a Rarefied Regime

Hang Song[1], Satyvir Singh[1,2], and Manuel Torrilhon

[1]Institute for Applied and Computational Mathematics, RWTH Aachen University, Germany
[2]Department of Mathematics, Graphic Era Deemed to be University, India
E-mail: song@acom.rwth-aachen.de; singh@acom.rwth-aachen.de; mt@acom.rwth-aachen.de

Abstract

This research conducts non-equilibrium flow simulations for polyatomic gases using the Grad-17 moment equations, highlighting their proficiency in capturing intricate flow behaviors across diverse shock dynamics cases. Comprehensive investigations of 1D sod shock tube problems and stable shock structures reveal that the Grad-17 model achieves higher accuracy in simulating non-equilibrium effects distinguished from conventional Euler and NSF models. Validation through comparisons with DSMC data demonstrates the Grad models' proficiency in resolving shock thickness, stress distributions, and heat flux in flows with pronounced non-equilibrium effects. Moreover, the study underscores the critical role of incorporating high-order moments in model development for rarefied flow regimes, where elevated Knudsen numbers exacerbate non-equilibrium effects.

Keywords: Polyatomic gas, discontinuous Galerkin, G17 moment equations

DOI: 10.1201/9788743806691-9

9.1 Introduction

Non-equilibrium phenomena in gases play a crucial role in scientific and engineering applications, especially in scenarios involving low-density flows or extreme conditions of high speed and temperature [1]. These effects arise when the molecular distribution of gas deviates significantly from equilibrium, which is commonly seen in areas such as hypersonic flight [2] and microfluidics [3]. The complexity increases for polyatomic gases, where internal energy modes, including rotational and vibrational states, must be accounted for. Understanding these phenomena is essential for predicting the performance of systems like re-entry vehicles, advanced propulsion technologies, and processes such as gas separation and chemical reaction modeling.

The Knudsen number (Kn) is a critical parameter for quantifying the impact of non-equilibrium effects on gas flow behavior. It is defined as the ratio of the gas molecule mean free path (λ) to a characteristic length scale (L), expressed as Kn $= \lambda/L$. Depending on its value, gas flows are classified into four regimes: continuum flow (Kn $\leqslant 0.001$), slip flow ($0.001 <$ Kn $\leqslant 0.1$), transition flow ($0.1 <$ Kn $\leqslant 10$), and free molecular flow (Kn > 10) [4]. At higher Kn values, the infrequency of molecular collisions amplifies non-equilibrium effects, which become more significant as the Mach number increases. These challenges make it difficult to construct reliable models, both theoretically and computationally [5].

The Boltzmann transport equation (BTE) is the foundation of understanding gas dynamics beyond the limits of local thermodynamic equilibrium [6]. Instead of focusing on macroscopic properties, BTE characterizes gas behavior through a probability distribution function that tracks particle velocities at different spatial and temporal points [7]. For polyatomic gases, internal energy modes were integrated into the equation [8]. Later, Bourgat et al. [9] refined this approach by introducing a model with a single continuous parameter to represent internal degrees of freedom. The non-linear nature of BTE makes solving it analytically or numerically challenging, especially when considering particle collisions and external forces. To address computational limitations, researchers often employ its linearized version [10].

A viable alternative to directly solving BTE is the direct simulation Monte Carlo (DSMC) method [11], which approximates gas behavior through particle-based simulations. Despite its effectiveness, DSMC becomes computationally demanding in the transition regime due to statistical noise, the need for large sample sizes, and increased collision frequencies.

In transitional gas regimes, macroscopic modeling offers a computationally efficient alternative to solving the Boltzmann equation. However, their accuracy is constrained by the Knudsen number. These models act as intermediaries between detailed molecular simulations and practical field equations, capturing critical non-equilibrium features. Among the most well-established methods for deriving macroscopic equations from BTE are the Chapman–Enskog expansion [12] and moment methods [13]. Nagnibeda and Kustova [14] employed the former to develop kinetic transport theory and explore relaxation processes in reacting gas mixtures under strongly non-equilibrium conditions.

Several enhancements of the popular Navior–Stokes–Fourier (NSF) framework have emerged within the BTE context, particularly through Chapman–Enskog expansions around equilibrium, leading to the Burnett and super-Burnett equations [12]. Meanwhile, moment methods provide an effective means of deriving macroscopic evolution equations directly from BTE by systematically incorporating higher-order terms. Grad [15] introduced this technique by expanding the distribution function, while Kogan [16] extended it through entropy maximization to generalize the Grad-13 moment equations. Subsequent developments were made by Zhdanov [17] and McCormack, [18] who independently formulated 17-moment equations for polyatomic gases, addressing slip velocity and temperature jump effects.

Recent research has focused on refining macroscopic equations to accommodate varying specific heats. Pavić-Čolić developed a Boltzmann collision operator within a continuous internal energy framework to describe polyatomic gases with temperature-dependent heat capacities [19]. Djordjić et al. [20] derived a Grad-17 moment system and transport coefficients for polyatomic gases based on a continuous internal energy variable. Additionally, the relationship between the collision kernel and molecular motion laws in collision processes is also being examined [21].

The Grad moment method extends kinetic theory beyond NSF equations to model non-equilibrium gas behavior. The G17 models, crucial for rarefied flows, incorporate higher-order moments to capture non-equilibrium effects. However, their numerical feasibility is largely unexplored. This work investigates polyatomic gas simulations using these models, focusing on complex shock dynamics. To assess their performance, a high-order discontinuous Galerkin solver is developed to solve Grad-type equations in various shock-driven flows, including 1D shock structures and shock tubes [22]. This study offers new insights into the effectiveness of these models in capturing non-equilibrium effects compared to lower-order and kinetic models.

9.2 Governing Equations

Numerically implementing the Boltzmann equation poses significant difficulties. Nonetheless, the Grad moment equations, obtained by expanding Hermite polynomials around the local Maxwellian distribution, offer a method to circumvent the necessity for intricate examination of molecular interactions and dynamics. These equations describe macroscopic properties or moments of the thermal velocity distribution function f, such as density, temperature, and higher-order terms. In polyatomic gases, the distribution function becomes more complex since internal degrees of freedom must be accounted for in addition to translational motion.

From a kinetic perspective, the state of a continuum-rarefied polyatomic gas is statistically described by a distribution function $f(\mathbf{x}, \mathbf{v}, I, t) \geq 0$, which depends on the position in physical space $\mathbf{x} \in \mathbb{R}^3$, particle velocity $\mathbf{v} \in \mathbb{R}^3$, microscopic internal energy $I \in \mathbb{R}_+$, and time $t \in \mathbb{R}_+$. In kinetic theory, moments are defined as integral projections of the distribution function over phase space. At this stage, we can define the following quantities: mass density: ρ, velocity: u_{i}, dynamic pressure: p, average pressure: $\overline{p}$, pressure difference: $p_{\mathrm{diff}} = p - \overline{p}$, pressure tensor: p_{ij}, stress tensor: σ_{ij}, translational heat flux: q_{i}, rotational heat flux: s_{i}, as shown in eqn (9.1).

$$\begin{pmatrix} \rho \\ \rho u_{\mathrm{i}} \\ p \\ \overline{p} \\ p_{\mathrm{diff}} \\ p_{\mathrm{ij}} \\ \sigma_{\mathrm{ij}} \\ q_{\mathrm{i}} \\ s_{\mathrm{i}} \\ q_{\mathrm{i}}^{\mathrm{tot}} \end{pmatrix} = \int_{\mathbb{R}^3 \times R_+} \begin{pmatrix} m \\ m v_{\mathrm{i}} \\ \frac{1}{3} m c^2 \\ \frac{2}{\delta+3}\left(\frac{1}{2} m c^2 + I\right) \\ \frac{1}{3} m c^2 - \frac{2}{\delta+3}\left(\frac{1}{2} m c^2 + I\right) \\ m c_{\mathrm{i}} c_{\mathrm{j}} \\ m c_{\mathrm{i}} c_{\mathrm{j}} - \frac{1}{3} m c^2 \delta_{\mathrm{ij}} \\ \frac{1}{2} m c^2 c_{\mathrm{i}} \\ I c_{\mathrm{i}} \\ \frac{1}{2} m c^2 c_{\mathrm{i}} + I c_{\mathrm{i}} \end{pmatrix} f \, d\mathbf{v} \, dI, \; i, j \in \{1, 2, 3\}, \tag{9.1}$$

$$\boldsymbol{U} = \begin{pmatrix} \rho \\ \rho u_{\mathrm{i}} \\ \rho |u|^2 + (\delta + 3)\overline{p} \\ \rho u_{\mathrm{i}} u_{\mathrm{j}} + p_{\mathrm{ij}} \\ (\rho |u|^2 + 3p) u_{\mathrm{i}} + 2 p_{\mathrm{ij}} u_{\mathrm{j}} + 2 q_{\mathrm{i}} \\ 2 s_{\mathrm{i}} + (\delta + 3)\overline{p} u_{\mathrm{i}} - 3 p u_{\mathrm{i}} \end{pmatrix}$$

$$
\boldsymbol{F}_{\mathrm{k}}=\begin{pmatrix}
\rho u_{\mathrm{k}} \\
\rho u_{\mathrm{i}} u_{\mathrm{k}}+p_{\mathrm{ik}} \\
(\rho|u|^2+(\delta+3)\overline{p})u_{\mathrm{k}}+2p_{\mathrm{jk}}u_{\mathrm{j}}+2q_{\mathrm{k}}+2s_{\mathrm{k}} \\
\rho u_{\mathrm{i}}u_{\mathrm{j}}u_{\mathrm{k}}+u_{\mathrm{i}}p_{\mathrm{jk}}+u_{\mathrm{j}}p_{\mathrm{ki}}+u_{\mathrm{k}}p_{\mathrm{ij}}+\frac{2}{5}(q_{\mathrm{i}}\delta_{\mathrm{jk}}+q_{\mathrm{j}}\delta_{\mathrm{ik}}+q_{\mathrm{k}}\delta_{\mathrm{ij}}) \\
(\rho|u|^2+3p)u_{\mathrm{i}}u_{\mathrm{k}}+2u_{\mathrm{i}}u_{\mathrm{j}}p_{\mathrm{jk}}+2u_{\mathrm{j}}u_{\mathrm{k}}p_{\mathrm{ij}}+|u|^2p_{\mathrm{ik}}+2u_{\mathrm{i}}q_{\mathrm{k}}+2u_{\mathrm{k}}q_{\mathrm{i}} \\
\quad+\frac{4}{5}(u_{\mathrm{i}}q_{\mathrm{k}}+u_{\mathrm{k}}q_{\mathrm{i}}+u_{\mathrm{j}}q_{\mathrm{j}}\delta_{\mathrm{ik}})+7\frac{\overline{p}}{\rho}p_{\mathrm{ik}}-5\frac{\overline{p}^2}{\rho}\delta_{\mathrm{ik}}+3\frac{\overline{p}}{\rho}p\delta_{\mathrm{ik}} \\
\frac{\overline{p}}{\rho}(\delta p_{\mathrm{ik}}-\delta p+\delta p\delta_{\mathrm{ik}}-3p\delta_{\mathrm{ik}}+3\overline{p}\delta_{\mathrm{ik}}) \\
\quad+2u_{\mathrm{i}}s_{\mathrm{k}}+2u_{\mathrm{k}}s_{\mathrm{i}}+u_{\mathrm{i}}u_{\mathrm{k}}((\delta+3)\overline{p}-3p)
\end{pmatrix}
$$

$$
\boldsymbol{S}=\begin{pmatrix}
0 \\
0 \\
0 \\
-\frac{\overline{p}}{\mu}(p_{\mathrm{ij}}-p\delta_{\mathrm{ij}})-\frac{\overline{p}}{\mu_{\mathrm{b}}}\frac{2\delta}{3\delta+9}(p-\overline{p})\delta_{\mathrm{ij}} \\
-\frac{\overline{p}}{\mu}(2u_{\mathrm{j}}p_{\mathrm{ij}}-2pu_{\mathrm{i}})-\frac{\overline{p}}{\mu_{\mathrm{b}}}\frac{2\delta}{3\delta+9}(p-\overline{p})(5u_{\mathrm{i}})-\frac{\overline{p}}{\mu}(2P_q^{(0)}q_{\mathrm{i}}+2P_q^{(1)}s_{\mathrm{i}}) \\
-\frac{\overline{p}}{\mu}(2P_s^{(0)}s_{\mathrm{i}}+2P_s^{(1)}q_{\mathrm{i}})+\frac{\overline{p}}{\mu_{\mathrm{b}}}\frac{2\delta}{3\delta+9}(p-\overline{p})(3u_{\mathrm{i}})
\end{pmatrix}
\tag{9.2}
$$

The Grad-17 moment (G17) equations, presented in eqn (9.2), are formulated from the Boltzmann transport equation (BTE) for polyatomic gases and incorporate supplementary moments to independently characterize the heat flux associated with translational and rotational energy [20, 21]. These equations capture the non-equilibrium dynamics of polyatomic gases by accounting for intricate internal degrees of freedom, including the translational heat flux (q_{i}) and rotational heat flux (s_{i}). As a result, the G17 equations for the respective densities are established. In these moment equations, the conservative variables $\boldsymbol{U}$ are defined as Ref. [20, 21].

Meanwhile, the production term $\boldsymbol{S}(\boldsymbol{U})$ embodies the influence of molecular collisions and relaxation processes. To derive $\boldsymbol{S}(\boldsymbol{U})$, we begin with the Boltzmann transport equation for polyatomic gases, employing a closure based on Grad's moment expansion. The moment equations are derived by integrating the Boltzmann equation with respect to weighted velocity moments of the distribution function. These source terms are obtained under the assumption of a BGK-type relaxation model, which simplifies the collision operator while retaining essential non-equilibrium phenomena beyond the Navier–Stokes–Fourier (NSF) equations. For a comprehensive derivation, readers are directed to Refs. [20, 21].

9.3 Numerical Method

This study employs Trixi.jl, a Julia-based framework, to perform moment equation simulations for non-equilibrium flows in polyatomic gases [23]. The framework specializes in addressing hyperbolic conservation laws with a modular and adaptable structure [24]. It employs advanced methods like discontinuous Galerkin spectral element method (DGSEM) to effectively manage systems with sharp gradients or discontinuities [25]. Stability and conservation are maintained through adaptive step sizing based on error or CFL conditions, as well as methods designed to preserve kinetic energy and entropy [26]. High-order basis functions, including Lobatto–Legendre polynomials, optimize accuracy while reducing computational costs. Moreover, multi-stage Runge–Kutta schemes improve efficiency in large-scale simulations. With Julia's parallel computing features, Trixi.jl provides excellent scalability, making it applicable to fields like fluid dynamics, aeroacoustics, electromagnetics, and astrophysics.

9.3.1 Spatial discretization in DG framework

Consider the 1D element $Q_l = [x_l - \frac{dx}{2}, x_l + \frac{dx}{2}]$. A mapping between Q_l and $\Omega = [-1, 1]$ is built through $x(\xi) = x_l + \frac{dx}{2}\xi, \xi \in [-1, 1]$. Then the original 1D hyperbolic equation, eqn (9.3), can be rewritten as eqn (9.4) on each element Q_l [23]:

$$\frac{\partial u}{\partial t} + \frac{\partial f}{\partial x} = 0, \tag{9.3}$$

$$\frac{dx}{2}\frac{\partial u}{\partial t} + \frac{\partial f}{\partial \xi} = 0. \tag{9.4}$$

To approximate the solution in each element, a polynomial of degree N is used to reduce the amount of calculations and space of information storage. Lagrange basis functions $\{l_{\mathrm{i}}\}_{\mathrm{j}=0}^{N}$ are used as polynomial basis, and the solution is approximated by eqn (9.5):

$$u(\xi, t) \mid_{Q_l} \approx u^{Q_l}(\xi, t) = \sum_{i=0}^{N} u_{\mathrm{i}}^{Q_l}(t) l_{\mathrm{i}}(\xi). \tag{9.5}$$

The $N+1$ Gauss–Lobatto nodes are chosen and are used for the Gaussian–Lobatto quadrature [27]. Due to the Kronecker property:

$$u^{Q_l}(\xi_{\mathrm{i}}, t) = u_{\mathrm{i}}^{Q_l}(t).$$

In this research, polynomial of degree $N = 3$. The nodes are $\{l_{\rm i}\} = \{-1, -\frac{1}{\sqrt{5}}, \frac{1}{\sqrt{5}}, 1\}$ $(i = 0, 1, 2, 3)$, and their corresponding weights are $\{w_{\rm i}\} = \{-\frac{1}{6}, -\frac{5}{6}, \frac{5}{6}, \frac{1}{6}\}$. Legendre-Gauss–Lobatto (LGL) mass matrix M is defined as $M_{\rm ij} = \delta_{\rm ij} w_{\rm j}$.

In the element Q_l, the terms of time derivatives and space derivatives are analyzed separately. Meanwhile, the original term of time derivatives can be expressed as eqn (9.6), and the time derivatives can be written as eqn (9.7):

$$\int_{-1}^{1}\left(\frac{dx}{2}\frac{\partial u(\xi,t)}{\partial t} + \frac{\partial f}{\partial \xi}\right) l_{\rm i}(\xi) d\xi = \frac{dx}{2}\int_{-1}^{1}\frac{\partial u(\xi,t)}{\partial t} l_{\rm i}(\xi) d\xi + \int_{-1}^{1}\frac{\partial f}{\partial \xi} l_{\rm i}(\xi) d\xi = 0. \tag{9.6}$$

$$\frac{dx}{2}\int_{-1}^{1}\frac{\partial u(\xi,t)}{\partial t} l_{\rm i}(\xi) d\xi \approx \frac{dx}{2}\int_{-1,N}^{1}\frac{\partial u(\xi,t)}{\partial t} l_{\rm i}(\xi) d\xi = \frac{dx}{2}\frac{\partial u(\xi_{\rm i},t)}{\partial t} w_{\rm i}. \tag{9.7}$$

For the term of space derivatives, according to spatial partial integration, the equation is shown as eqn (9.8):

$$\int_{-1}^{1}\frac{\partial f}{\partial \xi} l_{\rm i}(\xi) d\xi \approx [f l_{\rm i}]_{-1}^{1} - \int_{-1}^{1} f \frac{\partial l_{\rm i}}{\partial \xi} d\xi. \tag{9.8}$$

The first term on the right side is called the surface term. Because discontinous Galerkin method has the issue that the interfaces between two elements do not have unique values, it is necessary to define a numerical flux scheme f^*.

Then the surface term can be rewritten as eqn (9.9):

$$f_1^* = f_1^*(u_l, u_r). \tag{9.9}$$

u_l is the rightmost node of the left element, while u_r is the leftmost node of the right element. The surface term can be written as eqn (9.10):

$$[f^* l_{\rm i}]_{-1}^{1} = -f^* \mid_{\xi=-1} + f^* \mid^{\xi=+1}. \tag{9.10}$$

The second term on the right side is called the volume term, which can be expressed as eqn (9.11):

$$-\int_{-1}^{1} f \frac{\partial l_{\rm i}}{\partial \xi} d\xi = -\sum_{k=0}^{N} f_{\rm k} D_{\rm ki} w_{\rm k}. \tag{9.11}$$

$$D_{\mathrm{ki}} = \left(\frac{\partial l_{\mathrm{i}}}{\partial \xi} \bigg|_{\xi=\xi_{\mathrm{k}}} \right)_{\mathrm{ki}}$$

$$= \begin{pmatrix} -3.0 & 4.04508 & -1.54508 & 0.5 \\ -0.809017 & 0.0 & 1.11803 & -0.309017 \\ 0.309017 & -1.11803 & 0.0 & 0.809017 \\ -0.5 & 1.54508 & -4.04508 & 3.0 \end{pmatrix}.$$

We consider the so-called discontinuous Galerkin spectral element method (DGSEM) with collocation. This approach is based on a nodal DG approximation, employing $N+1$ Gauss–Lobatto nodes ξ_{i} in the interval $[-1,1]$, along with corresponding interpolation weights w_{i}. These weights are utilized for both numerical integration and interpolation using the Lagrange polynomial basis l_{i} of degree N. The Lagrange functions are constructed with these nodes, ensuring they satisfy the Kronecker property at the Gauss–Lobatto points. The weak formulation of the DGSEM for a single element is given by eqn (9.12):

$$\frac{dx}{2}\underline{u_t} = -M^{-1}B\underline{f}^* + M^{-1}D^T M\underline{f}, \tag{9.12}$$

where $\underline{u} = (u_0, u_1, \ldots, u_N)^T \in \mathbb{R}^{N+1}$ represents the collected pointwise evaluation of u at the discretization nodes. The nodal values of the flux function f are determined using collocation in $\underline{f}$, since $\underline{f}_{\mathrm{j}} = f(\underline{u}_{\mathrm{j}})$. Additionally, the numerical flux is computed as $f^* = f^*(u^-, u^+)$. $B = diag(-1, 0, 0, ..., 0, 1)$, and the discretization method has the property $MD + (MD)^T = B$, called summation-by-parts (SBP) property [28].

According to Kopriva, the numerical framework applies surface flux and volume flux, shown as eqn (9.13) [29]:

$$\frac{dx}{2}\underline{u_t} = -M^{-1}B\underline{f}^*_{\mathrm{surface}} - (2D - M^{-1}B)\underline{f}_{\mathrm{volume}}, \tag{9.13}$$

$$f_{\mathrm{ik}} = f(u(\xi_{\mathrm{i}}, t), u(\xi_{\mathrm{k}}, t)). \tag{9.14}$$

In this research, the Local Lax–Friedrichs scheme is select as surface flux and volume flux. The Local Lax–Friedrichs scheme [30], known for its stabilizing effect, is utilized to address the numerical discontinuity, as shown in eqn (9.15):

$$f_{\mathrm{LocalLaxFriedrichs}} = \frac{1}{2}(f(u^-) + f(u^+)) - \frac{1}{2}\lambda(f(u^+) - f(u^-)), \tag{9.15}$$

λ is the eigenvalue of the local flow field. In this research, it is selected as $\max\{|u_i| + c\}$. c is the sound of speed, which is commonly defined as $c = \sqrt{\gamma\frac{p}{\rho}}$.

9.3.2 Time discretization

The time integral used in this research is a five-stage fourth-order Runge–Kutta low-storage scheme [31]. Compared with the classical fourth-order method, it requires only 2N storage locations; it is considerably more efficient and accurate than existing schemes. For the time-based equations eqn (9.16), the relationship between two time steps $U_0 = U(t_0)$ and $U_1 = U(t_0 + h)$ is shown as eqn (9.17):

$$\frac{dU}{dt} = F(t, U(t)); \quad U(t_0) = U_0, \tag{9.16}$$

$$k_1 = F(t_0, U^0); \quad k_i = F(t_0 + c_i h, U^0 + h\sum_{j=1}^{i-1} a_{i,j}k_j)(i = 2, 3, ..., M);$$

$$U^1 = U^0 + h\sum_{j=1}^{M} b_j k_j. \tag{9.17}$$

To reduce the storage of time integration, the original Runge–Kutta method is developed by Williamson [32] and Fyfe [33], as shown in eqn (9.18). Only arrays $\{U_j\}$ and $\{dU_j\}$ are kept in storage.

$$dU_j = A_j dU_{j-1} + hF(U_j); \quad U_j = U_{j-1} + B_j dU_j. \tag{9.18}$$

The relationship between two sets of coefficients $\{A_j\}$ and $\{B_j\}$ and the original coefficients is as shown in eqn (9.19):

$$\begin{aligned} &B_j = a_{j+1,j}(j \neq M); \quad B_M = b_M; \quad A_j = (b_{j-1} - B_{j-1})/b_j(j \neq 1, b_j \neq 0) \\ &A_j = (a_{j+1,j-1} - c_j)/B_j \quad (j \neq 1, b_j = 0) \end{aligned} \tag{9.19}$$

The exact values of $\{A_j\}$, $\{B_j\}$ are shown as Ref. [31].

9.4 Results and Disscussion

This section presents an in-depth analysis of non-equilibrium flow simulations for polyatomic gases, including Euler, NSF, and G17 systems. Two test cases: 1D shock structure and shock tube are examined. The study explores flow field properties and non-equilibrium effects to assess parameter influences and uncover the mechanisms governing these behaviors.

9.4.1 Sod shock tube

The shock tube scenario involves two initially motionless gases with distinct characteristics, such as density, temperature, and pressure, separated by a diaphragm. Once the diaphragm is abruptly removed, the resulting pressure disparity induces shock waves and expansion fans that propagate through the medium. This setup is extensively employed to study shock wave mechanics, gas dynamics under extreme conditions, and interactions between shock waves and material boundaries [34].

To validate the computational model, an 80 mm shock tube filled with nitrogen gas at a uniform temperature of 273.15 K is examined. The tube is divided into two regions, with the right-side gas exhibiting a mean free path of 1 mm. On the left side, the gas possesses five times the density and pressure of the right. The system's evolution is analyzed at 4×10^{-5} seconds, capturing shock waves, contact discontinuities, and expansion waves for comparison with DSMC results.

For rarefied gas flows, the DSMC approach serves as the benchmark for validating G17 models. As a direct solver of the Boltzmann equation, DSMC avoids moment truncation and continuum approximations, making it the reference standard for rarefied flow simulations. It accurately models key non-equilibrium effects, including velocity slip, thermal transpiration, and high Knudsen number transport – phenomena that the NSF and Grad moment methods fail to fully capture. Rigorously validated against experimental data, DSMC remains the preferred solver for rarefied gas studies, particularly in hypersonic regimes and shock wave structures. This study adopts the variable soft sphere (VSS) model to describe collisions.

Figure 9.1 presents the density and stress distributions in the 1D shock tube scenario, comparing Euler, NSF, and G17 models against DSMC data. The Euler and NSF formulations struggle to capture non-equilibrium behavior, especially the stress near the shock front and contact surface. In contrast, the G17 equations exhibit strong agreement with DSMC, whereas the NSF-based stress predictions show substantial deviation.

To evaluate the impact of the Knudsen number (Kn) on the solutions of various moment models, including Euler and G17, two Kn values (Kn $=$ $0.005, 0.1$) are analyzed. The parameter $\overline{p}/\mu$ in the Grad moment framework

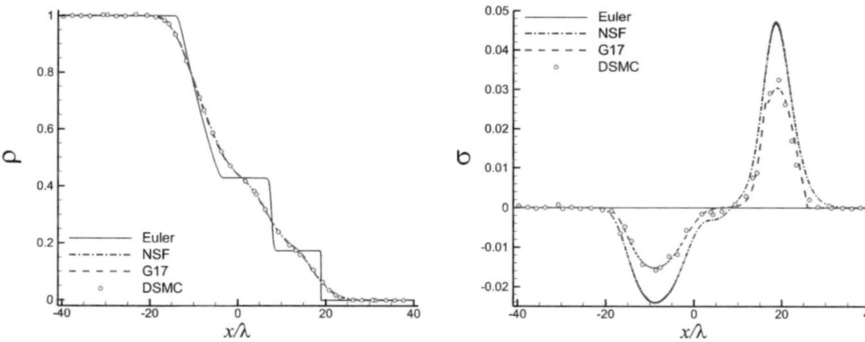

Figure 9.1 Validation of 1D sod shock problem: profiles of density (ρ), stress (σ) solutions with the Euler, NSF, and G17 results.

is replaced by ρ/Kn, ensuring all variables remain dimensionless for consistency. The initial conditions for G17 models in the 1D shock tube case are defined as eqn (9.20):

$$(\rho, u, \overline{p}, p, \sigma, q, s) = \begin{cases} (5.0, 0.0, 5.0, 5.0, 0.0, 0.0, 0.0), & \text{if } x \leq 0 \\ (1.0, 0.0, 1.0, 1.0, 0.0, 0.0, 0.0), & \text{if } x > 0 \end{cases}. \tag{9.20}$$

Figure 9.2 compares key physical quantities – density (ρ), velocity (v), average pressure ($\overline{p}$), pressure difference (p_{diff}), stress (σ), and total energy heat flux (q^{tot}) – in a 1D shock tube problem at $t = 1$. The analysis employs the Euler and G17 models for two Knudsen numbers ($\mathrm{Kn} = 0.005, 0.1$). The findings reveal the limitations of the Euler equations, which assume local thermodynamic equilibrium and struggle to account for non-equilibrium effects, particularly at $\mathrm{Kn} = 0.1$.

At low Knudsen numbers ($\mathrm{Kn} = 0.005$), both models exhibit similar trends. However, as Kn increases, discrepancies among them become more pronounced, with the G17 model providing the most realistic representation of non-equilibrium effects, such as pressure differences, stress, and heat flux. Notably, at higher Kn, reduced molecular collision frequency slows momentum and energy transfer. As a result, discontinuities like shock fronts become less distinct, while rarefaction waves smooth out due to increased dissipation effects:

$$R = \frac{1}{\overline{p}}\sqrt{\frac{1}{2}\sigma_{\text{ij}}^2 + \frac{\delta}{\delta + 5}\mathrm{Pr}\frac{\rho}{\overline{p}}(q_{\text{i}} + s_{\text{i}})^2}, \tag{9.21}$$

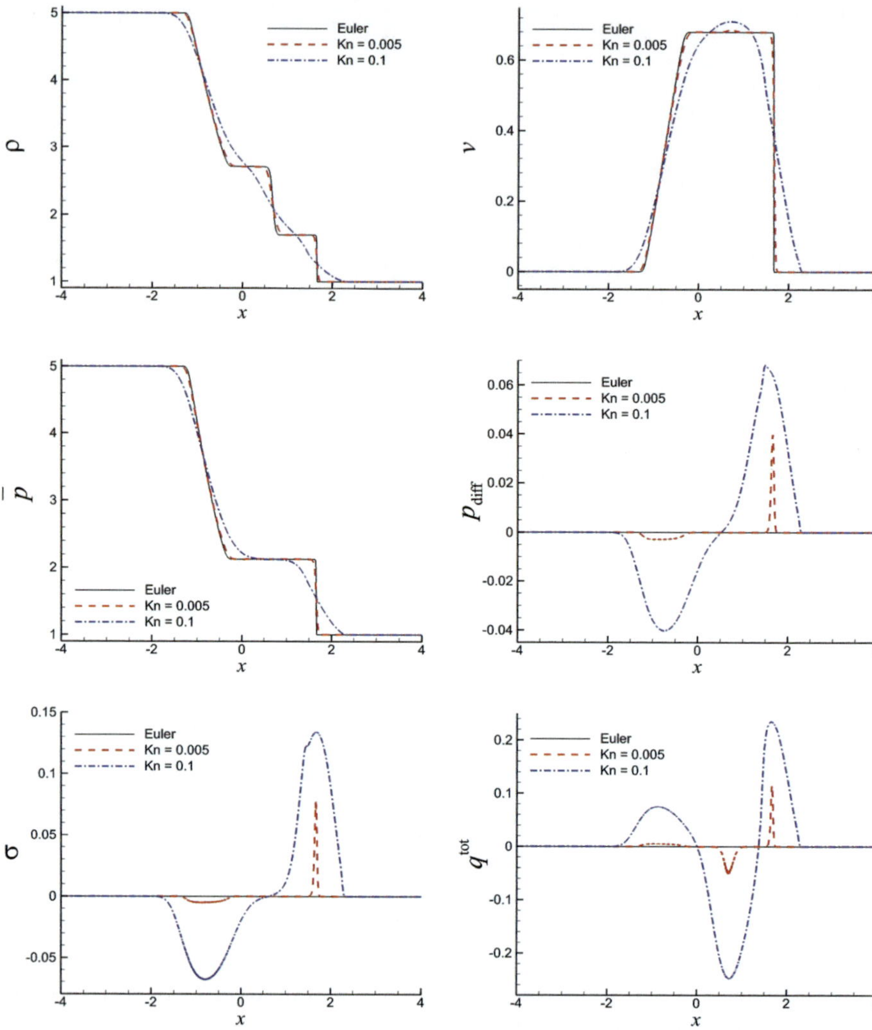

Figure 9.2 1D sod shock problem: profiles of density (ρ), velocity (v), average pressure ($\overline{p}$), pressure difference (p_{diff}), stress (σ), and heat flux of total (q^{tot}) energy solutions with the Euler, G17 results at Kn $= 0.005, 0.1$.

Figure 9.3 explores the evolution of stress-heat flux ($\sigma - q$), and degree of thermal non-equilibrium (R), which is defined as the Rayleigh–Onsager dissipation function eqn (9.21) in a 1D shock tube for various moment models at Kn $= 0.005$ and Kn $= 0.1$ when $t = 1$. The correlation between stress and heat flux is depicted in Figure 9.3(a). At Kn $= 0.005$, all models exhibit

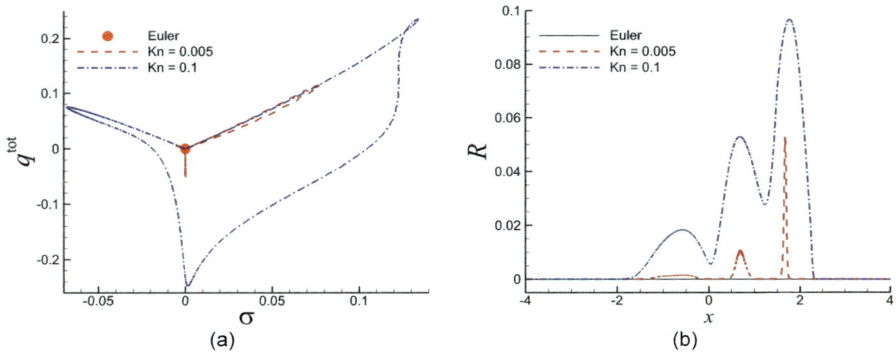

Figure 9.3 1D sod shock problem: profiles of stress-heat flux ($\sigma - q^{tot}$) and degree of non-equilibrium (R) solutions with the Euler, G17 results at Kn $= 0.005, 0.1$.

comparable trends, reflecting an almost equilibrium state. The G17 model, which accounts for higher-order moments, captures flow interactions more precisely. However, as Kn increases to 0.1, deviations become more evident due to intensified non-equilibrium effects.

In Figure 9.3(b), the non-equilibrium parameter R is presented. At Kn $= 0.005$, R remains minimal, indicating a system close to equilibrium. Conversely, at Kn $= 0.1$, R rises substantially, particularly around the shock front. The Euler model, based on equilibrium assumptions, is unable to reflect these variations, while the G17 model demonstrates superior in capturing non-equilibrium behavior. For an adequate description of rarefied flows and transition regime phenomena, high-order G17 equations are crucial.

9.4.2 1D shock wave

This test case involves nitrogen gas for non-equilibrium flow simulations. In the pre-shock region (left side), the mean free path is 1 millimeter, inflow Mach number is 1.2, and the temperature is 273.15 K. Reference values are taken from Bird [11].

Figure 9.4 presents the normalized physical variable profiles obtained from the Euler, NSF and G17 models, alongside DSMC results at Ma $= 1.2$. Since Euler equations assume inviscid flow, they predict an abrupt discontinuity, whereas the NSF and Grad models capture the finite shock thickness, yielding a smoother transition. Both the G17 equations and the NSF model exhibit strong performance in predicting density (ρ), velocity (v), and average pressure ($\overline{p}$), with G17 demonstrating superior consistency

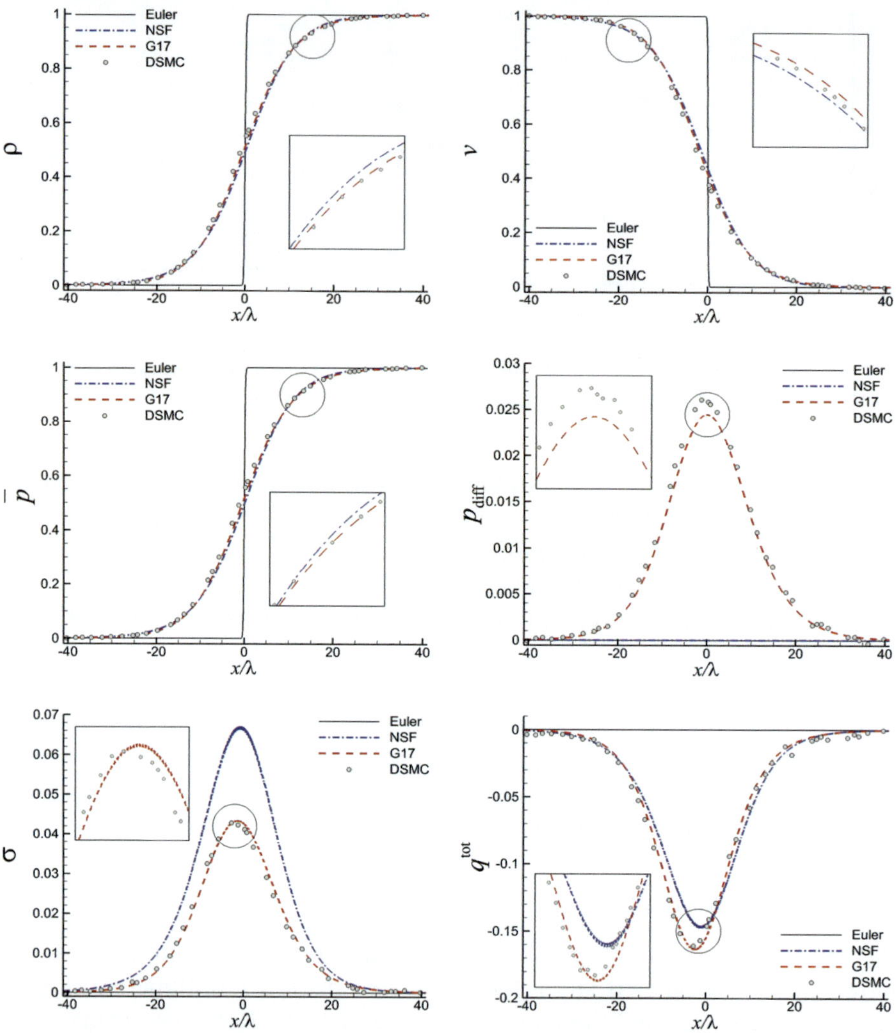

Figure 9.4 1D shock structure problem: profiles of normalized density (ρ), normalized velocity (v), normalized average pressure ($\overline{p}$), pressure difference (p_{diff}), normalized stress (σ), and normalized heat flux of total (q^{tot}) energy solutions with the Euler, NSF, G17 models and DSMC results at Ma $= 1.2$.

with DSMC results, particularly in the shock layer, where non-equilibrium effects are more pronounced. However, the NSF equations lose precision in estimating stress (σ), pressure difference (p_{diff}), and total heat flux (q^{tot}), whereas G17 continues to deliver reliable and accurate results.

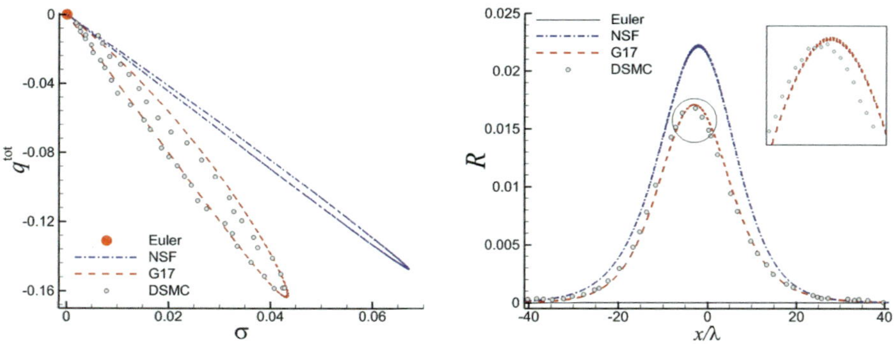

Figure 9.5 1D shock structure problem: profiles of stress-heat flux ($\sigma - q^{\text{tot}}$) and degree of non-equilibrium (R) solutions with the Euler, NSF, G17 models and DSMC results at Ma = 1.2.

Figure 9.5 presents the results of 1D shock structures at Ma $=$ 1.2, comparing normalized stress versus total energy heat flux ($\sigma - q$) and the non-equilibrium intensity (R) across various models and DSMC data. The G17 model accurately captures steeper gradients and dissipation within shock structures, surpassing traditional models in depicting the intricate nonlinear and non-equilibrium dynamics of high-speed flows.

9.5 Conclusion

This study examines non-equilibrium flow simulations based on the G17 moment equations for polyatomic gases, highlighting their ability to accurately model complex flow behavior in diverse shock dynamics scenarios. Through detailed examinations of 1D shock structures and shock tubes, the research reveals that the G17 model offers superior performance in predicting non-equilibrium effects compared to the traditional Euler and NSF models. Comparison with DSMC data verifies that the Grad models accurately capture shock thickness, stress, and heat flux in high-speed flows, where non-equilibrium effects are substantial. Furthermore, the impact of the Knudsen number underscores the requirements for higher-order moment models, particularly in low-density regions where non-equilibrium effects are more significant.

This work aimed to develop and utilize the G17 moment equations for modeling non-equilibrium flows in polyatomic gases. These models present significant improvements over conventional kinetic models, though further refinement is still possible. Exploring even higher-order moment systems could lead to more precise predictions and allow for better representation of subtle flow behavior under extreme conditions. Future studies will aim to report additional findings on this topic.

References

[1] N. Singh and T. E. Schwartzentruber. 'Aerothermodynamic correlations for high-speed flow'. *J. Fluid Mech.*, 821:421–439, 2017.

[2] S. Singh, A. Karchani, T. Chourushi, and R. S. Myong. 'A three-dimensional modal discontinuous Galerkin method for the second-order Boltzmann-Curtiss-based constitutive model of rarefied and microscale gas flows'. *J. Comput. Phys.*, 457:111052, 2022.

[3] G. E. Karniadakis, A. Beskok, and M. Gad el Hak. 'Micro flows: fundamentals and simulation'. *Appl. Mech. Rev.*, 55(4):B76, 2002.

[4] Y. Gao, Z. Hou, X. Wu, and P. Xu. The impact of sample size on transport properties of carbon-paper and carbon-cloth gdls: Direct simulation using the lattice boltzmann model. *International Journal of Heat and Mass Transfer*, 118:1325–1339, 2018.

[5] M. Torrilhon. 'Modeling nonequilibrium gas flow based on moment equations'. *Annu. Rev. Fluid Mech.*, 48(1):429–458, 2016.

[6] L. Boltzmann. *Weitere Studien über das Wärmegleichgewicht unter Gasmolekülen*. Aus der kk Hof- und Staatsdruckerei, 1872.

[7] C. Cercignani. 'Theory and application of the Boltzmann equation'. 1975.

[8] W. Chang and G. E. Uhlenbeck. 'Transport phenomena in polyatomic gases'. Tech. rep., 1951.

[9] J.-F. Bourgat, L. Desvillettes, P. Le Tallec, and B. Perthame. 'Microreversible collisions for polyatomic gases and Boltzmann's theorem'. *Eur. J. Mech. B, Fluids*, 13(2):237–254, 1994.

[10] C. Cercignani. *The Boltzmann Equation*. Springer, 1988.

[11] G. A. Bird. *Molecular Gas Dynamics and the Direct Simulation of Gas Flows*. Oxford Univ. Press, 1994.

[12] S. Chapman and T. G. Cowling. 'The mathematical theory of non-uniform gases'. *Cambridge Univ. Press*, 12, 1970.

[13] H. Grad. 'On the kinetic theory of rarefied gases'. *Commun. Pure Appl. Math.*, 2(4):331–407, 1949.

[14] E. Nagnibeda and E. Kustova. *Non-equilibrium Reacting Gas Flows: Kinetic Theory of Transport and Relaxation Processes*. Springer, 2009.

[15] H. Grad. 'The profile of a steady plane shock wave'. *Commun. Pure Appl. Math.*, 5(3):257–300, 1952.

[16] A. M. Kogan. 'Derivation of Grad's type equations and study of their relaxation properties by the method of maximization of entropy'. *J. Appl. Math. Mech.*, 29(1):130–142, 1965.

[17] V. M. Zhdanov. 'The kinetic theory of a polyatomic gas'. *Sov. Phys. JETP*, 26:1187, 1968.

[18] F. J. McCormack. 'Kinetic Equations for Polyatomic Gases: The 17-Moment Approximation'. *Phys. Fluids*, 11(12):2533–2543, 1968.

[19] M. Pavić-Čolić and S. Simić. 'Kinetic description of polyatomic gases with temperature-dependent specific heats'. *Phys. Rev. Fluids*, 7(8):083401, 2022.

[20] V. Djordjić, G. Oblapenko, M. Pavić-Čolić, and M. Torrilhon. 'Boltzmann collision operator for polyatomic gases in agreement with experimental data and DSMC method'. *Continuum Mech. Thermodyn.*, 35(1):103–119, 2023.

[21] V. Djordjić, M. Pavić-Čolić, and M. Torrilhon. '14-and 17-moment systems for polytropic gases-Comparison regarding transport coefficients'. In *AIP Conf. Proc.*, volume 2996. AIP, 2024.

[22] H. Song, S. Singh, and M. Torrilhon. Non-equilibrium flow simulations based on grad-14 and grad-17 moment equations for polytropic gases. *Physics of Fluids*, 37(3), 2025.

[23] H. Ranocha, M. Schlottke-Lakemper, A. R. Winters, E. Faulhaber, J. Chan, and G. Gassner. 'Adaptive numerical simulations with Trixi.jl: A case study of Julia for scientific computing'. *Proc. JuliaCon Conf.*, 1(1):77, 2022.

[24] M. Schlottke-Lakemper, A. R. Winters, H. Ranocha, and G. J. Gassner. 'A purely hyperbolic discontinuous Galerkin approach for self-gravitating gas dynamics'. *J. Comput. Phys.*, 442:110467, 2021.

[25] S. Hennemann, A. M. Rueda-Ramírez, F. J. Hindenlang, and G. J. Gassner. 'A provably entropy stable subcell shock capturing approach for high order split form DG for the compressible Euler equations'. *J. Comput. Phys.*, 426:109935, 2021.

[26] H. Ranocha, M. Schlottke-Lakemper, J. Chan, A. M. Rueda-Ramírez, A. R. Winters, F. Hindenlang, and G. J. Gassner. 'Efficient implementation of modern entropy stable and kinetic energy preserving discontinuous Galerkin methods for conservation laws'. *ACM Trans. Math. Softw.*, 49(4):1–30, 2023.

[27] A. T. Patera. 'A spectral element method for fluid dynamics: Laminar flow in a channel expansion'. *J. Comput. Phys.*, 54(3):468–488, 1984.

[28] G. J. Gassner. 'A skew-symmetric discontinuous Galerkin spectral element discretization and its relation to SBP-SAT finite difference methods'. *SIAM J. Sci. Comput.*, 35(3):A1233–A1253, 2013.

[29] D. A. Kopriva. *'Implementing Spectral Methods for Partial Differential Equations: Algorithms for Scientists and Engineers'*. Springer, 2009.

[30] K. O. Friedrichs and P. D. Lax. 'Systems of conservation equations with a convex extension'. *Proc. Natl. Acad. Sci.*, 68(8):1686–1688, 1971.

[31] M. H. Carpenter and C. A. Kennedy. 'Fourth-order 2N-storage Runge–Kutta schemes'. Tech. rep., 1994.

[32] J. H. Williamson. 'Low-storage Runge–Kutta schemes'. *J. Comput. Phys.*, 35(1):48–56, 1980.

[33] D. J. Fyfe. 'Economical evaluation of Runge–Kutta formulae'. *Math. Comput.*, 20(95):392–398, 1966.

[34] S. Singh, H. Song, and M. Torrilhon. 'Modal discontinuous Galerkin simulations for Grad's 13 moment equations: Application to Riemann problem in continuum-rarefied flow regime'. *J. Comput. Theor. Transp.*, 53(6):398–422, 2024.

10

Single Formula for Each Velocity Component in the Incompressible Turbulent Boundary Layer over a Rotating Broad Cone and Disk

Oleh G. Shvydkyi

Independent researcher, Ukraine
E-mail: oleg.shv.jw@gmail.com

Abstract

At present, there is no single formula for the streamwise velocity component in the turbulent boundary layer over a flat plate. In the viscous sub-layer, the velocity changes linearly; then there is a region with a logarithmic law of velocity change (so-called log law region). At the junction of these regions is the so-called buffer region, where the velocity obviously changes according to a certain Kolmogorov power law. Thus, for each of these regions, its own distinctive formula is used. This paper presents a method for obtaining a single formula for each velocity component in the incompressible turbulent boundary layer over a rotating broad cone and disk (the disk is considered as the limiting position of the broad cone when the polar angle of the cone wall tends to zero). The obtained formula for the azimuthal velocity component in the boundary layer over a disk (analogous to the streamwise velocity component in the boundary layer above a flat plate) works in the viscous sublayer, the buffer region, and at the beginning of the log law region. The obtained formulas for the radial and axial velocity components in the boundary layer over a disk can be applied to the entire thickness of the turbulent boundary layer. With appropriate adjustment of the calibration coefficients, all three

DOI: 10.1201/9788743806691-10

formulas for the velocity components in the turbulent boundary layer over a disk can also be used to describe the laminar boundary layer.

Keywords: Tunnel mathematics, functions of a complex variable, steady Navier–Stokes equations, turbulence modeling, turbulent boundary layers, incompressible flow, broad cone, disk, rotation

10.1 Introduction

Consider the streamwise velocity profile for a turbulent boundary layer developing over an unbounded flat plate [1, 2, 6, 8, 9, 14], as shown in Figure 10.1. Similarly, the azimuthal velocity profile in a turbulent boundary layer formed above a rotating disk, which is known as the Karman turbulent boundary layer (KTBL) [5], exhibits a comparable structure in the reference frame rotating with the disk [5, 6, 7, 14]. As observed in Figure 10.1, this profile resembles a table fork shape.

The steep blue section of the profile within the range $10^0 < \xi < 10^1$ corresponds to the viscous sublayer. The green linear segment, resembling the handle of the fork and characterized by $\xi > 10^1$, corresponds to the logarithmic (log-law) region. The intermediate gray segment, in the range $5 < \xi < 30$, represents the buffer region, where the Kolmogorov power law is applicable [1, 9]. The values along the ξ-axis are presented on a decimal logarithmic scale.

Here, the dimensionless wall-normal coordinate ξ is defined as

$$\xi = \frac{zv_*}{\nu}, \tag{10.1}$$

where v_* denotes the friction velocity, which is typically on the order of the turbulence-induced fluctuating velocity [1, 2, 5, 9]. The friction velocity is also involved in computing the surface shear stress, σ, expressed as

$$\sigma = \rho v_*^2, \tag{10.2}$$

where ρ is the fluid density, ν is the kinematic viscosity, and z denotes the wall-normal coordinate.

Figure 10.2 presents a plot similar to the one shown in Figure 10.1, but with a linear scale used along the abscissa axis instead of a decimal logarithmic scale. The blue curve follows the linear velocity law:

$$v = \frac{v_*^2 z}{\nu}, \tag{10.3}$$

Upon reaching the boundary of the viscous sublayer (at approximately $z \approx 4 \times 10^{-5}$ m), the velocity profile transitions to the green logarithmic curve:

$$v = v_* \left(2.5 \log \frac{z v_*}{\nu} + 5.1\right). \tag{10.4}$$

To describe the streamwise velocity distribution in a turbulent boundary layer, we can also use empirical power-law formulations. The $\frac{1}{7}$-power velocity distribution law [2, 3] is given by

$$v = 8.74\, v_* \left(\frac{z v_*}{\nu}\right)^{1/7}, \tag{10.5}$$

while the $\frac{1}{6}$-power velocity distribution law [3, 4] is expressed as

$$v = 7.8\, v_* \left(\frac{z v_*}{\nu}\right)^{1/6}. \tag{10.6}$$

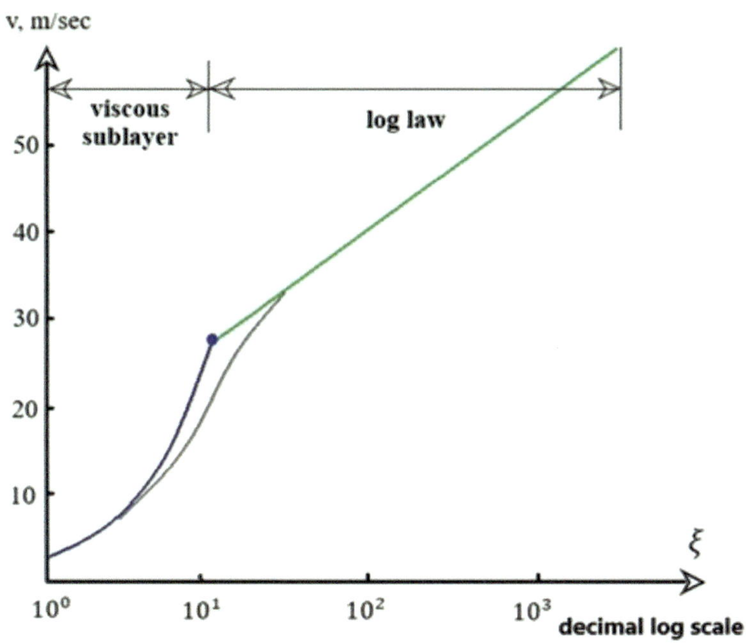

Figure 10.1 Streamwise velocity plot in the form of a table fork profile for a turbulent boundary layer over an unbounded flat plate. The blue part of the plot with $10^0 < \xi < 10^1$ corresponds to the viscous sublayer; the green part with $\xi > 10^1$ corresponds to the log-law region; the gray part of the plot with $5 < \xi < 30$ corresponds to the buffer region. Values along the ξ-axis are plotted on a *decimal log scale*.

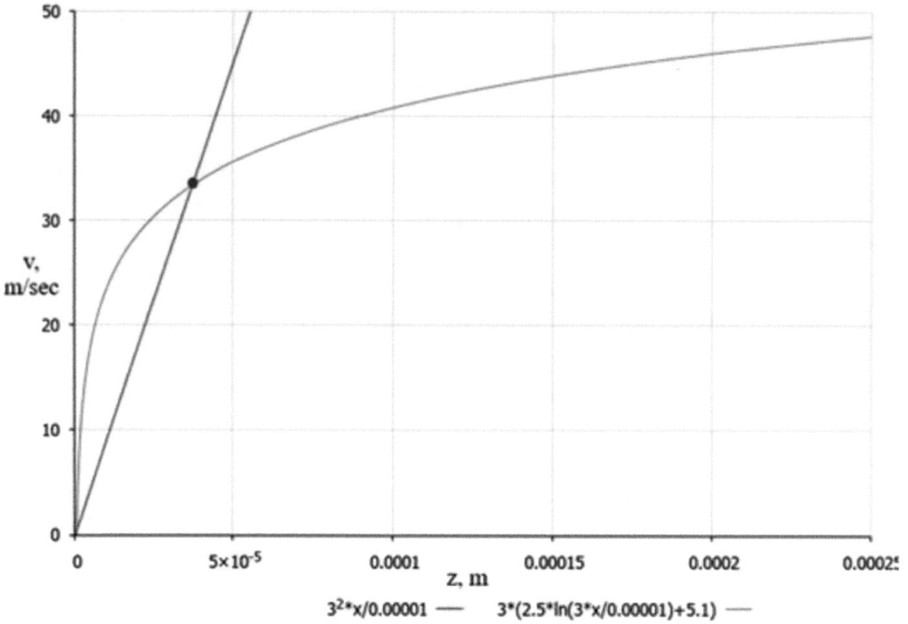

Figure 10.2 Plot corresponding to the one shown in Figure 10.1, but without using the decimal log scale on the abscissa axis. First, the blue line follows the linear law (3); upon reaching the boundary of the viscous sublayer ($z \approx 4 \times 10^{-5}$ m), the profile transitions to the green logarithmic curve (4). Here, $v_* = 3$ m/sec and $\nu = 0.00001$ m^2/sec (air under normal conditions).

In Figure 10.3, the velocity profiles corresponding to eqn (10.5) and (10.6), which were originally derived for turbulent flow of liquids in pipes, are superimposed on those shown in Figure 10.2 for comparison.

As observed, the velocity profiles corresponding to relations (10.5) and (10.6) effectively describe the streamwise velocity distribution in the turbulent boundary layer, beyond the viscous sublayer. Notably, the framework of algebraic tunnel mathematics [25] leads to an expression for the radial velocity component in a turbulent boundary layer over a rotating disk. This expression is remarkably similar to relation eqn (10.6), and is given by

$$v_{\mathrm{r}} \sim \rho^{z/6\delta}, \tag{10.7}$$

where $\frac{z}{\delta} \sim 1$ within the turbulent boundary layer, and δ denotes the boundary layer thickness.

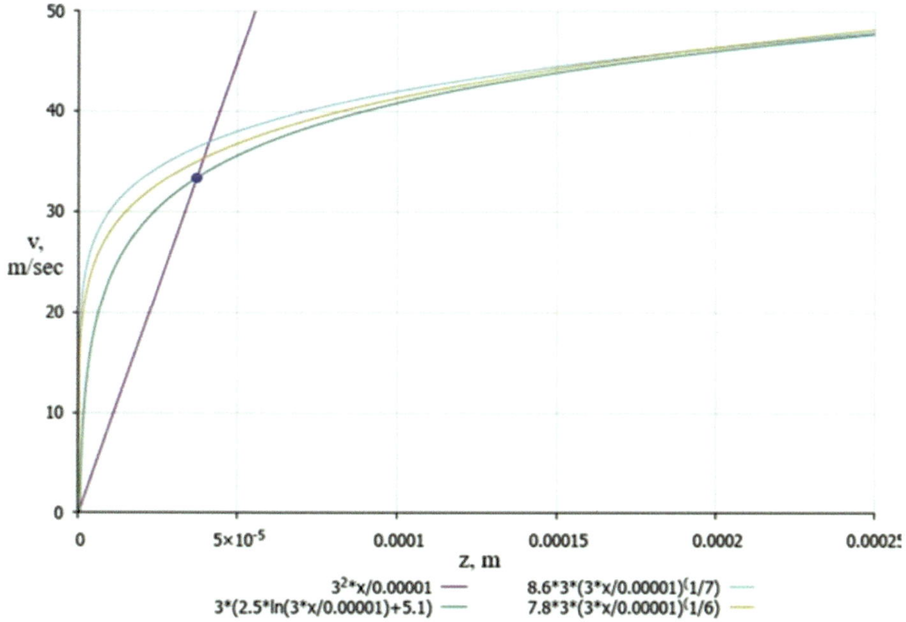

Figure 10.3 Plots for relations (10.5) and (10.6) added to those shown in Figure 10.2: the cyan curve corresponds to eqn (10.5), and the orange curve corresponds to eqn (10.6).

The analytical models that aim to describe the velocity distribution in a turbulent boundary layer are generally quite sophisticated. For instance, in the context of atmospheric surface layers, an autoregressive moving average (ARMA) model is utilized [9]. This model incorporates both the historical behavior and random disturbances of the streamwise velocity fluctuations characteristic of boundary-layer turbulence. Despite the complexity of the mathematical procedures involved, the theoretical formulation ultimately yields a logarithmic trace that aligns closely with the logarithmic law described by relation (10.4). This agreement provides strong validation for the ARMA model in representing turbulent boundary-layer behavior.

The processes occurring in the viscous sublayer play a crucial role in determining how velocity increases in the outer region of the turbulent boundary layer [10]. Consequently, considerable research efforts have been dedicated to modeling the viscous sublayer [11–13]. It is understood that the dynamics within the viscous sublayer are more closely aligned with solid-state physics than with conventional fluid or gas mechanics. Specifically, the viscous sublayer is influenced by the interaction between the wall's electrostatic potential field and the molecular electric moments of the surrounding liquid or gas. A detailed understanding of this interaction is essential for developing suitable mathematical frameworks to accurately describe the behavior within the viscous sublayer. From a theoretical standpoint, it may be possible to significantly reduce the thickness of the viscous sublayer – and consequently the entire boundary layer – by embedding atoms with predefined electric moments into the crystal lattice of the wall. Such a modification could lead to notable simplifications in the mathematical modeling, while also causing a substantial reduction in drag force due to the elimination of skin friction.

It should be noted that relations (10.3) and (10.4), which form the basis of the so-called law of the wall (LoW), are effective for modeling an equilibrium turbulent boundary layer – also known as the equilibrium wall model (EWM) – under zero pressure gradient conditions [1, 2]. However, recent studies [10, 15, 16] have reported suboptimal performance of the LoW-based EWM in 3D turbulent boundary layers.

In the present context, we consider a flat plate geometry, for which the turbulent velocity profile is assumed to be unbounded [1, 2]. This geometrical assumption is what gives rise to the logarithmic velocity law in eqn (10.4). Nonetheless, for a turbulent boundary layer over a flat plate, the velocity profile may deviate from a strictly logarithmic form. For instance, an exponential velocity profile, which varies linearly for small values of the argument, may be more appropriate in the viscous sublayer region (consistent with eqn (10.3)).

Consequently, considerable efforts have been directed toward refining the loW and its associated models [17–19]. In particular, the work by [20] processed experimental data and proposed the following modified damping function:

$$D_{\mathrm{k}} = 1 - \exp\left(-\frac{\xi}{A_{\mathrm{k}}^{+}}\right), \quad A_{\mathrm{k}}^{+} = \frac{25.0}{cP^{+} + 1}, \tag{10.8}$$

where the damping function D_{k} is incorporated into the eddy viscosity scaling:

$$v_t = kzv_* D_{\mathrm{k}}.$$

Here, k is the von Kármán constant; ξ and v_* are defined in eqn (10.1); P^+ denotes the normalized pressure gradient (with respect to the wall shear stress σ in eqn (10.2)); $\delta_\nu = \nu/v_*$ is the viscous length scale; and c is an empirical constant.

It is important to note that expressions of the form $1 \pm \exp\left(-\frac{z}{r}\right)$ naturally emerge when applying the apparatus of tunnel mathematics, even in the context of a laminar boundary layer over a rotating disk [22, 23]. To determine the velocity profile in a turbulent boundary layer under the influence of an arbitrary pressure gradient, it is essential to adopt an appropriate velocity scaling. A recently proposed velocity scaling [21] is given as follows:

$$\xi^* = \int_0^z \frac{F(\xi^*)}{F'(\xi^*)} \left[\frac{1}{\sigma^+} \frac{\partial \sigma^+}{\partial z} - \frac{1}{1+g} \frac{\partial g}{\partial z} + \frac{F'(\xi_0)}{F(\xi_0)} \frac{d\xi_0}{dz} \right] dz, \tag{10.9}$$

$$V^* = \int_0^v \frac{v}{\sigma_{xz}} \frac{F(\xi^*)}{F'(\xi^*)} \left[\frac{1}{\sigma^+} \frac{\partial \sigma^+}{\partial z} - \frac{1}{1+g} \frac{\partial g}{\partial z} + \frac{F'(\xi_0)}{F(\xi_0)} \frac{d\xi_0}{dz} \right] dv, \tag{10.10}$$

where the transformed velocity V^* satisfies the classical law of the wall (LoW) as a function of ξ^*. Here, $\sigma^+ = \sigma_{xz}/\sigma$, and the function $F(\alpha) = 1 + k\alpha(1 - \exp(-\alpha/A))^2$ characterizes the total shear stress behavior in the inner layer of an equilibrium turbulent boundary layer. The derivative of F is denoted by F', and g represents the Lagrangian integral of the total shear stress, given by

$$g = \frac{\nu F(\xi_0)}{v_*^2} \int_0^t \frac{\partial^2 \sigma_{xz}}{\partial z^2} \, dT. \tag{10.11}$$

The parameter g accounts for the memory or history effects [10]. In the absence of a pressure gradient, the transformation reduces to the classical form where $\xi^* = \xi$, $V^* = \nu/v_*$, thus recovering the original relations (10.1), (10.3), and (10.4). Under pressure gradient conditions, this transformation provides a generalized description of the mean flow within turbulent boundary layers.

The quantity g accounts for the history effect [10]. In the absence of a pressure gradient, the transformation reduces to the classical law of the wall (LoW), yielding $\xi^* = \xi$ and $V^* = \nu/v_*$, as seen in relations (10.1), (10.3), and (10.4). Under the influence of a pressure gradient, however, the transformation provides a more comprehensive description of the mean velocity profile in boundary layers. As can be seen from eqn (10.9), the integrand of ξ^* contains functions that are themselves dependent on the transformed variable ξ^*, rather than solely on the original scaling variable ξ. This recursive dependence adds complexity to the integration process and demands substantial computational resources. A similar challenge arises when addressing the velocity profile in a turbulent boundary layer over a rotating broad cone and disk within the framework of tunnel mathematics. In this context, one encounters analogous complications. Nevertheless, the apparatus of tunnel mathematics provides tools to tackle this problem using conventional analytical techniques.

10.2 Theory

10.2.1 Broad cone – disk system

We adopt a modified spherical coordinate system [27], in which the polar angle θ is measured from the xy-plane (rather than from the z-axis, as is standard). This convention is illustrated in Figure 10.4. To convert from the conventional spherical coordinate system, we apply the following transformations:

$$\begin{aligned} v_\theta &\to -v_\theta, \\ \theta &\to \frac{\pi}{2} - \theta, \\ \partial_\theta &\to -\partial_\theta. \end{aligned} \tag{10.12}$$

The broad cone–disk system is depicted in Figure 10.4.

10.2.2 Main equations of tunnel mathematics

Tunnel mathematics equations applied to the components of the vector velocity field in the Cartesian coordinate system are given as follows [22]:

$$\begin{aligned} &\frac{\partial u}{\partial x} + \frac{(uw)y}{(x^2+y^2)^{3/2}}\left(2+\left(\frac{y}{x}\right)^2\right) + \frac{1}{\sqrt{x^2+y^2}}\frac{\partial(vw)}{\partial x} \\ &= -\frac{\partial v}{\partial y} + \frac{(vw)x}{(x^2+y^2)^{3/2}}\left(2+\left(\frac{x}{y}\right)^2\right) - \frac{1}{\sqrt{x^2+y^2}}\frac{\partial(uw)}{\partial y} \\ &= -\frac{2y}{\sqrt{x^2+y^2}}\frac{\partial u}{\partial z} + \frac{1}{x}\frac{\partial(uw)}{\partial z} + \frac{1}{y}\frac{\partial(vw)}{\partial z} - \frac{2y}{(x^2+y^2)}\frac{\partial(vw)}{\partial z}, \end{aligned} \tag{10.13}$$

$$\begin{aligned} &\frac{\partial v}{\partial x} + \frac{(uw)y}{(x^2+y^2)^{3/2}} - \frac{1}{\sqrt{x^2+y^2}}\frac{\partial(uw)}{\partial x} \\ &= -\frac{\partial u}{\partial y} + \frac{(uw)x}{(x^2+y^2)^{3/2}} - \frac{1}{\sqrt{x^2+y^2}}\frac{\partial(vw)}{\partial y} \\ &= -\frac{2y}{\sqrt{x^2+y^2}}\frac{\partial v}{\partial z} + \frac{2y}{(x^2+y^2)}\frac{\partial(vw)}{\partial z}, \end{aligned} \tag{10.14}$$

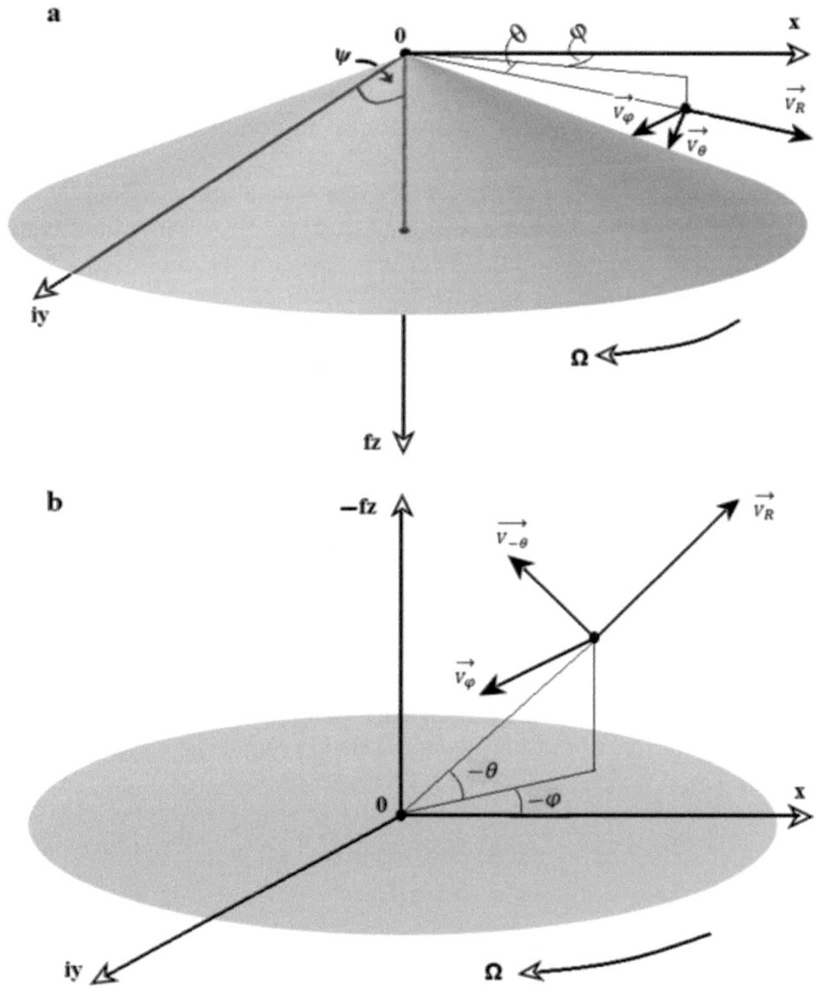

Figure 10.4 (a) Broad cone: ψ – half-apex angle of cone, φ – azimuthal angle, θ – polar angle, $V_\mathrm{R}, V_\theta, V_\varphi$ – velocity field components in a modified spherical coordinate system (we measure the polar angle θ from the xy-plane, not from the z-axis, as is usually done). Ω – angular velocity of rotation, (b) Disk.

$$\frac{1}{x}\frac{\partial(uw)}{\partial x}+\frac{1}{y}\frac{\partial(vw)}{\partial x}=-\frac{i}{y}\frac{\partial(vw)}{\partial y}-\frac{i}{x}\frac{\partial(uw)}{\partial y}$$
$$=-\frac{\partial u}{\partial z}-i\frac{\partial v}{\partial z}+\frac{i}{\sqrt{x^2+y^2}}\frac{\partial(uw)}{\partial z}-\frac{1}{\sqrt{x^2+y^2}}\frac{\partial(vw)}{\partial z}. \quad (10.15)$$

Eqn (10.13) and (10.14) are of an algebraic nature; however, eqn (10.15) is of a *complex nature* and constitutes the core of the main theorem in tunnel mathematics [22]. It is evident that for $w = 0$, eqn (10.13) and (10.14) reduce to the ordinary planar Cauchy–Riemann conditions [22, 26], while eqn (10.15) disappears.

Functions u, v, w in eqn (10.13)–(10.15) represent the components of the velocity field in the Cartesian coordinate system. These relations connect the velocity field components in the Cartesian and the modified spherical coordinate system defined in eqn (10.12) [27]:

$$u = v_x = \cos\varphi\,(v_{\mathrm{R}}\cos\theta - v_\theta\sin\theta) - v_\varphi\sin\varphi, \tag{10.16}$$

$$v = v_y = \sin\varphi\,(v_{\mathrm{R}}\cos\theta - v_\theta\sin\theta) + v_\varphi\cos\varphi, \tag{10.17}$$

$$w = v_{\mathrm{z}} = v_{\mathrm{R}}\sin\theta + v_\theta\cos\theta. \tag{10.18}$$

There also exist inverse transformation formulas:

$$v_{\mathrm{R}} = \cos\theta\,(u\cos\varphi + v\sin\varphi) + w\sin\theta, \tag{10.19}$$

$$v_\theta = -\sin\theta\,(u\cos\varphi + v\sin\varphi) + w\cos\theta, \tag{10.20}$$

$$v_\varphi = -u\sin\varphi + v\cos\varphi. \tag{10.21}$$

Additionally, the standard transformation rules between the Cartesian and the modified spherical coordinate system (10.12) remain valid [27]:

$$x = R\cos\theta\cos\varphi, \tag{10.22}$$

$$y = R\cos\theta\sin\varphi, \tag{10.23}$$

$$z = R\sin\theta. \tag{10.24}$$

Recall that we work in the modified spherical coordinate system defined in eqn (10.12) [27], that is *we measure the polar angle θ from the xy-plane (not from the z-axis, as is usually the case)* (see Figure 10.4). Using eqn (10.16)–(10.18), we derive the following relations:

$$(uw) = \cos\varphi\left(\left(v_{\mathrm{R}}^2 - v_\theta^2\right)\frac{\sin 2\theta}{2} + v_{\mathrm{R}}v_\theta\cos 2\theta\right) - v_\varphi\sin\varphi\,(v_{\mathrm{R}}\sin\theta + v_\theta\cos\theta)\,, \tag{10.25}$$

$$(vw) = \sin\varphi\left(\left(v_{\mathrm{R}}^2 - v_\theta^2\right)\frac{\sin 2\theta}{2} + v_{\mathrm{R}}v_\theta\cos 2\theta\right) + v_\varphi\cos\varphi\,(v_{\mathrm{R}}\sin\theta + v_\theta\cos\theta)\,. \tag{10.26}$$

For our broad cone–disk system, we consider the following limiting case.

Broad cone: $v_{\mathrm{R}}, v_\theta \ll v_\varphi$. Taking into account eqn (10.22) and (10.23), we can derive from (10.25) and (10.26) the following simplified condition for a broad cone:

$$(uw)x = -(vw)y. \tag{10.27}$$

We will get the solution for the rotating disk as a limiting transition for the broad cone.

Besides, the use of tunnel mathematics has certain advantages because it allows for straightforward computation of Laplacians of functions. For example, the following relations hold in tunnel mathematics:

$$P = u(x, y, z) + iv(x, y, z) + fw(x, y, z), \tag{10.28}$$

where P is a spatial function of a complex variable; the functions u, v, w are real-valued functions of the variables x, y, z; and f is a special operator defined in tunnel mathematics [22]:

$$\Delta u = 0, \tag{10.29}$$

$$\Delta v = 0, \tag{10.30}$$

Relations (10.29) and (10.30) are the standard Laplace equations in the planar theory of complex variables [22, 26].

$$\Delta(uw) = \frac{(vw)x}{y^3} - \frac{(y^2 - x^2)}{xy} C_{\mathrm{i}} - C_{\mathrm{r}}, \tag{10.31}$$

$$\Delta(vw) = \frac{(uw)y}{x^3} - \frac{(y^2 - x^2)}{xy} C_{\mathrm{r}} + C_{\mathrm{i}}, \tag{10.32}$$

where C_{r} and C_{i} are constants (which may depend on the z-coordinate), and are defined in the following relations:

$$\frac{\partial(uw)}{\partial x} + i\frac{\partial(uw)}{\partial y} = (C_{\mathrm{r}} + iC_{\mathrm{i}})x, \tag{10.33}$$

$$\frac{\partial(vw)}{\partial x} + i\frac{\partial(vw)}{\partial y} = -(C_{\mathrm{r}} + iC_{\mathrm{i}})y. \tag{10.34}$$

Eqn (10.33) and (10.34) correspond to the first equality in eqn (10.15). In eqn (10.29)–(10.32), $\Delta = \frac{\partial^2}{\partial x^2} + \frac{\partial^2}{\partial y^2}$ is the Laplace operator applied on planes parallel to the xy-plane.

Relations (10.31) and (10.32) indicate that, on these working platforms, the Laplacians of functions depend linearly on the functions themselves. This significantly simplifies the solution of differential equations. Given that Laplacians appear in numerous physical contexts – such as the Schrödinger equation in quantum mechanics, the Lamé equation in elasticity theory, and the eikonal equation in optics – tunnel mathematics shows promising potential for becoming an essential tool in theoretical physics.

10.3 Results and Discussion

10.3.1 Broad cone

Using the first equalities in eqn (10.13) and (10.14), along with condition (10.27) for the broad cone $(uw)x = -(vw)y$, we compute the first derivatives of the functions (uw) and (vw) with respect to ∂x and ∂y:

$$\frac{\partial(uw)}{\partial x} = -(uw)\left(\frac{x}{x^2+y^2} + \frac{x}{y^2} - \frac{1}{x}\right), \tag{10.35}$$

$$\frac{\partial(vw)}{\partial x} = -(vw)\left(\frac{x}{x^2+y^2} + \frac{x}{y^2} - \frac{2}{x}\right), \tag{10.36}$$

$$\frac{\partial(uw)}{\partial y} = -(uw)\left(\frac{y}{x^2+y^2} + \frac{y}{x^2} - \frac{2}{y}\right), \tag{10.37}$$

$$\frac{\partial(vw)}{\partial y} = -(vw)\left(\frac{y}{x^2+y^2} + \frac{y}{x^2} - \frac{1}{y}\right). \tag{10.38}$$

To determine the derivatives with respect to ∂z, we apply the planar Laplacian operator to eqn (10.15):

$$\Delta\left(\frac{1}{x}\frac{\partial(uw)}{\partial x} + \frac{1}{y}\frac{\partial(vw)}{\partial x}\right) = -\frac{\partial\Delta u}{\partial z} - \Delta\left(\frac{1}{\sqrt{x^2+y^2}} \cdot \frac{\partial(vw)}{\partial z}\right). \tag{10.39}$$

Using the harmonicity of function u in eqn (10.29), and applying tunnel mathematics relations for the planar Laplacians in eqn (10.31) and (10.32), we obtain the required derivative with respect to ∂z:

$$\frac{\partial(uw)}{\partial z} = -(uw)\left(-\sqrt{x^2+y^2} \cdot \frac{\beta_2}{\beta_3} + \frac{1}{\beta_3} \cdot \frac{(x^2+y^2)^{3/2}}{(xy)^2} \cdot \left(\frac{y}{x}C_\mathrm{i} + C_\mathrm{r}\right)\right.$$
$$\left. \cdot \frac{1}{(uw)} - \frac{1}{\beta_3} \cdot \left(\frac{(y^2-x^2)}{xy} \cdot \frac{\partial C_\mathrm{r}}{\partial z} - \frac{\partial C_\mathrm{i}}{\partial z}\right) \cdot \frac{1}{(uw)}\right) \tag{10.40}$$

where:

$$\beta_2 = -\frac{1}{y^2}\left(\left(\frac{x}{y^2}\right)^2 + \frac{4}{y^2} + \frac{2}{x^2}\right) + \frac{2}{x^2+y^2}\left(\frac{1}{y^2} + \frac{6}{x^2+y^2}\right), \quad (10.41)$$

$$\beta_3 = \frac{1}{(xy)^3}\left((x^2+y^2)^2 - x^2\left(4(x^2+y^2) - x^2\left(8 - \frac{7x^2}{x^2+y^2}\right)\right)\right). \quad (10.42)$$

In a similar way, we find:

$$\frac{\partial(vw)}{\partial z} = -(vw)\left(-\sqrt{x^2+y^2}\cdot\frac{\beta_2}{\beta_3} - \frac{1}{\beta_3}\cdot\frac{(x^2+y^2)^{3/2}}{(xy)^2}\cdot\left(C_\mathrm{i} + \frac{x}{y}C_\mathrm{r}\right)\right.$$
$$\left.\cdot\frac{1}{(vw)} + \frac{x}{y\beta_3}\cdot\left(\frac{(y^2-x^2)}{xy}\cdot\frac{\partial C_\mathrm{r}}{\partial z} - \frac{\partial C_\mathrm{i}}{\partial z}\right)\cdot\frac{1}{(vw)}\right). \quad (10.43)$$

The presence of constants C_r and C_i in eqn (10.40) and (10.43) will later provide us with the opportunity for the necessary mathematical maneuver in order to satisfy the conditions of harmonicity (10.29) and (10.30) for the velocity components.

Now, using eqn (10.35)–(10.38), (10.40), and (10.43), we derive the expressions for the required functions (uw) and (vw):

$$(uw) = -B_0B_1\frac{xy^2}{x^2+y^2}\cdot\frac{b_1^2}{a_1a_2^2}\exp\left(1 - \frac{1}{2}\left(\frac{x^2+y^2}{xy}\right)^2\right.$$
$$+ z\sqrt{x^2+y^2}\cdot\frac{\beta_2}{\beta_3} - \frac{(x^2+y^2)^{3/2}}{\beta_3(xy)^2}\int\left(\frac{y}{x}C_\mathrm{i} + C_\mathrm{r}\right)\frac{dz}{(uw)}$$
$$\left.+\frac{1}{\beta_3}\int\left(\frac{(y^2-x^2)}{xy}\frac{\partial C_\mathrm{r}}{\partial z} - \frac{\partial C_\mathrm{i}}{\partial z}\right)\frac{dz}{(uw)}\right). \quad (10.44)$$

$$(vw) = B_0B_2\frac{x^2y}{x^2+y^2}\cdot\frac{b_1^2}{a_1^2a_2}\exp\left(1 - \frac{1}{2}\left(\frac{x^2+y^2}{xy}\right)^2\right.$$
$$+ z\sqrt{x^2+y^2}\cdot\frac{\beta_2}{\beta_3} + \frac{(x^2+y^2)^{3/2}}{\beta_3(xy)^2}\int\left(C_\mathrm{i} + \frac{x}{y}C_\mathrm{r}\right)\frac{dz}{(vw)}$$
$$\left.-\frac{x}{y\beta_3}\int\left(\frac{(y^2-x^2)}{xy}\frac{\partial C_\mathrm{r}}{\partial z} - \frac{\partial C_\mathrm{i}}{\partial z}\right)\frac{dz}{(vw)}\right). \quad (10.45)$$

where $B_0, B_1, B_2, a_1, a_2, b_1$ are constants defined by:

$$[B_0] = [B_1] = [B_2] = \frac{m}{\text{sec}}, \quad [a_1] = [a_2] = [b_1] = m.$$

Using these results, we can construct the components of the vector velocity field in the Cartesian coordinate system. In order to satisfy condition (10.27), that is $(uw)x = -(vw)y$, we select from eqn (10.44) and (10.45) the components of analytical functions [26] in the following manner:

$$u = -B_1 \frac{y}{a_2} \exp(\Sigma), \tag{10.46}$$

$$v = B_2 \frac{x}{a_1} \exp(\Sigma). \tag{10.47}$$

where:

$$\Sigma = \frac{1}{\beta_3} \int \left(-\frac{(x^2+y^2)^{3/2}}{(xy)^2} \left(\frac{y}{x} C_{\mathrm{i}} + C_{\mathrm{r}} \right) + \frac{(y^2-x^2)}{xy} \frac{\partial C_{\mathrm{r}}}{\partial z} - \frac{\partial C_{\mathrm{i}}}{\partial z} \right) \frac{dz}{(uw)}$$
$$= \frac{1}{\beta_3} \int (\cdots) \frac{dz}{(uw)}, \tag{10.48}$$

where in turn:

$$(\cdots) = \left(-\frac{(x^2+y^2)^{3/2}}{(xy)^2} \left(\frac{y}{x} C_{\mathrm{i}} + C_{\mathrm{r}} \right) + \frac{(y^2-x^2)}{xy} \frac{\partial C_{\mathrm{r}}}{\partial z} - \frac{\partial C_{\mathrm{i}}}{\partial z} \right). \tag{10.49}$$

It is easy to see from eqn (10.46) and (10.47) that the following equalities for constants hold:

$$B_1 = B_2; \quad a_1 = a_2. \tag{10.50}$$

The functions u and v in eqn (10.46) and (10.47) must satisfy the conditions of harmonicity (10.29) and (10.30), respectively. We will address this later. For now, we construct the function w, and from it derive the radial and polar components of the velocity for a broad cone, since these components turn out to be independent of the functions u and v.

10.3.1.1 Radial and polar components of the velocity

After selecting the functions u and v from eqn (10.44) and (10.45), the following expression is obtained for the function w:

$$w = B_0 \frac{xy}{x^2+y^2} \cdot \frac{b_1^2}{a_1^2} \exp\left(1 - \frac{1}{2} \left(\frac{x^2+y^2}{xy} \right)^2 + z\sqrt{x^2+y^2} \cdot \frac{\beta_2}{\beta_3} \right), \tag{10.51}$$

where we assume the following equality for the constants:

$$b_1^2 = 2a_1^2. \tag{10.52}$$

If we rewrite the last exponential term in eqn (10.51) as

$$z\sqrt{x^2+y^2}\cdot\frac{\beta_2}{\beta_3} = zxy\sqrt{x^2+y^2}\cdot\frac{\beta_2'}{\beta_3'}, \tag{10.53}$$

where

$$\beta_2' = -\left(\frac{x^2}{y^2}+\left(1+\frac{x^2}{y^2}\right)\left(2+\frac{x^2}{y^2}\right)\right)+\frac{2x^2}{x^2+y^2}+\frac{12(xy)^2}{(x^2+y^2)^2}, \tag{10.54}$$

$$\beta_3' = (x^2+y^2)^2 - x^2\left(4(x^2+y^2)-x^2\left(8-\frac{7x^2}{x^2+y^2}\right)\right), \tag{10.55}$$

then the function w takes exactly the same form as that in the article [23], and we can utilize the results therein.

Using eqn (10.19), (10.20), and (10.22)–(10.24), we can obtain the radial and polar components of the vector velocity field in the modified (10.12) spherical coordinate system

$$v_{\mathrm{R}} = 2B_0B_3\sin\theta\exp\left(1-\frac{1}{2}B_3^{-2}+B_3B_6\tan\theta,\right). \tag{10.56}$$

$$v_\theta = 2B_0B_3\cos\theta\exp\left(1-\frac{1}{2}B_3^{-2}+B_3B_6\tan\theta\right), \tag{10.57}$$

where the constants are assigned the following expressions

$$B_3 = \frac{\sin 2\varphi}{2}, \tag{10.58}$$

$$B_6 = \frac{-B_4+2B_5(1+6(1-B_5))}{1-B_5(4-B_5(8-7B_5))}, \tag{10.59}$$

$$B_4 = \cot^2\varphi + \left(1+\cot^2\varphi\right)\left(2+\cot^2\varphi\right), \tag{10.60}$$

$$B_5 = \cos^2\varphi. \tag{10.61}$$

It is seen from eqn (10.56) and (10.57) that the components v_{R} and v_θ do not depend on R. This is clearly a consequence of the simplifications we have adopted, particularly those arising from relation (10.27). Since the physical problems under consideration possess axial symmetry, we are free to

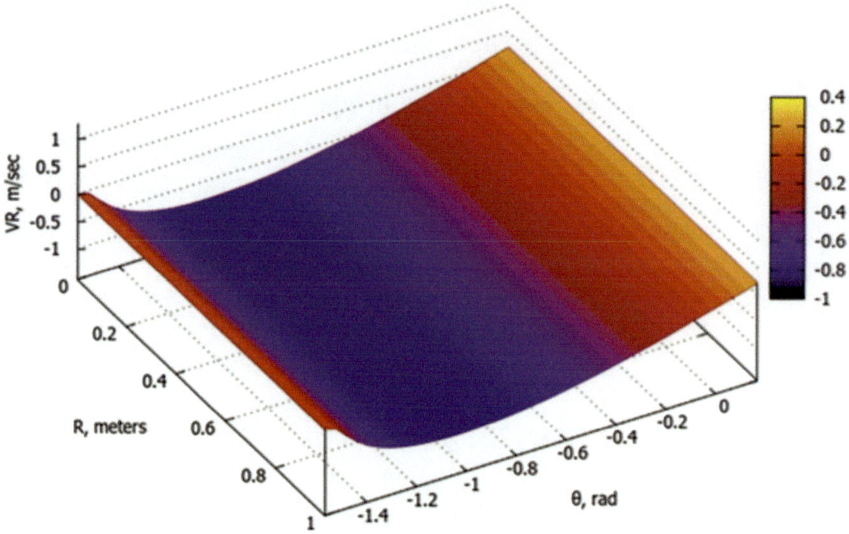

Figure 10.5 Plot for v_{R} component in eqn (10.56) at $\psi = 80°$ ($\theta_{\mathrm{w}} = 10°$ (0.174 rad)) and $B_0 = 200\,\dfrac{m}{\sec}$, $B_3 = 0.3$; $B_6 = 0.5$.

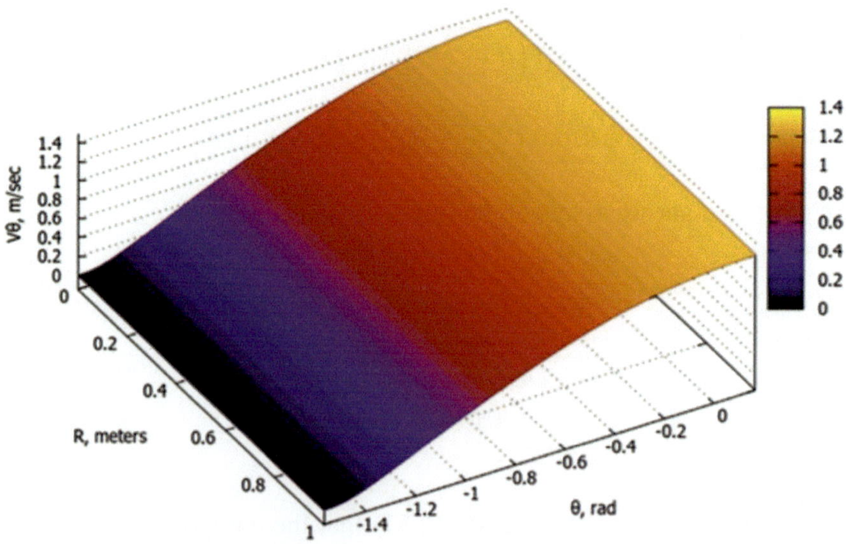

Figure 10.6 Plot for v_θ component in eqn (10.57) at $\psi = 80°$ ($\theta_{\mathrm{w}} = 10°$ (0.174 rad)) and $B_0 = 200\,\dfrac{m}{\sec}$, $B_3 = 0.3$; $B_6 = 0.5$.

assign values to the constants B_3 through B_6 that yield physically meaningful solutions.

In Figures 10.5 and 10.6, we present plots of the radial and polar components of the vector velocity field as expressed in eqn (10.56) and (10.57). Recall that the angle θ_{w} is the angle corresponding to the wall of cone; for a broad cone Figure 10.4(a) we assume that $\theta_{\mathrm{w}} = 10° = 0.174\,\mathrm{rad}$ ($\psi = 80°$). As can be seen from Figures. 10.5 and 10.6, the simplifications we made lead to the impossibility of meeting the non-slip boundary conditions for a broad cone (especially, for v_θ component). However, as we will see below, this situation will not affect the fulfillment of similar boundary conditions for a rotating disk.

10.3.1.2 Azimuthal components of the velocity

The azimuthal component of the velocity is determined by expression (10.21); therefore, we need to find the form of the functions u and v in (10.46) and (10.47). Imposing the harmonicity conditions (10.29) and (10.30) on these expressions and applying the well-known formula for the Laplacian of the product of two functions:

$$\Delta(uw) = u\Delta w + w\Delta u + 2\vec{\nabla} u \vec{\nabla} w, \tag{10.62}$$

we come to the conclusion that the following equality must be satisfied:

$$x\frac{\partial}{\partial y}\exp(\Sigma) - y\frac{\partial}{\partial x}\exp(\Sigma) = \begin{vmatrix} x & y \\ \frac{\partial}{\partial x}\exp(\Sigma) & \frac{\partial}{\partial y}\exp(\Sigma) \end{vmatrix} = 0. \tag{10.63}$$

It is easy to see that the determinant in (10.63) is the z-component of a certain vector

$$\vec{M} = \vec{R} \times \vec{\nabla}\exp(\Sigma) = \begin{vmatrix} \vec{i} & \vec{j} & \vec{k} \\ x & y & z \\ \frac{\partial}{\partial x}\exp(\Sigma) & \frac{\partial}{\partial y}\exp(\Sigma) & \frac{\partial}{\partial z}\exp(\Sigma) \end{vmatrix}, \tag{10.64}$$

where

$$\vec{R} = \vec{R}(x, y, z); \tag{10.65}$$

$$\vec{\nabla}\exp(\Sigma) = \left(\vec{i}\frac{\partial}{\partial x} + \vec{j}\frac{\partial}{\partial y} + \vec{k}\frac{\partial}{\partial z}\right)\exp(\Sigma). \tag{10.66}$$

Obviously, the vector $\vec{M}$ represents the moment of a certain force acting in the region of the turbulent boundary layer. The fact that the z-component of this vector is equal to zero may indicate a conservation law of some other quantity, probably related to the angular acceleration of the vortex motion in the turbulent boundary layer.

To satisfy relation (10.63), the constants C_i and C_r in (10.48) (and they can be functions of z, see relations (10.31) and (10.32)) must take the following form

$$C_\mathrm{i} = C_{0\mathrm{i}} \exp\left(-\frac{\delta_\mathrm{i}}{\gamma_\mathrm{i}} z\right), \tag{10.67}$$

$$C_\mathrm{r} = C_{0\mathrm{r}} \exp\left(\frac{\delta_\mathrm{r}}{\gamma_\mathrm{r}} z\right), \tag{10.68}$$

where $C_{0\mathrm{i}}$ and $C_{0\mathrm{r}}$ are constants; we will also introduce the following denotations

$$F_\mathrm{i} = F_\mathrm{i}(z) = -\frac{\delta_\mathrm{i}}{\gamma_\mathrm{i}} = -\frac{(x^2+y^2)^{3/2}}{(xy)^2}\left(\frac{y}{x} + \frac{1}{\gamma_\mathrm{i}}\left(1 - \frac{3y^2}{x^2}\right)\right), \tag{10.69}$$

$$F_\mathrm{r} = F_\mathrm{r}(z) = \frac{\delta_\mathrm{r}}{\gamma_\mathrm{r}} = \frac{(x^2+y^2)^{3/2}}{(xy)^2}\left(\gamma_\mathrm{i} - \frac{2(y^2-x^2)}{xy}\right) \Bigg/ \left(\gamma_\mathrm{i}\frac{(y^2-x^2)}{xy} - \left(2 + \frac{x^4+y^4}{(xy)^2}\right)\right), \tag{10.70}$$

$$\gamma_\mathrm{i} = \frac{\beta_7}{\beta_4} + \frac{y}{x}, \tag{10.71}$$

$$\beta_7 = 2xy\left(4(x^2+y^2) - x^2\left(2 + \frac{7(2y^2-x^2)}{x^2+y^2}\right)\right), \tag{10.72}$$

$$\beta_4 = \beta_3' = (x^2+y^2)^2 - x^2\left(4(x^2+y^2) - x^2\left(8 - \frac{7x^2}{x^2+y^2}\right)\right). \tag{10.73}$$

The denotations we adopted in (10.69) and (10.70) show that the solutions we obtain later for the velocity profile will work only on certain spatial surfaces. Using the accepted denotations, we can differentiate (10.67) and (10.68) with respect to z:

$$\frac{\partial C_\mathrm{i}}{\partial z} = C_\mathrm{i}\left(F_\mathrm{i} + z\frac{\partial F_\mathrm{i}}{\partial z}\right), \tag{10.74}$$

$$\frac{\partial C_\mathrm{r}}{\partial z} = C_\mathrm{r}\left(F_\mathrm{r} + z\frac{\partial F_\mathrm{r}}{\partial z}\right). \tag{10.75}$$

Suppose the expressions in brackets on the right-hand side of (10.74) and (10.75) are equal to the constants d_i and d_r, respectively:

$$z\frac{\partial F_\mathrm{i}}{\partial z} + F_\mathrm{i} - d_\mathrm{i} = 0, \tag{10.76}$$

$$z\frac{\partial F_\mathrm{r}}{\partial z} + F_\mathrm{r} - d_\mathrm{r} = 0. \tag{10.77}$$

Having solved the differential eqn (10.76) and (10.77), we can determine the specific forms of the functions F_i and F_r:

$$F_\mathrm{i} = -\frac{d_{0\mathrm{i}}}{z} + d_\mathrm{i}, \tag{10.78}$$

$$F_\mathrm{r} = -\frac{d_{0\mathrm{r}}}{z} + d_\mathrm{r}, \tag{10.79}$$

where $d_{0\mathrm{i}}$ and $d_{0\mathrm{r}}$ are constants.

Substituting eqn (10.78) and (10.79) into (10.67) and (10.68), we obtain expressions for the functions C_i and C_r:

$$C_\mathrm{i} = C_{0\mathrm{i}} \exp(F_\mathrm{i} z) = C_{0\mathrm{i}} \exp(-d_{0\mathrm{i}}) \exp(z d_\mathrm{i}), \tag{10.80}$$

$$C_\mathrm{r} = C_{0\mathrm{r}} \exp(F_\mathrm{r} z) = C_{0\mathrm{r}} \exp(-d_{0\mathrm{r}}) \exp(z d_\mathrm{r}). \tag{10.81}$$

We can now rewrite eqn (10.74) and (10.75) in the following simplified form:

$$\frac{\partial C_\mathrm{i}}{\partial z} = d_\mathrm{i} C_\mathrm{i}; \tag{10.82}$$

$$\frac{\partial C_\mathrm{r}}{\partial z} = d_\mathrm{r} C_\mathrm{r}. \tag{10.83}$$

We will also write down the values of the second derivatives:

$$\frac{\partial^2 C_\mathrm{i}}{\partial z^2} = d_\mathrm{i}^2 C_\mathrm{i}, \tag{10.84}$$

$$\frac{\partial^2 C_\mathrm{r}}{\partial z^2} = d_\mathrm{r}^2 C_\mathrm{r}. \tag{10.85}$$

Now we have everything we need to integrate eqn (10.48) and find the functions u and v in eqn (10.46) and (10.47). Considering the form of the function (uw) in eqn (10.44), we can rewrite eqn (10.48) as follows:

$$\Sigma = \frac{1}{\beta_3 D_1} \int \frac{(\cdots)dz}{\exp\left(z\sqrt{x^2 + y^2 \frac{\beta_2}{\beta_3}}\right) \exp(\Sigma)}, \tag{10.86}$$

where

$$D_1 = -B_0 B_1 \frac{xy^2}{x^2+y^2} \frac{b_1^2}{a_1^2 a_2^2} \exp\left(1 - \frac{1}{2}\left(\frac{x^2+y^2}{xy}\right)^2\right), \quad (10.87)$$

and $(\cdots)$ is taken from eqn (10.49).

Taking the differential of eqn (10.86) and transferring $\exp(\Sigma)$ to the left-hand side, we obtain:

$$\exp(\Sigma)\, d\Sigma = \frac{1}{\beta_3 D_1} \frac{(\cdots)dz}{\exp\left(z\sqrt{x^2+y^2}\frac{\beta_2}{\beta_3}\right)}. \quad (10.88)$$

We see that the left-hand side of eqn (10.88) is integrated elementarily; the right-hand side is integrated by parts:

$$\exp(\Sigma) = \frac{1}{\beta_3 D_1}\left(-\frac{\beta_3}{\beta_2\sqrt{x^2+y^2}}\right)\left((\cdots)\exp\left(-z\sqrt{x^2+y^2}\frac{\beta_2}{\beta_3}\right)\right.$$
$$\left. - \int \exp\left(-z\sqrt{x^2+y^2}\frac{\beta_2}{\beta_3}\right) d(\cdots)\right). \quad (10.89)$$

Choosing the constants d_{i} and d_{r} in eqn (10.82)–(10.85) to be quite small, we can neglect the first- and second-order derivatives in $(\cdots)$. Then we arrive at the expression:

$$\exp(\Sigma) = \frac{1}{\beta_2 D_1}\frac{x^2+y^2}{(xy)^2}\exp\left(-z\sqrt{x^2+y^2}\frac{\beta_2}{\beta_3}\right)\left(\frac{y}{x}C_{\mathrm{i}} + C_{\mathrm{r}}\right). \quad (10.90)$$

Substituting eqn (10.90) into (10.46) and (10.47), we find the following expressions for the functions u and v:

$$u = (\beta_2 w)^{-1}\frac{x^2+y^2}{(xy)^2}\left(\frac{y}{x}C_{\mathrm{i}} + C_{\mathrm{r}}\right), \quad (10.91)$$

$$v = -(\beta_2 w)^{-1}\frac{x^2+y^2}{(xy)^2}\left(C_{\mathrm{i}} + \frac{x}{y}C_{\mathrm{r}}\right), \quad (10.92)$$

where w is taken from eqn (10.51).

For relation eqn (10.27) to hold for a broad cone ($(uw)x = -(vw)y$), we must require that the function C_{r} vanish, that is that the constant $d_{0\mathrm{r}}$ in eqn (10.81) be positive and sufficiently large. Substituting eqn (10.91) and (10.92) into eqn (10.21), we find the general form for the azimuthal component of the velocity over a broad cone:

$$v_\varphi = -(\beta_2 w)^{-1}(B_3 R\cos\theta)^{-2}\frac{C_\mathrm{i}}{\sqrt{B_5}}, \tag{10.93}$$

where B_3 is taken from eqn (10.58); B_5 from eqn (10.61); C_i from eqn (10.80).

After imposing the boundary condition:

$$v_\varphi = \Omega R\cos\theta_\mathrm{w} \quad \text{at } \theta = \theta_\mathrm{w}, \quad \text{Figure 10.4(a)} \tag{10.94}$$

we obtain the final form for the azimuthal component of the velocity:

$$v_\varphi = \frac{\Omega R}{\cos\theta_\mathrm{w}}\cos^2\theta \exp\left(B_3 B_6(\tan\theta_\mathrm{w} - \tan\theta) + d_\mathrm{i} R(\sin\theta - \sin\theta_\mathrm{w})\right), \tag{10.95}$$

where θ_w is the angle corresponding to the wall of the cone Figure 10.4(a); B_6 is taken from eqn (10.59).

In Figure 10.7, we show the plot of the azimuthal components of the vector velocity field in eqn (10.95). Figure 10.8 shows the surface on which solution (10.95) works; this surface is constructed based on the form of

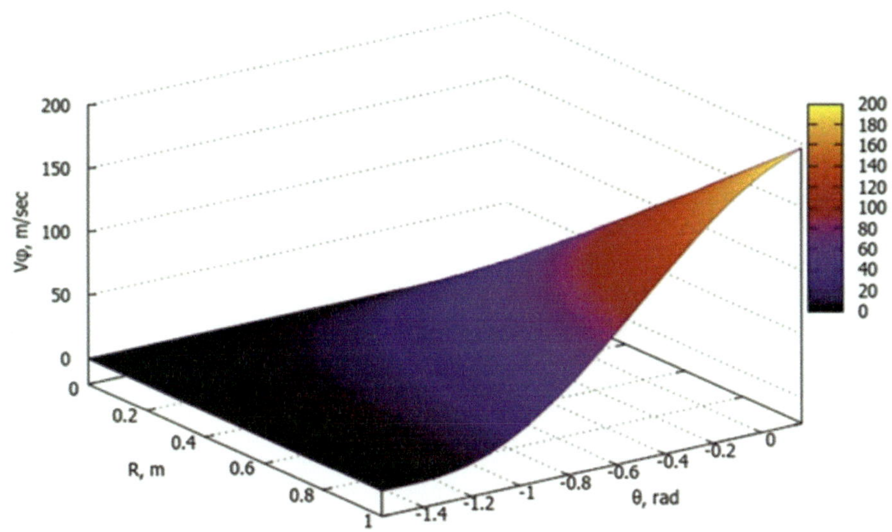

Figure 10.7 Plot for v_φ component in eqn (10.95) at $\psi = 80°$ ($\theta_\mathrm{W} = 10°$ (0.174 rad)); $\Omega = 200\,\frac{\mathrm{rad}}{\mathrm{sec}}$, $B_3 = -1$; $B_6 = 0.5$; $d_\mathrm{i} = 0.1$.

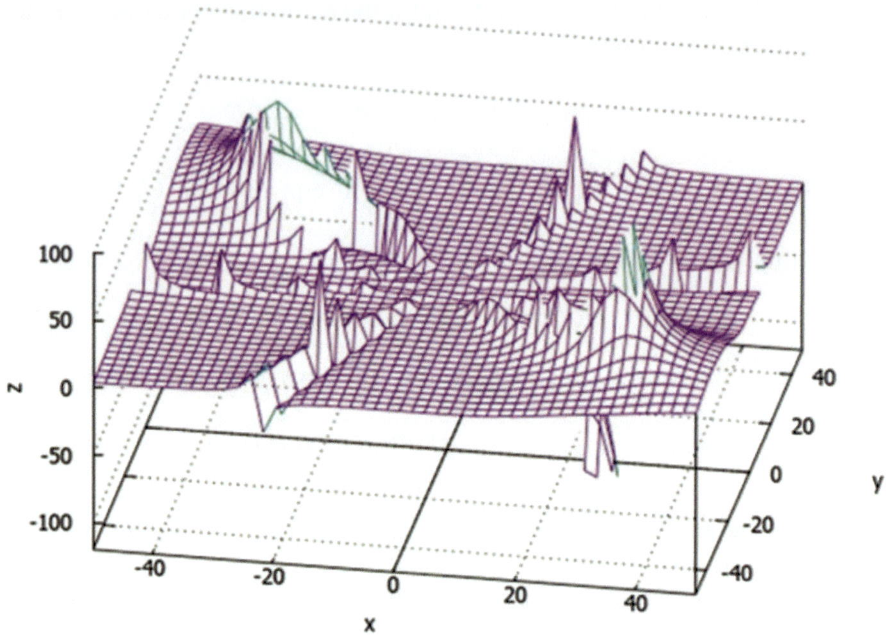

Figure 10.8 Plot of the surface on which solution (10.95) works; this surface is constructed based on the form of the function $F_{\mathrm{i}} = F_{\mathrm{i}}(z)$ in eqn (10.78) and (10.69).

the function $F_{\mathrm{i}} = F_{\mathrm{i}}(z)$ in eqn (10.78) and (10.69). As can be seen, this surface is mostly close to zero, but in some places it can also take non-zero values, and with different rates of increase. Due to this behavior of the surface, a favorable opportunity is created for fitting the modeling curve using calibration coefficients.

10.3.2 Disk

The solution for a rotating disk can be obtained as a limiting case of the broad cone configuration (see Figure 10.4). To achieve this, we introduce the following transformations into the broad cone solution, effectively transitioning to cylindrical coordinates:

$$\theta_{\mathrm{W}} \to 0, \tag{10.96}$$

$$\theta \to -\theta, \tag{10.97}$$

$$v_{\mathrm{R}} \to v_{\mathrm{R}}, \tag{10.98}$$

$$v_\theta \to v_{-\theta}, \tag{10.99}$$

$$v_\varphi \to v_\varphi, \tag{10.100}$$

$$R\cos\theta \to r, \tag{10.101}$$

$$-R\sin\theta \to -z, \tag{10.102}$$

$$\cos\theta \to \frac{r}{R} = \frac{r}{\sqrt{r^2+z^2}}, \tag{10.103}$$

$$\sin\theta \to \frac{z}{R} = \frac{z}{\sqrt{r^2+z^2}}, \tag{10.104}$$

$$\tan\theta \to \frac{z}{r}, \tag{10.105}$$

$$R^2 = r^2 + (-z)^2 = r^2 + z^2, \tag{10.106}$$

$$v_\mathrm{r} = v_\mathrm{R}\cos\theta + v_{-\theta}\sin\theta, \tag{10.107}$$

$$-v_\mathrm{z} = -v_\mathrm{R}\sin\theta + v_{-\theta}\cos\theta. \tag{10.108}$$

In examining the velocity profile within the turbulent boundary layer above a rotating disk, special attention must be given to the high-quality studies by the Swedish school [5–8]. These references are cited to support and validate the correctness of the current results.

10.3.2.1 Radial and axial (wall-normal) components of the velocity

Substituting relations (10.56) and (10.57) for the radial and polar components of the vector velocity field for a broad cone into relations (10.107) and (10.108) for a disk, we obtain expressions for v_r and $-v_\mathrm{z}$ (the minus sign accounts for the axial direction since the z-axis in Figure 10.4 is directed downward). These describe the radial and axial components of velocity for a rotating disk in cylindrical coordinates:

$$v_\mathrm{r} = 2B_0B_3\frac{r(-z)}{r^2+z^2}\exp\left(1-\frac{1}{2}B_3^{-2} - B_3B_6\frac{(-z)}{r}\right)$$
$$\left(1+\exp\left(2B_3B_6\frac{(-z)}{r}\right)\right), \tag{10.109}$$

$$-v_\mathrm{z} = -2B_0B_3\frac{z^2}{r^2+z^2}\exp\left(1-\frac{1}{2}B_3^{-2} - B_3B_6\frac{(-z)}{r}\right)$$
$$\left(1-\left(\frac{r}{z}\right)^2\exp\left(2B_3B_6\frac{(-z)}{r}\right)\right). \tag{10.110}$$

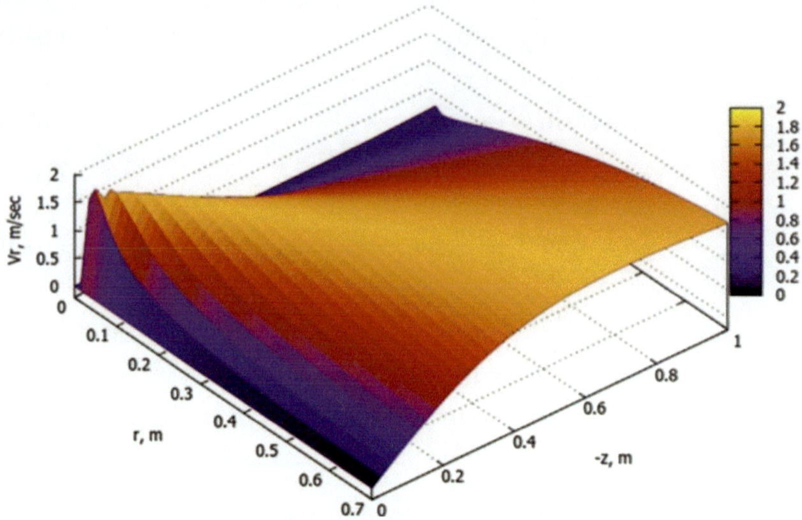

Figure 10.9 Plot for v_r component of the fluid vector velocity field over a rotating disk in eqn (10.109) at $B_0 = 200\,\frac{\text{m}}{\text{sec}}$, $B_3 = 0.33$, and $B_6 = 0.1$.

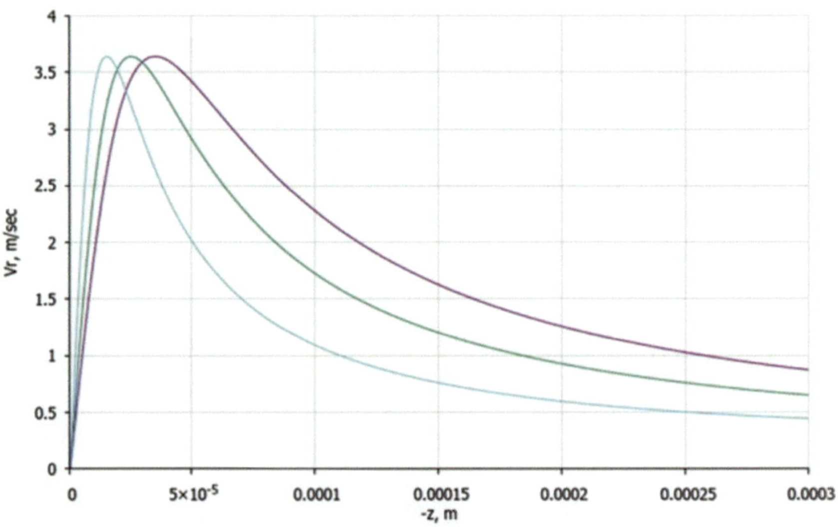

Figure 10.10 Plot for v_r component of the fluid vector velocity field over a rotating disk in eqn (10.109) at $B_0 = 200\,\frac{\text{m}}{\text{sec}}$, $B_3 = 0.33$, and $B_6 = 0.1$, scaled to the area of turbulent boundary layer at different values of r-coordinate: the blue curve at $r = 0.7\,\text{m}$; the green curve at $r = 0.5\,\text{m}$; and the cyan curve at $r = 0.3\,\text{m}$.

Figures 10.9–10.12 present plots of the radial and axial (wall-normal) velocity components defined in eqn (10.109) and (10.110). Figure 10.10 illustrates the 2D distribution of the radial velocity component from

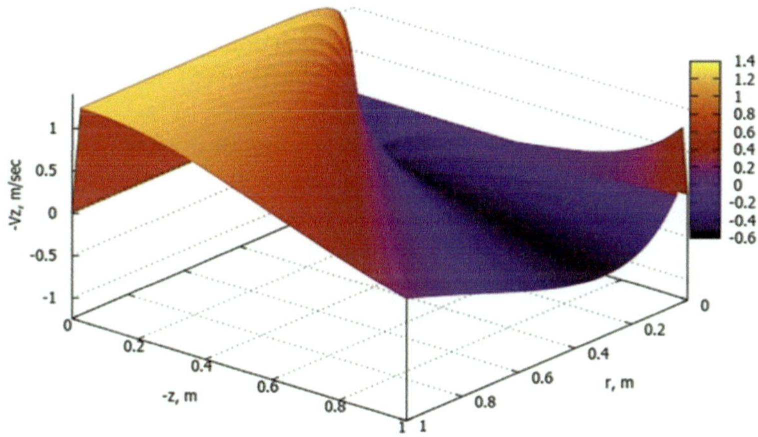

Figure 10.11 Plot for $-v_z$ component of the fluid vector velocity field over a rotating disk in eqn (10.110) at $B_0 = 200\,\frac{\mathrm{m}}{\mathrm{sec}}$, $B_3 = 0.3$, and $B_6 = 0.5$.

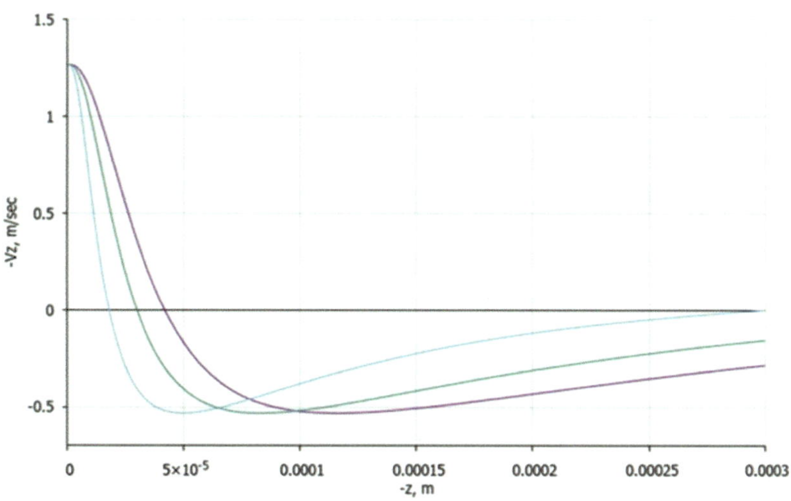

Figure 10.12 Plot for $-v_z$ component of the fluid vector velocity field over a rotating disk in eqn (10.110) at $B_0 = 200\,\frac{\mathrm{m}}{\mathrm{sec}}$, $B_3 = 0.3$, and $B_6 = 0.5$, scaled to the area of turbulent boundary layer at different values of r-coordinate: the blue curve at $r = 0.7$ m; the green curve at $r = 0.5$ m; and the cyan curve at $r = 0.3$ m.

eqn (10.109), scaled to the region of the turbulent boundary layer for various r-coordinate values. These plots show strong agreement with those presented in the reference paper [6]. When comparing the plots, it is essential to note that in [6], the dimensionless coordinate $z^+ = \xi$ from eqn (10.1) is plotted on the abscissa in a logarithmic scale. The consistency of the profiles can be observed from the location of their peaks, which correspond to the boundary of the viscous sublayer – specifically, at $z^+ = 10^1$ in [6], and approximately $z \approx 4 \times 10^{-5}$ m in Figure 10.10.

Figure 10.12 shows the 2D plots of the axial (wall-normal) component in eqn (10.110) scaled to the area of the turbulent boundary layer for different r-coordinate values. The literature generally does not provide results for the axial velocity component in a turbulent boundary layer above a disk. Since we have nothing to compare with, we simply describe our results in Figure 10.12. In the logarithmic region of the turbulent boundary layer, the gas moves toward the rotating disk, gradually increasing its velocity. Upon reaching the boundary of the buffer region ($z \approx 1 \times 10^{-4}$ m), the velocity of such movement begins to decrease and disappears completely at the boundary of the viscous sublayer ($z \approx 4 \times 10^{-5}$ m). In the viscous sublayer, the gas moves away from the disk with a velocity that decreases almost linearly.

10.3.2.2 Azimuthal components of the velocity

The azimuthal component of the velocity in the turbulent boundary layer above the disk is of particular interest because, in the reference frame associated with the disk, it is analogous to the streamwise velocity component in the turbulent boundary layer above a flat plate. Substituting relations (10.96)-(10.97) and (10.100)–(10.106) into (10.95), we obtain expressions for v_φ component for a rotating disk:

$$v_\varphi = \Omega \frac{r^2}{\sqrt{r^2 + z^2}} \exp\left((-z)\left(\frac{B_3 B_6}{r} + d_{\mathrm{i}}\right)\right). \tag{10.111}$$

In Figures. 10.13 and 10.14, we show plots of the azimuthal component of the velocity in eqn (10.111) in the area of the turbulent boundary layer. Figure 10.14 shows the 2D plot of the azimuthal component of the velocity in eqn (10.111) in the area of the turbulent boundary layer at $r = 0.7$ m. To make a comparison with Figure 10.2, plots corresponding to the linear law (10.3) and the logarithmic law (10.4) were added to the plot of the azimuthal component. We see that the quality of the modeling up to $z \approx 3 \times 10^{-4}$ m (which corresponds to the beginning of the log law region) is quite good. Further, significant deviations are observed, probably related to the simplifications we

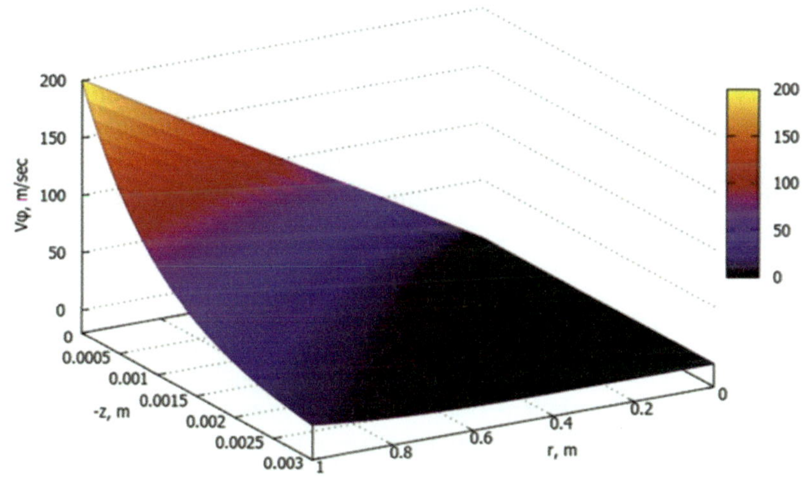

Figure 10.13 Plot for v_φ component of the fluid vector velocity field over a rotating disk in eqn (10.111) at $\Omega = 200\,\frac{\text{rad}}{\text{sec}}$, $B_3 = -2000$, $B_6 = 0.5$, and $d_\text{i} = 0.1$ in the area of turbulent boundary layer.

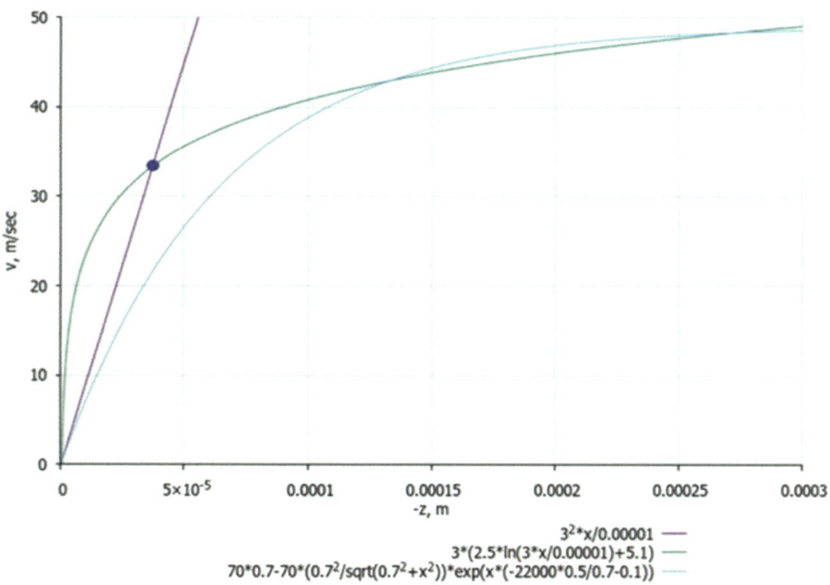

Figure 10.14 Plot for v_φ component of the fluid vector velocity field over a rotating disk in eqn (10.111) in the reference frame associated with the disk at $\Omega = 70\,\frac{\text{rad}}{\text{sec}}$, $B_3 = -22000$, $B_6 = 0.5$, $d_\text{i} = -0.1$, and $r = 0.7\,\text{m}$ in the area of turbulent boundary layer (the cyan curve); the blue line corresponds to the linear law (10.3); the green curve corresponds to the log law region (10.4).

adopted in modeling e.g., when we equated the expressions in brackets on the right-hand side of (10.74) and (10.75) to constants). One way or another, it is necessary to further develop this method in order to improve the quality of the modeling. It remains to add that similar plots for the azimuthal component of velocity can be obtained for any value of the r-coordinate; moreover, the smaller the value of r, the greater the angular velocity Ω, and the smaller the constant B_3 should be.

10.4 Conclusions

In the tunnel mathematics space, single formulas for each velocity component in the incompressible turbulent boundary layer over a rotating broad cone and disk were obtained. The formula for the azimuthal velocity component in the boundary layer over a disk (analogous to the streamwise velocity component in the boundary layer above a flat plate) works in the viscous sublayer, the buffer region, and at the beginning of the log law region. Further improvement of the theory is necessary to make this formula workable over the entire thickness of the turbulent boundary layer. The obtained formulas for the radial and axial velocity components in the boundary layer over a disk can be applied to the entire thickness of the turbulent boundary layer. With appropriate adjustment of the calibration coefficients, all three formulas for the velocity components in the turbulent boundary layer over a disk can be sensitive to changes in the pressure gradient and are suitable for describing a laminar boundary layer.

The method of tunnel mathematics allows us to work with coordinates and velocities in their pure form and not to apply specific scaling. Here, simple mathematical scaling is allowed, when, for example, it is necessary to transfer the solution to the boundary layer region. Due to this, the mathematical apparatus used is more transparent and simple; also, when reading formulas and plots, the reader does not need to constantly keep in mind a large number of additional physical parameters, which, moreover, can also change from point to point (for example, the fluctuating velocity used to scale the axial coordinate can change with a change in height above the disk).

References

[1] LD Landau. Fluid mechanics, course of theoretical physics, vol. 6. *Addison-Wesley Publishing Co.*, 1959.

[2] H Schlichting. Boundary-layer theory, 7th ednmcgraw-hill, 1979.

[3] Jaroslav Štigler. Analytical velocity profile in tube for laminar and turbulent flow. *Engineering Mechanics*, 21(6):371–379, 2014.

[4] Bruce R Munson, Donald F Young, and Theodore H Okiishi. *Fundamentals of fluid mechanics*. Wiley, 1990.

[5] P Henrik Alfredsson, Kentaro Kato, and RJ Lingwood. Flows over rotating disks and cones. *Annual Review of Fluid Mechanics*, 56(1):45–68, 2024.

[6] Ellinor Appelquist, Philipp Schlatter, P Henrik Alfredsson, and RJ Lingwood. Turbulence in the rotating-disk boundary layer investigated through direct numerical simulations. *European Journal of Mechanics-B/Fluids*, 70:6–18, 2018.

[7] Shintaro Imayama, Rebecca J Lingwood, and P Henrik Alfredsson. The turbulent rotating-disk boundary layer. *European Journal of Mechanics-B/Fluids*, 48:245–253, 2014.

[8] Philipp Schlatter and Ramis Örlü. Assessment of direct numerical simulation data of turbulent boundary layers. *Journal of Fluid Mechanics*, 659:116–126, 2010.

[9] Fei-Chi Zhang, Jin-Han Xie, Song Xi Chen, and Xiaojing Zheng. Robust relation of streamwise velocity autocorrelation in atmospheric surface layers based on an autoregressive moving average model. *Journal of Fluid Mechanics*, 981:A20, 2024.

[10] Xiang IA Yang, Peng ES Chen, Wen Zhang, and Robert Kunz. Predictive near-wall modelling for turbulent boundary layers with arbitrary pressure gradients. *Journal of Fluid Mechanics*, 993:A1, 2024.

[11] Ugo Piomelli and Elias Balaras. Wall-layer models for large-eddy simulations. *Annual review of fluid mechanics*, 34(1):349–374, 2002.

[12] Georgi Kalitzin, Gorazd Medic, Gianluca Iaccarino, and Paul Durbin. Near-wall behavior of rans turbulence models and implications for wall functions. *Journal of Computational Physics*, 204(1):265–291, 2005.

[13] Alexander J Smits, Beverley J McKeon, and Ivan Marusic. High–reynolds number wall turbulence. *Annual Review of Fluid Mechanics*, 43(1):353–375, 2011.

[14] Ludwig Prandtl. 7. bericht über untersuchungen zur ausgebildeten turbulenz. *ZAMM-Journal of Applied Mathematics and Mechanics/Zeitschrift für Angewandte Mathematik und Mechanik*, 5(2):136–139, 1925.

[15] Christoffer Hansen, Xiang IA Yang, and Mahdi Abkar. A pod-mode-augmented wall model and its applications to flows at non-equilibrium conditions. *Journal of Fluid Mechanics*, 975:A24, 2023.

[16] Adrián Lozano-Durán, Marco G Giometto, George Ilhwan Park, and Parviz Moin. Non-equilibrium 3D boundary layers at moderate reynolds numbers. *Journal of Fluid Mechanics*, 883:A20, 2020.
[17] Matthew Subrahmanyam, Zan Xu, Brian Cantwell, and Juan J Alonso. A new wall-stress model for large-eddy simulations. In *AIAA SCITECH 2023 Forum*, page 0285, 2023.
[18] Mitchell Fowler, Tamer A Zaki, and Charles Meneveau. A lagrangian relaxation towards equilibrium wall model for large eddy simulation. *Journal of Fluid Mechanics*, 934:A44, 2022.
[19] Xinyi LD Huang and Xiang IA Yang. A bayesian approach to the mean flow in a channel with small but arbitrarily directional system rotation. *Physics of Fluids*, 33(1), 2021.
[20] William Morrow Kays, Michael E Crawford, and Bernhard Weigand. *Convective heat and mass transfer*, volume 4. McGraw-Hill New York, 1980.
[21] Peng ES Chen, Wen Wu, Kevin P Griffin, Yipeng Shi, and Xiang IA Yang. A universal velocity transformation for boundary layers with pressure gradients. *Journal of Fluid Mechanics*, 970:A3, 2023.
[22] Oleh Shvydkyi. Tensor foundations of tunnel mathematics. 2023.
[23] G Shvydkyi et al. Sharp cone-broad cone-disk: Analytical solutions in the tunnel mathematics space to the steady navier–stokes equations in the area of boundary layer for incompressible symmetric flows entrained by these rotating bodies. *Acceleron Aerospace Journal*, 3(6):624–647, 2024.
[24] Souayeh, B., & Ramesh, K. (Eds.). (2025). Advances in Computational Fluid Dynamics (1st ed.). Chapter 1. Steady Navier–Stokes Solutions in Rotating Disk Boundary Layers. CRC Press. https://doi.org/10.1201/9781003611820
[25] OG Shvydkyi. Analytical solution to the steady navier–stokes equation for a compressible fluid entrained by a rotating disk of finite radius in the area of boundary layer by means tunnel mathematics (subsonic and hypersonic cases). 2022.
[26] Mikhail Lavrentiev and Boris Vladimirovich Shabat. Méthodes de la théorie des fonctions d'une variable complexe. 1972.
[27] Tomás Soler and Jen-Yu Han. Orthogonal curvilinear coordinates [and more] for engineers and physical scientists, 560. *Chicago: Excel Book Writing*, 2024.

11

Optimization Techniques in Heat Transfer and Fluid Dynamics

Pourya Bazyar[1] and Ehsan Sheidaee[2]

[1]Department of Mechanical Engineering and Production Management, Hamburg University of Applied Science, Germany
[2]Biosystems Engineering Department, TarbiatModares University, Iran
E-mail: pourya.bazyar@haw-hamburg.de; sheidaee.ehsan@modares.ac.ir

Abstract

Optimization techniques play a crucial role in enhancing the efficiency and performance of heat transfer and fluid dynamic systems. This chapter presents a comprehensive review of advanced optimization methodologies applied to the design and analysis of heat exchangers (HE), computational fluid dynamics (CFD), computational heat transfer (CHT), and topology optimization (TO) methods. A significant focus is placed on the parametric design optimization of various heat exchanger configurations, including shell-and-tube, plate-fin, and fin-tube heat exchangers. Traditional trial-and-error methods, often tedious and time-consuming, lack the guarantee of optimal solutions. Therefore, advanced optimization algorithms have been increasingly adopted to achieve superior performance, addressing complex structural and physical parameters effectively. This review consolidates existing research efforts, providing a valuable resource for industrial designers and researchers exploring parameter optimization of heat exchangers. Topology optimization has emerged as a powerful numerical approach for optimizing thermal systems, particularly in the context of conductive, convective, and conjugate heat transfer. Over the past two decades, various TO methodologies have been developed, yet they remain scattered across the literature, with limited comparative studies assessing their effectiveness. Furthermore, recent advancements in computational techniques have significantly influenced the

DOI: 10.1201/9788743806691-11

evolution of CFD and CHT. These numerical tools, widely used in engineering design and analysis, require a deep understanding of mechanics and numerical methods to maximize their potential. Finite element methods (FEM), coupled with applied case studies, have proven instrumental in solving heat conduction, incompressible viscous flow, and convection heat transfer problems. An emerging approach in conjugate heat transfer optimization integrates TO with a Lagrangian optimization framework. This method aims to design more efficient coolant systems by minimizing pressure losses and maximizing heat transfer. The ability to define and achieve an optimal balance between heat transfer enhancement and pressure losses presents a significant advancement in thermal system optimization.

Keywords: Heat exchangers, computational fluid dynamics, CFD, computational heat transfer, CHT, topology optimization, TO, thermal system design

11.1 Introduction

FEA and optimization techniques are essential for improving the efficiency and performance of heat transfer and fluid flow systems in different variety of research areas [1–6], which are widely used in industries such as aerospace, automotive, and energy. The design of turbine cooling systems, heat exchangers, and other thermal management components requires a careful balance between maximizing heat transfer and minimizing pressure losses [7]. Traditional optimization methods, including shape and size optimization, have been commonly used; however, the advent of topology optimization (TO) has introduced new possibilities, especially with the development of additive manufacturing [8]. TO allows for the design of highly efficient and complex geometries that were previously unachievable with conventional manufacturing techniques [9]. The challenges associated with these optimization problems arise due to their non-linear constraints and multi-objective nature, which require advanced computational techniques [10]. Various optimization algorithms, such as genetic algorithms, particle swarm optimization, and artificial bee colony methods, have been developed to tackle these challenges [11]. This chapter explores the application of these advanced optimization techniques in thermal-fluid systems and highlights the growing role of computational intelligence in solving complex heat transfer problems. Traditional methods, including shape and size optimization, have been extensively utilized; however, the emergence of topology optimization (TO) has revolutionized the field, enabling the design of highly complex geometries optimized for maximum heat transfer efficiency. With the advancement of additive manufacturing, TO has gained increased significance, as it allows the production of intricate structures that would otherwise be impossible using conventional manufacturing methods [12]. The complexity of these optimization problems, often involving multiple constraints and objectives, necessitates the use of advanced computational algorithms [13]. This chapter provides an overview of these advanced optimization strategies and their applications in improving the efficiency and effectiveness of heat transfer systems. The optimization of heat transfer and fluid dynamics systems is fundamental to improving the efficiency and reliability of thermal management solutions across various engineering applications. Industries such as aerospace, energy, and electronics rely on efficient cooling techniques to maintain optimal operating conditions and prevent thermal failure. The design of components like turbine cooling channels, heat exchangers, and microfluidic devices requires a careful trade-off between heat transfer enhancement and pressure loss minimization [14]. While traditional shape and size optimization methods have been widely used, recent advancements in topology optimization (TO) have opened new opportunities for designing innovative cooling structures with improved performance. The integration of TO with additive manufacturing has further expanded design possibilities, enabling the fabrication of intricate geometries tailored for efficient heat dissipation [15]. Given the complexity of these optimization problems, sophisticated computational algorithms are essential to address non-linear constraints and multi-objective functions [10]. This chapter delves into these state-of-the-art optimization methodologies and their transformative impact on heat transfer applications.

11.2 Background and Importance of Optimization in Heat Transfer and Fluid Dynamics

Heat transfer and fluid dynamics play a fundamental role in engineering applications, influencing the performance and efficiency of various systems. From power generation to aerospace engineering, effective thermal management ensures the longevity, safety, and energy efficiency of mechanical and electronic components [16]. Optimizing heat transfer and fluid flow processes is critical to achieving superior performance while reducing operational costs.

11.2.1 Significance of heat transfer and fluid dynamics in engineering applications

Heat transfer and fluid dynamics govern the movement of heat and fluids in a wide range of engineering systems. Effective thermal management is crucial in various fields, including:

- *Aerospace engineering:* Aircraft and spacecraft require advanced cooling mechanisms to manage high thermal loads. Turbine cooling, fuel system heat exchangers, and spacecraft thermal control systems rely on optimized heat transfer techniques [17].
- *Automotive industry:* Engine cooling, exhaust gas heat recovery, and HVAC (heating, ventilation, and air conditioning) systems use fluid dynamics to enhance efficiency and fuel economy [18].
- *Power generation:* Steam turbines, condensers, and nuclear reactors depend on optimized heat transfer processes to improve energy conversion and operational reliability [19].
- *Electronics and microprocessors:* Electronic devices require efficient cooling solutions, such as heat sinks and liquid cooling, to prevent overheating and ensure stable operation [20].

11.2.2 Importance of optimization in heat transfer and fluid flow

- *Improving efficiency and performance:* Optimization techniques help engineers design systems that achieve maximum heat transfer rates while minimizing energy consumption and pressure losses. Optimizing the geometry of heat exchangers can significantly enhance thermal performance while reducing material usage [21].
- *Cost-effectiveness:* By improving heat transfer efficiency, optimization reduces operational costs, energy consumption, and material waste [22]. This is particularly important in industries like power generation, where even small efficiency improvements translate to significant cost savings.
- *Environmental benefits:* Enhanced heat transfer efficiency leads to lower energy consumption and reduced greenhouse gas emissions. Optimizing cooling systems in industrial processes can decrease water and energy usage, contributing to sustainable engineering practices [23].

11.2.3 Applications of optimization in various industries

- *Aerospace industry:* Aircraft engines and spacecraft systems operate under extreme thermal conditions. Figure 11.1 illustrates an optimized turbine blade cooling system, demonstrating how advanced heat transfer techniques enhance efficiency and lifespan (Table 11.1) [24].

- *Automotive industry:* In internal combustion engines, heat exchangers and radiators are optimized to achieve maximum heat dissipation with minimal airflow resistance [34]. Figure 11.2 showcases a CFD-simulated optimized cooling system for an electric vehicle battery pack.
- *Power generation:* Thermal power plants rely on optimized cooling towers, steam condensers, and turbine heat exchangers to improve efficiency and reliability [35]. By employing computational optimization methods, power plants can maximize heat recovery and minimize fuel consumption (Table 11.2).
- *Electronics cooling:* Miniaturized electronic components require highly efficient heat dissipation techniques [36]. Optimized heat sinks and liquid cooling systems, significantly improve device performance and reliability.
- *Results:* Optimization techniques in heat transfer and fluid dynamics are essential for improving efficiency, performance, and cost-effectiveness across multiple industries. Advances in computational fluid dynamics

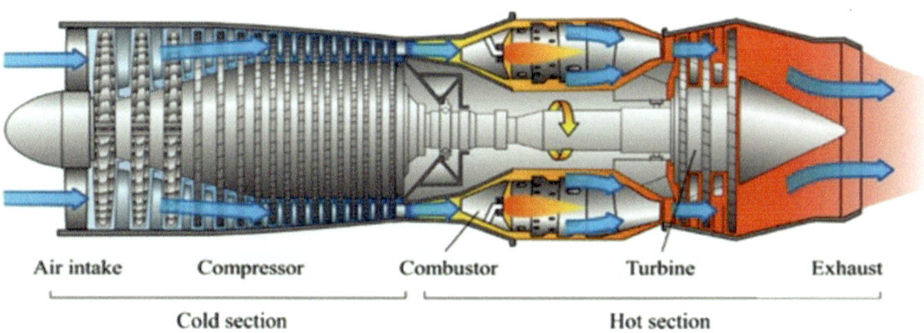

Figure 11.1 Gas turbine blade [24].

Table 11.1 Optimization strategies in electronics cooling

Method	Application	Benefit
Microchannel heat sinks [25]	CPU and GPU cooling [26]	Increased heat dissipation [27]
Phase-change cooling [28]	High-performance computing [29]	Improved temperature control [30]
AI-based thermal design [31]	Server cooling systems [32]	Energy efficiency [33]

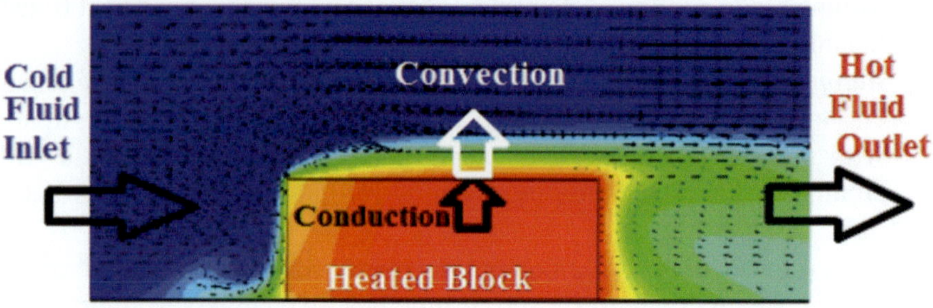

Figure 11.2 Example of a CFD simulated turbulent flow in a heat exchange.

Table 11.2 Optimization strategies in electronics cooling

Optimization method	Application area	Advantages	Challenges
Computational fluid dynamics (CFD)[37]	Heat exchangers, engine cooling[22]	Accurate simulations, cost-effective[38]	Requires high computational power[30]
Topology optimization[39]	Additive manufacturing[40]	Optimized material usage, complex geometries[41]	Limited by manufacturing constraints[22]
Genetic algorithms[42]	Electronics cooling[43]	Efficient in large solution spaces[44]	Computationally expensive[45]
Experimental prototyping[27]	Aerospace, automotive[46]	Real-world validation[47]	Time-consuming, expensive[48]

(CFD), artificial intelligence (AI), and additive manufacturing continue to drive innovation in thermal management solutions. By integrating these optimization strategies, engineers can develop highly efficient, sustainable, and cost-effective heat transfer systems.

11.3 Evolution of Computational Methods in Optimization

The optimization of heat transfer and fluid dynamics has undergone significant evolution due to advancements in computational methods. Initially reliant on experimental and analytical approaches, engineers have increasingly adopted computational techniques such as computational fluid dynamics (CFD), finite element method (FEM), and conjugate heat transfer (CHT) simulations [49]. More recently, artificial intelligence (AI) and machine learning (ML) have been integrated into the optimization process, enabling more efficient and intelligent solutions. This section explores the historical development, current advancements, and future trends in computational optimization methods.

11.3.1 Historical perspective: From experimental to computational approaches

Traditionally, heat transfer and fluid flow optimization relied heavily on empirical testing and analytical methods. Experimental techniques involved building physical prototypes and conducting controlled tests, while analytical methods used mathematical models to approximate fluid flow and thermal behavior (Table 11.3) [50].

Experimental methods provided reliable results but were costly and time-consuming. Analytical methods improved efficiency but were constrained by mathematical complexity. The emergence of computational methods bridged this gap by allowing simulations under various conditions without extensive physical testing.

11.3.2 Computational fluid dynamics (CFD)

CFD revolutionized heat transfer and fluid flow optimization by providing detailed insights into complex flow behaviors. By solving the Navier–Stokes equations numerically, CFD enables the analysis of heat dissipation, turbulence, and fluid-structure interactions [51].

11.3.3 Key advantages of CFD:

- Detailed visualization: Provides flow field visualization, including velocity profiles and temperature distributions.
- Parametric studies: Allows engineers to test multiple configurations without physical modifications.
- Cost and time efficiency: Reduces reliance on experimental prototypes.

Table 11.3 Traditional versus computational methods in optimization

Feature	Experimental methods	Analytical methods	Computational methods
Accuracy	High (but limited to test conditions)	Moderate (simplified models)	Very high (detailed simulations)
Cost	Expensive	Low	Moderate to high
Time requirement	Long	Short	Short to moderate
Flexibility	Limited to physical constraints	Limited to solvable equations	High (various conditions simulated)

Despite its advantages, CFD requires significant computational power and expertise to ensure accurate modeling and meshing techniques.

11.3.4 Finite element method (FEM)

The finite element method (FEM) is another widely used numerical technique in optimization. While CFD focuses on fluid flow, FEM primarily addresses structural and thermal analysis (Figure 11.3) [52]. FEM discretizes a system into smaller elements, solving for temperature distributions and stress responses in heat exchangers, turbine blades, and microchannel networks (Table 11.4).

11.3.5 Applications of FEM in optimization

- Thermal stress analysis: Determines the effects of temperature gradients on material integrity.

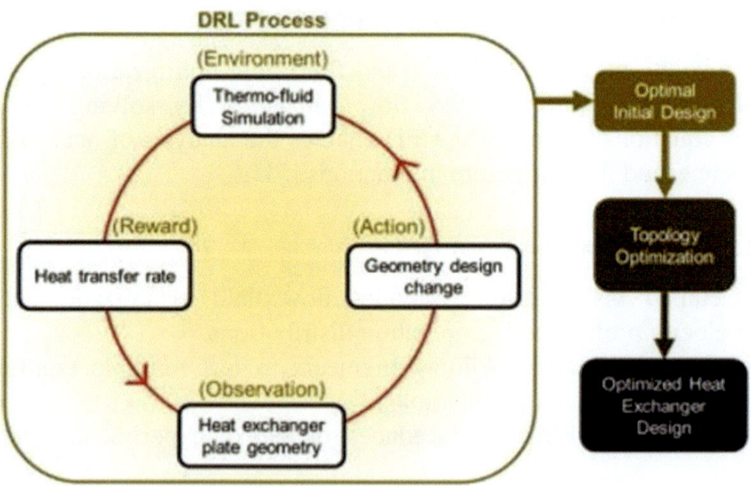

Figure 11.3 AI-driven topology optimization of a heat exchanger [53].

Table 11.4 Comparison of CFD and FEM in heat transfer optimization

Feature	CFD	FEM
Application	Fluid flow and heat transfer	Structural and Thermal Analysis
Governing equations	Navier–Stokes	Heat conduction and stress equations
Primary use case	Optimizing flow and heat dissipation	Optimizing material distribution and thermal stress

- Conjugate heat transfer (CHT) simulations: Combines fluid and solid heat transfer modeling.
- Topology optimization: Improves component shapes to maximize performance.

11.3.6 Conjugate heat transfer (CHT) modeling:

CHT modeling integrates both CFD and FEM techniques to provide a more comprehensive optimization approach [54]. It solves the heat transfer between solid and fluid domains simultaneously, making it particularly useful in high-temperature applications such as gas turbines and electronic cooling systems (Figure 11.4).

11.3.7 Benefits of CHT modeling

- More accurate predictions: Simultaneously solves for conduction, convection, and radiation.
- Enhanced design insights: Helps in optimizing coolant flow and thermal gradients.
- Reduces design failures: Ensures structural integrity under high thermal loads.

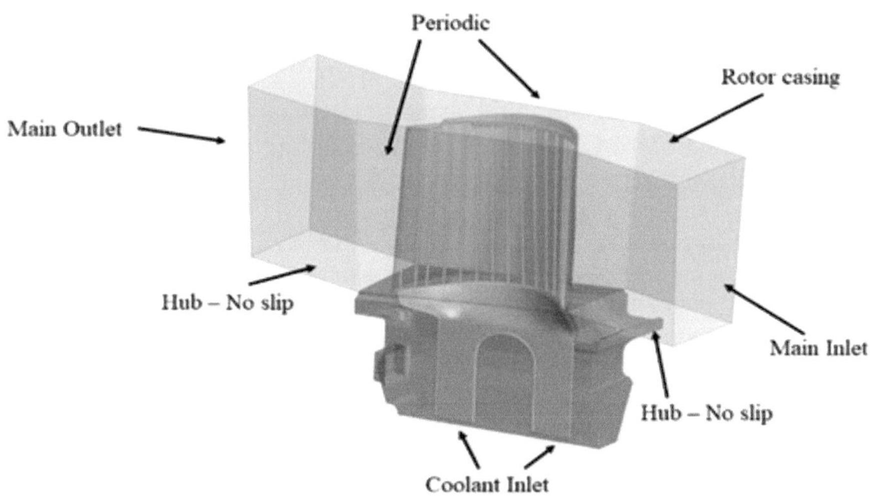

Figure 11.4 Conjugate heat transfer analysis of a turbine blade [55].

11.3.8 The role of AI and machine learning in optimization

Recent advancements in AI and ML have transformed heat transfer and fluid dynamics optimization. Traditional methods rely on iterative simulations, but AI-driven techniques can predict optimal solutions more efficiently (Table 11.5).

11.3.9 Applications of AI in optimization

- Surrogate modeling: Uses ML algorithms to approximate complex CFD and FEM simulations.
- Automated design optimization: AI-driven algorithms optimize heat exchanger geometries, fin structures, and coolant flow paths.
- Predictive maintenance: AI models forecast component failures based on temperature and flow data.

11.3.10 Future trends in computational optimization

With the continuous advancements in computing power, optimization methods will become more efficient and widely applicable. Some key trends include:

- Real-time optimization: Advanced computing resources will enable real-time simulations for heat transfer applications [56].
- Integration of AI with CFD/FEM: Hybrid models will reduce computational costs while maintaining accuracy [57].
- Quantum computing for fluid dynamics: Quantum-based simulations could revolutionize CFD computations, enabling previously impossible optimizations [58].

11.3.11 Results

The evolution of computational methods has significantly improved the efficiency and accuracy of heat transfer and fluid dynamics optimization. While traditional CFD, FEM, and CHT methods remain essential, AI and

Table 11.5 AI versus traditional computational methods in optimization

Feature	Traditional CFD/FEM	AI-based approaches
Computational cost	High	Lower (after training)
Optimization speed	Slow	Fast
Accuracy	High	High (if trained properly)
Flexibility	Requires manual adjustments	Learns and adapts automatically

machine learning are ushering in a new era of intelligent optimization. As these technologies continue to develop, engineering designs will become more cost-effective, efficient, and capable of handling increasingly complex applications.

11.4 Challenges in Heat Transfer and Fluid Dynamics Optimization

Optimizing heat transfer and fluid dynamics systems is a crucial challenge in various engineering applications [59]. The complexity arises due to non-linearities, multi-objective constraints, material limitations, and the need to balance performance with economic and environmental factors [60]. This section explores the challenges in optimizing these systems, highlighting difficulties in designing efficient heat exchangers, cooling systems, and pressure drop minimization strategies.

11.4.1 Complexity of thermal and fluid system optimization

Heat transfer and fluid flow processes are governed by nonlinear equations such as the Navier–Stokes equations, Fourier's law of heat conduction, and the convection-diffusion equation [61]. These nonlinearities create difficulties in predicting and optimizing system behavior accurately. Additionally, the multi-objective nature of optimization – where factors like efficiency, cost, size, and energy consumption must be simultaneously considered – adds another layer of complexity (Table 11.6) [62].

11.4.2 Difficulties in achieving optimal heat exchanger designs

Heat exchangers are essential in various industrial applications, but their optimization is challenging due to conflicting objectives such as maximizing heat transfer while minimizing pressure drop and material usage [66]. Traditional

Table 11.6 Key nonlinear equations in heat transfer and fluid dynamics

Equation	**Description**
Navier–Stokes equations [63]	Governs fluid flow behavior in terms of velocity and pressure fields
Fourier's law [64]	Defines the conduction heat transfer rate proportional to the temperature gradient
Convection–diffusion equation [65]	Describes the combined effect of heat conduction and convective transport

design methods rely on empirical correlations and trial-and-error approaches, which are often inefficient and time-consuming [67]. Modern optimization techniques, including topology optimization and computational fluid dynamics (CFD)-based design, offer solutions. However, computational expense and convergence issues remain significant obstacles. Manufacturing constraints,

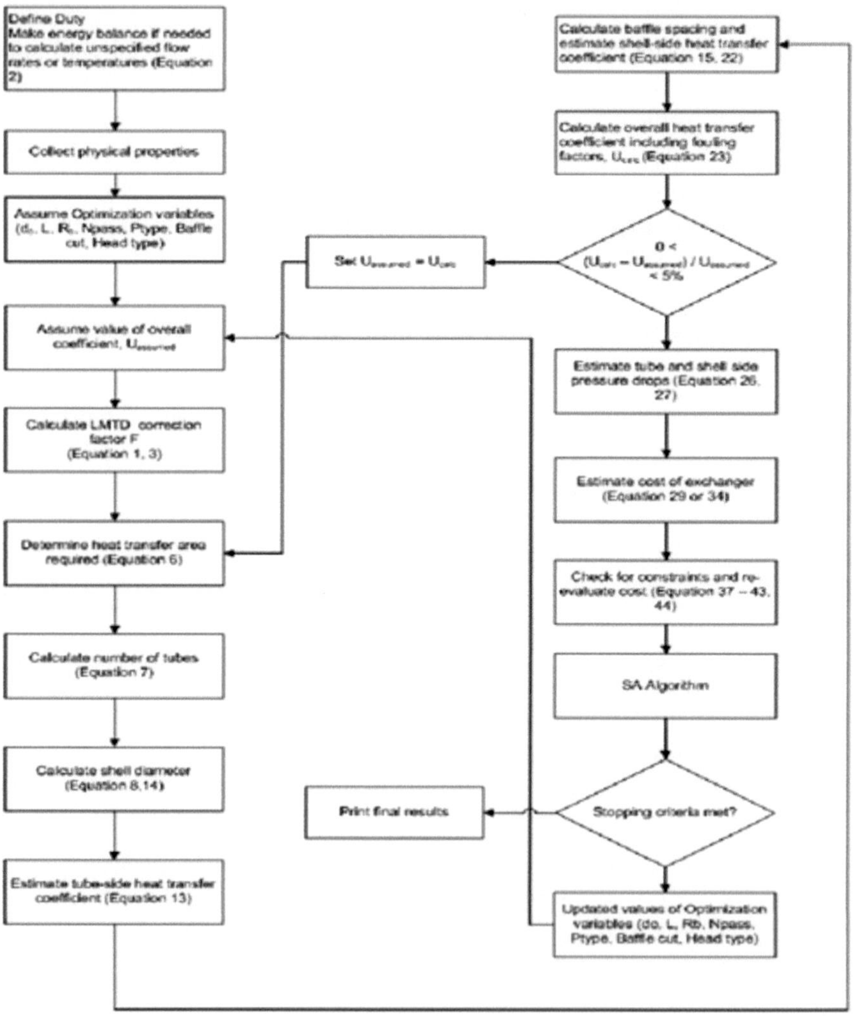

Figure 11.5 Typical heat exchanger optimization trade-offs [68].

such as limitations in material properties and fabrication techniques, further restrict the design space (Figure 11.5).

11.4.3 Challenges in efficient cooling system optimization

Cooling systems, especially in high-performance applications such as gas turbines, data centers, and automotive engines, must ensure efficient heat dissipation while maintaining structural integrity and minimizing energy consumption [69]. The primary challenges include:

- Non-uniform heat distribution: Leads to thermal stresses and material degradation [70].
- Flow maldistribution: Causes inefficient cooling due to uneven fluid velocity profiles [71].
- Material compatibility: Requires balancing thermal conductivity, corrosion resistance, and mechanical strength [72].

New cooling strategies, such as microchannel cooling and phase-change cooling, provide promising solutions but require advanced modeling and optimization techniques to determine the most effective configurations (Table 11.7).

11.4.4 Addressing pressure drop and energy efficiency

Minimizing pressure drop is a fundamental goal in fluid system design. Excessive pressure losses lead to increased pumping power requirements, reducing overall system efficiency [73]. Achieving an optimal balance between pressure drop and heat transfer effectiveness is a challenging task, particularly in systems with complex geometries [74].

Optimization approaches such as shape optimization, topology optimization, and artificial intelligence (AI)-driven design help improve flow performance while maintaining desired thermal characteristics. However, these methods require extensive computational resources and robust algorithms to ensure convergence to an optimal solution (Figure 11.6).

Table 11.7 Comparison of traditional and advanced cooling techniques

Cooling technique	Efficiency	Cost	Complexity
Traditional finned heat sink	Moderate	Low	Low
Microchannel cooling	High	Moderate	High
Phase-change cooling	Very high	High	Very high

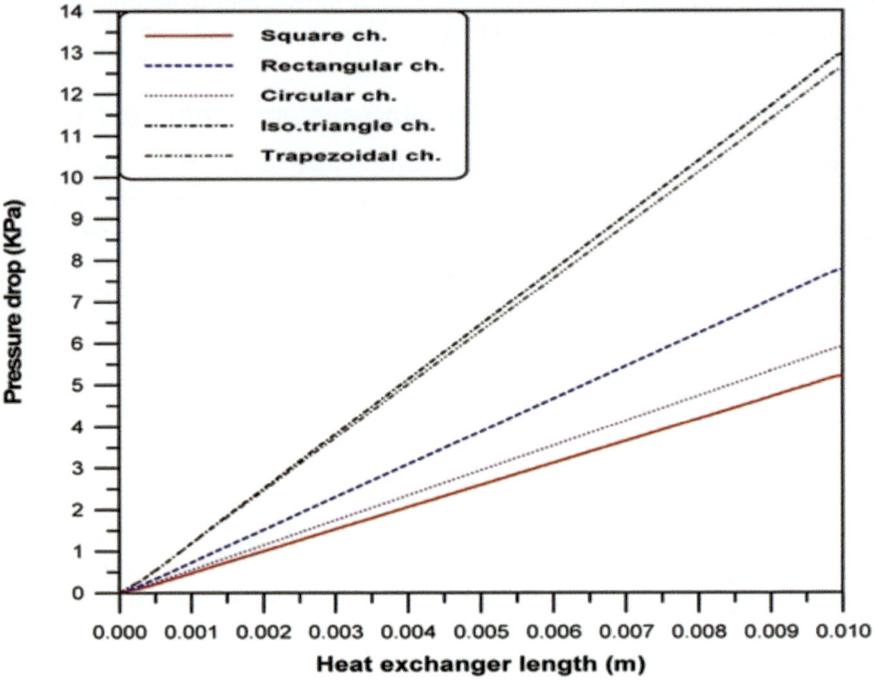

Figure 11.6 Effect of flow channel geometry on pressure drop and heat transfer [75].

11.4.5 Limitations of traditional trial-and-error methods

Historically, engineers have relied on empirical correlations and experimental testing to optimize thermal and fluid systems. While these methods provide insights into system behavior, they are time-consuming, costly, and often fail to identify truly optimal solutions. Some key limitations include:

- Limited design space exploration: Traditional methods focus on iterative improvements rather than global optimization.
- High experimental costs: Physical testing of multiple prototypes is expensive and labor-intensive.
- Lack of adaptability: Changes in operating conditions or system configurations require extensive redesign efforts.

11.4.6 The need for advanced optimization algorithms

To overcome the limitations of traditional methods, advanced computational optimization techniques have emerged (Table 11.8). These include:

Table 11.8 Comparison of advanced optimization techniques

Optimization technique	Strengths	Weaknesses
Genetic algorithm (GA) [77]	Good for complex, non-linear problems	Requires tuning of parameters
Particle swarm optimization (PSO) [78]	Fast convergence in many cases	May get trapped in local optima
Topology optimization (TO) [79]	Generates novel and efficient designs	Computationally expensive
Machine Learning-Based [80]	Can predict optimal designs quickly	Requires extensive training data

- Genetic algorithms (GA): Mimic natural selection to explore optimal design configurations [76].
- Particle swarm optimization (PSO): Uses swarm intelligence to find the best design parameters.
- Topology optimization (TO): Provides innovative structural layouts optimized for specific performance objectives.
- Machine learning-based optimization: Leverages AI to predict optimal configurations based on historical data and simulations.

11.5 Conclusion

Optimizing heat transfer and fluid dynamics systems involves overcoming challenges related to nonlinearities, multi-objective constraints, material limitations, and traditional trial-and-error methods. While advanced computational techniques such as CFD, topology optimization, and AI-driven methods provide promising solutions, they also present challenges in terms of computational cost and implementation complexity. Future research should focus on integrating these techniques with high-performance computing (HPC) and additive manufacturing to further enhance optimization capabilities.

11.5.1 Future work

- Development of hybrid optimization approaches combining AI and traditional methods.
- Improved computational efficiency through parallel computing and cloud-based simulations (see Table 11.9).
- Exploration of novel materials and manufacturing techniques to enable innovative thermal management solutions.

Table 11.9 Comparison of parametric optimization in different applications

Application	Key parameter	Optimization goal
Heat exchangers	Fin spacing	Maximizing heat transfer efficiency
Pipe networks	Diameter	Reducing pressure loss
Turbine cooling	Channel geometry	Minimizing thermal gradients

11.5.2 Parametric optimization

Parametric optimization involves adjusting predefined design parameters to find the best possible solution within given constraints. This method relies on numerical simulations and iterative calculations to refine geometries and operational parameters.

11.5.3 Advantages of parametric optimization

- Enables precise control over design variables.
- Efficiently integrates with computational fluid dynamics (CFD) and finite element methods (FEM).
- Provides rapid convergence to optimal solutions.

11.5.3.1 Applications in heat transfer and fluid dynamics

- Design of heat exchanger fins for maximum thermal efficiency.
- Optimization of pipe diameters in cooling and heating systems.
- Enhancement of turbine blade cooling techniques.

11.5.4 Topology optimization

Topology optimization (TO) is a more advanced method that optimizes the material distribution within a given design space to maximize performance. Unlike parametric optimization, TO does not rely on predefined shapes but instead evolves structures based on physics-driven principles (Figure 11.7).

11.5.4.1 Advantages of topology optimization

- Produces innovative and highly efficient designs.
- Allows for the reduction of material usage while maintaining structural integrity.
- Improves thermal and fluid flow performance in complex systems.

11.5.4.2 Applications in heat transfer and fluid dynamics

- Design of porous structures for heat exchangers.

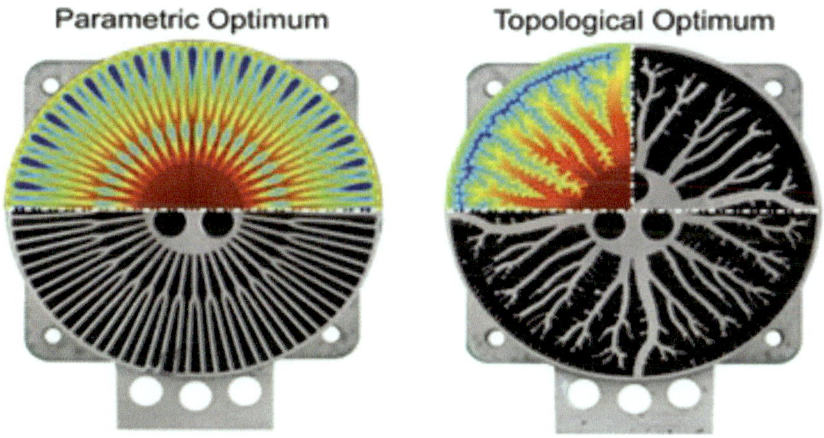

Figure 11.7 Example of a topology optimized heat exchanger design (heat sink).

- Optimization of cooling channels in gas turbine blades.
- Development of lightweight and efficient radiators.

11.5.5 Evolutionary algorithms

Evolutionary algorithms (EAs) are optimization techniques inspired by natural selection, where potential solutions evolve over successive generations to achieve better performance.

11.5.5.1 Types of evolutionary algorithms

- Genetic algorithms (GA): Uses selection, crossover, and mutation to evolve solutions.
- Particle swarm optimization (PSO): Models solution optimization based on swarm behavior.
- Ant colony optimization (ACO): Mimics ant foraging behavior to optimize path-based problems.

11.5.5.2 Advantages of evolutionary algorithms

- Capable of handling complex, multi-objective optimization problems.
- Does not require gradient information, making it suitable for non-differentiable problems.
- Can find global optima where traditional methods may get stuck in local minima (see Table 11.10).

Table 11.10 Comparison of evolutionary algorithms in heat transfer applications

Algorithm	Strengths	Common applications
Genetic algorithm	Handles large design spaces	Heat sink optimization
Particle swarm optimization	Fast convergence	Fluid flow optimization
Ant colony optimization	Suitable for network-based problems	Pipe routing optimization

11.5.5.3 Applications in heat transfer and fluid dynamics

- Optimization of microchannel heat sinks.
- Finding optimal flow distribution in heat exchangers.
- Improving aerodynamics in cooling systems.

11.5.6 Influence of additive manufacturing on optimization

The rise of additive manufacturing (AM) has dramatically influenced optimization strategies by removing many of the constraints imposed by traditional manufacturing processes.

11.5.6.1 Benefits of AM in optimized designs

- Enables fabrication of highly complex geometries.
- Reduces material waste by using only what is necessary.
- Allows for integration of multi-functional components in a single build.

Advanced optimization techniques such as parametric optimization, topology optimization, and evolutionary algorithms have revolutionized heat transfer and fluid dynamics applications. The integration of these methods with additive manufacturing further enhances design flexibility and efficiency. As computational power increases and AI-driven approaches become more prevalent, the future of optimization in engineering is set to reach new heights, enabling innovative and highly efficient designs.

11.6 Discussion

In this chapter, the importance of optimization in heat transfer and fluid dynamics has been discussed, emphasizing its role in enhancing efficiency, performance, and cost-effectiveness in various engineering applications. Industries such as aerospace, automotive, power generation, and electronics have been shown to benefit significantly from these techniques, as optimization enables better thermal management, reduced energy consumption, and improved system reliability. The necessity of optimization in heat exchangers, cooling systems, and other thermal-fluid applications has been highlighted, with a focus on the challenges associated with designing effective and efficient components. The evolution of computational methods in optimization has also been explored. It has been shown how the transition from traditional experimental and analytical techniques to advanced computational approaches, such as computational fluid dynamics (CFD), finite element method (FEM), and conjugate heat transfer (CHT), has revolutionized engineering design. These computational techniques have allowed complex problems to be solved with greater accuracy and efficiency. Additionally, the increasing integration of artificial intelligence (AI) and machine learning in optimization processes has been examined. Through AI-driven methods, design solutions have been enhanced by automating the optimization process, reducing computational time, and improving accuracy in predicting optimal configurations. A comprehensive overview of advanced optimization techniques has been provided, covering parametric optimization, topology optimization, and evolutionary algorithms. Each approach has been described in terms of its advantages over conventional methods. Parametric optimization has been discussed as a method where predefined design variables are adjusted to achieve optimal performance. Topology optimization has been introduced as a powerful approach that enables the generation of highly efficient designs by distributing material within a given space while respecting constraints. Evolutionary algorithms, including genetic algorithms and particle swarm optimization, have been explained as nature-inspired techniques capable of exploring a vast design space to identify optimal solutions. The influence of emerging technologies, such as additive manufacturing, on optimization strategies has been addressed. It has been shown that additive manufacturing has eliminated many traditional manufacturing constraints, enabling the realization of highly complex and optimized structures that were previously unattainable. A structured outline of the chapter has been presented, explaining how each section delves into specific optimization methods and their applications. The logical progression of the chapter has been designed to provide a clear understanding of how optimization techniques are applied in real-world engineering problems. Examples of various optimization case studies have been included, demonstrating their effectiveness in improving thermal-fluid systems. Tables and figures have been utilized throughout the chapter to enhance comprehension and provide visual representations of key concepts, methodologies, and comparative analyses. It has been emphasized that the application of optimization techniques in heat transfer and fluid dynamics continues to evolve. As new computational methods and advanced manufacturing technologies emerge, further improvements in thermal management and energy efficiency are expected. Research in the field has been shown to be ongoing, with continuous efforts being made to refine optimization algorithms, improve computational accuracy, and expand the applicability of these techniques to new and challenging engineering problems. The integration of AI and machine learning is anticipated to play an increasingly significant role in optimization processes, further automating design improvements and enabling faster and more efficient engineering solutions. In conclusion, the chapter has provided a detailed examination of optimization in heat transfer and fluid dynamics, the evolution of computational methods, and the impact of advanced optimization techniques on modern engineering applications. The necessity of optimization in various industries has been demonstrated, and the advantages of computational approaches over traditional methods have been established. Through the discussion of parametric optimization, topology optimization, and evolutionary algorithms, a deeper understanding of how modern techniques contribute to the design of more efficient and high-performance systems has been provided. Additionally, the role of additive manufacturing in shaping the future of optimization has been highlighted, showcasing how manufacturing constraints are being overcome to produce highly complex and efficient designs. Future advancements in optimization methodologies are expected to lead to even greater improvements in heat transfer and fluid dynamics applications. With the continuous development of computational power, AI-driven approaches, and additive manufacturing technologies, optimization strategies will become more sophisticated and accessible. Engineers and researchers will be able to develop more efficient, cost-effective, and innovative solutions to meet the growing demands of various industries. The insights presented in this chapter serve as a foundation for further exploration and application of optimization techniques in heat transfer and fluid dynamics, contributing to the advancement of engineering practices and technological progress.

References

[1] P. Bazyar et al. An overview of selected material properties in finite element modeling of the human femur. *Biomechanics*, 3(1):124–135, 2023.

[2] P. Bazyar et al. Finite element analysis of small-scale head of combine harvester for harvesting fine-grain products. *International Journal of Advanced Biological and Biomedical Research*, 8(4):340–358, 2020.

[3] P. Bazyar et al. Small-scale head of combine for harvesting sesame. *Agricultural Engineering*, 22(4):02, 2019.

[4] P. Bazyar and E. Sheidaee. Design and simulating lattice structures in the fe analysis of the femur bone. *Bioprinting*, 37:e00326, 2024.

[5] P. Bazyar and A. Baumgart. Effects of additional mechanisms on the performance of workshop crane. Technical report, 2022.

[6] S. Gholami et al. Effective parameters in the separation of pectin by vacuum membrane filtering. Technical report, 2023.

[7] J.J. KlemeÅą et al. Heat transfer enhancement, intensification and optimisation in heat exchanger network retrofit and operation. *Renewable and Sustainable Energy Reviews*, 120:109644, 2020.

[8] L. Meng et al. From topology optimization design to additive manufacturing: Today's success and tomorrow's roadmap. *Archives of Computational Methods in Engineering*, 27:805–830, 2020.

[9] Y. Zhang et al. Printing, folding and assembly methods for forming 3d mesostructures in advanced materials. *Nature Reviews Materials*, 2(4):1–17, 2017.

[10] I. Rahimi et al. A review on constraint handling techniques for population-based algorithms: from single-objective to multi-objective optimization. *Archives of Computational Methods in Engineering*, 30(3):2181–2209, 2023.

[11] V. Selvi and R. Umarani. Comparative analysis of ant colony and particle swarm optimization techniques. *International Journal of Computer Applications*, 5(4):1–6, 2010.

[12] G. Singh, A. Mehta, and H. Vasudev. Sustainability of additive manufacturing: a comprehensive review. *Progress in Additive Manufacturing*, 9(6):2249–2272, 2024.

[13] G. Ausiello et al. *Complexity and approximation: Combinatorial optimization problems and their approximability properties*. Springer Science & Business Media, 2012.

[14] S. Li et al. A state-of-the-art overview on the developing trend of heat transfer enhancement by single-phase flow at micro scale. *International Journal of Heat and Mass Transfer*, 143:118476, 2019.

[15] S. Yang and Y.F. Zhao. Additive manufacturing-enabled design theory and methodology: a critical review. *The International Journal of Advanced Manufacturing Technology*, 80:327–342, 2015.

[16] Y.-G. Lv et al. Thermal management technologies used for high heat flux automobiles and aircraft: A review. *Energies*, 15(21):8316, 2022.

[17] A.J. Juhasz. *Analysis and numerical optimization of gas turbine space power systems with nuclear fission reactor heat sources.* PhD thesis, Cleveland State University, 2005.

[18] W. Lin and B. Sunden. Vehicle cooling systems for reducing fuel consumption and carbon dioxide: literature survey. In *SAE 2010 International Powertrains, Fuels & Lubricants Meeting*, 2010.

[19] M. Jaszczur et al. Hydrogen production using high temperature nuclear reactors: Efficiency analysis of a combined cycle. *International Journal of Hydrogen Energy*, 41(19):7861–7871, 2016.

[20] V. Bianco, M. De Rosa, and K. Vafai. Phase-change materials for thermal management of electronic devices. *Applied Thermal Engineering*, 214:118839, 2022.

[21] D.T. Crane and G.S. Jackson. Optimization of cross flow heat exchangers for thermoelectric waste heat recovery. *Energy conversion and management*, 45(9-10):1565–1582, 2004.

[22] H. Cho, A.D. Smith, and P. Mago. Combined cooling, heating and power: A review of performance improvement and optimization. *Applied Energy*, 136:168–185, 2014.

[23] P.E. Ohenhen et al. Sustainable cooling solutions for electronics: A comprehensive review: Investigating the latest techniques and materials, their effectiveness in mechanical applications, and associated environmental benefits. *World Journal of Advanced Research and Reviews*, 21(1):957–972, 2024.

[24] P. Primiceri et al. Photo-induced ignition phenomenon of carbon nanotubes by xenon pulsed light: Ignition tests analysis, automotive and new potential applications, future developments. *Journal of applied research and technology*, 15(6):609–623, 2017.

[25] A. Husain and K.-Y. Kim. Shape optimization of micro-channel heat sink for micro-electronic cooling. *IEEE Transactions on Components and Packaging Technologies*, 31(2):322–330, 2008.

[26] B. Ramakrishnan et al. Cpu overclocking: A performance assessment of air, cold plates, and two-phase immersion cooling. *IEEE Transactions on Components, Packaging and Manufacturing Technology*, 11(10):1703–1715, 2021.

[27] A. Michalak et al. A thermal management design methodology for advanced power electronics utilizing genetic optimization and additive manufacturing techniques. In *2020 19th IEEE Intersociety Conference on Thermal and Thermomechanical Phenomena in Electronic Systems (ITherm)*. IEEE, 2020.

[28] T.-J. Lu. Thermal management of high power electronics with phase change cooling. *International journal of heat and mass transfer*, 43(13):2245–2256, 2000.

[29] O. Sarood et al. "cool" load balancing for high performance computing data centers. *IEEE Transactions on Computers*, 61(12):1752–1764, 2012.

[30] Q. Zhang et al. A survey on data center cooling systems: Technology, power consumption modeling and control strategy optimization. *Journal of Systems Architecture*, 119:102253, 2021.

[31] W.M. Shaban et al. Optimizing photovoltaic thermal solar systems efficiency through advanced artificial intelligence driven thermal management techniques. *Applied Thermal Engineering*, 247:123029, 2024.

[32] W. He et al. Performance optimization of server water cooling system based on minimum energy consumption analysis. *Applied Energy*, 303:117620, 2021.

[33] Q. Chang et al. Optimization control strategies and evaluation metrics of cooling systems in data centers: A review. *Sustainability*, 16(16):7222, 2024.

[34] M.H. Bargal et al. Thermohydraulic performance augmentation and heat transfer enhancement of automotive radiators using nano-coolants: a critical review. *Journal of Thermal Analysis and Calorimetry*, pages 1–53, 2025.

[35] R.A. Chaplin. Thermal power plants. In *Encyclopedia of Life Support Systems*. 2009.

[36] Z.-Q. Yu, M.-T. Li, and B.-Y. Cao. A comprehensive review on microchannel heat sinks for electronics cooling. *International Journal of Extreme Manufacturing*, 6(2):022005, 2024.

[37] C.-J. Zhao et al. A review of computational fluid dynamics for forced-air cooling process. *Applied Energy*, 168:314–331, 2016.

[38] L. Ventola et al. Unshrouded plate fin heat sinks for electronics cooling: Validation of a comprehensive thermal model and cost optimization in semi-active configuration. *energies*, 9(8):608, 2016.

[39] X. Chen et al. Multi-objective topology optimization design of silicon carbide metal oxide semiconductor field effect transistors power module liquid-cooled heatsink for electric vehicles. *Applied Thermal Engineering*, 254:123861, 2024.

[40] K.L. Kirsch and K.A. Thole. Experimental investigation of numerically optimized wavy microchannels created through additive manufacturing. In *Turbo Expo: Power for Land, Sea, and Air*. American Society of Mechanical Engineers, 2017.

[41] E.M. Dede. Optimization and design of a multipass branching microchannel heat sink for electronics cooling. *Journal of Electronic Packaging*, 134(4):041001, 2012.

[42] G. Liu, J.-J. Zhou, and J. Wang. Coefficients identification in electronic system cooling simulation through genetic algorithm. *Computers & Structures*, 80(1):23–30, 2002.

[43] K.K. Bodla, J.Y. Murthy, and S.V. Garimella. Optimization under uncertainty for electronics cooling design. In *ENCYCLOPEDIA OF THERMAL PACKAGING: Thermal Packaging Tools*, pages 233–265. World Scientific, 2015.

[44] G. Bramerdorfer et al. Modern electrical machine design optimization: Techniques, trends, and best practices. *IEEE Transactions on Industrial Electronics*, 65(10):7672–7684, 2018.

[45] A. Singh, S. Rangarajan, and B. Sammakia. Utilizing ragone framework for optimized phase change material-based heat sink design in electronic cooling applications. *International Journal of Heat and Mass Transfer*, 227:125518, 2024.

[46] Y. Gai et al. Cooling of automotive traction motors: Schemes, examples, and computation methods. *IEEE Transactions on Industrial Electronics*, 66(3):1681–1692, 2018.

[47] H.-J. Lin et al. Quantifying carbon emissions in cold chain transport: A real-world data-driven approach. *Transportation Research Part D: Transport and Environment*, 142:104679, 2025.

[48] I.O. Alade et al. Modeling thermal conductivity enhancement of metal and metallic oxide nanofluids using support vector regression. *Advanced Powder Technology*, 29(1):157–167, 2018.

[49] A. Afzal et al. Steady and transient state analyses on conjugate laminar forced convection heat transfer. *Archives of Computational Methods in Engineering*, 27(1):135–170, 2020.

[50] Y. Jaluria. Thermal processing of materials: from basic research to engineering. In *Heat Transfer Summer Conference*, 2003.

[51] M. Badshah, S. Badshah, and S. Jan. Comparison of computational fluid dynamics and fluid structure interaction models for the performance prediction of tidal current turbines. *Journal of Ocean Engineering and Science*, 5(2):164–172, 2020.

[52] B. Meyghani et al. A comparison of different finite element methods in the thermal analysis of friction stir welding (fsw). *Metals*, 7(10):450, 2017.

[53] G. Lee et al. Design optimization of heat exchanger using deep reinforcement learning. *International Communications in Heat and Mass Transfer*, 159:107991, 2024.

[54] C. Sun et al. A multi-time scales semi-decoupled cht (coupled heat transfer) model and its application on piston transient heat transfer simulation. *Applied Thermal Engineering*, 229:120548, 2023.

[55] L. Laveneziana et al. Conjugate heat transfer analysis of the aero-thermal impact of different feeding geometries for internal cooling in lifetime extension processes for heavy-duty gas turbines. *Energies*, 15(9):3022, 2022.

[56] X. Guillaud et al. Applications of real-time simulation technologies in power and energy systems. *IEEE Power and Energy Technology Systems Journal*, 2(3):103–115, 2015.

[57] H. Li et al. A review of the numerical strategies for solving ship hydroelasticity based on cfd-fem technology. *Ships and Offshore Structures*, 19(11):1912–1930, 2024.

[58] S.S. Bharadwaj and K.R. Sreenivasan. An introduction to algorithms in quantum computation of fluid dynamics. Technical Report 58, STO EN-AVT-377, 2022.

[59] M.S. Rahman. Computational fluid dynamics for predicting and controlling fluid flow in industrial equipment. *European Journal of Advances in Engineering and Technology*, 11(9):1–9, 2024.

[60] H. Alcocer-García et al. Multi-objective optimization of intensified processes for the purification of levulinic acid involving economic and environmental objectives. *Chemical Engineering and Processing-Process Intensification*, 136:123–137, 2019.

[61] G.F. Naterer. *Advanced heat transfer*. CRc Press, 2021.

[62] C. Diakaki, E. Grigoroudis, and D. Kolokotsa. Towards a multi-objective optimization approach for improving energy efficiency in buildings. *Energy and buildings*, 40(9):1747–1754, 2008.

[63] P. Constantin and C. FoiaÅ§. *Navier–stokes equations*. University of Chicago press, 1988.

[64] F. Bonetto and L. Rey-Bellet. Fourier law. Technical Report 20006, Atlanta, GA, 2006.

[65] M. Stynes. Steady-state convection-diffusion problems. *Acta Numerica*, 14:445–508, 2005.

[66] A. Fawaz et al. Topology optimization of heat exchangers: A review. *Energy*, 252:124053, 2022.

[67] S. An et al. Multifunctional metasurface design with a generative adversarial network. *Advanced Optical Materials*, 9(5):2001433, 2021.

[68] N.M. Khalfe, K.S. Lahiri, and K.S. Wadhwa. Simulated annealing technique to design minimum cost exchanger. *Chemical Industry and Chemical Engineering Quarterly*, 17(4):409–427, 2011.

[69] D. Wu and R. Wang. Combined cooling, heating and power: A review. *progress in energy and combustion science*, 32(5-6):459–495, 2006.

[70] C. MarugÃąn-Cruz et al. Heat transfer and thermal stresses in a circular tube with a non-uniform heat flux. *International Journal of Heat and Mass Transfer*, 96:256–266, 2016.

[71] M. Habib et al. Evaluation of flow maldistribution in air-cooled heat exchangers. *Computers & Fluids*, 38(3):677–690, 2009.

[72] A.R.J. Hussain et al. Review of polymers for heat exchanger applications: Factors concerning thermal conductivity. *Applied Thermal Engineering*, 113:1118–1127, 2017.

[73] F. Yeaple. *Fluid power design handbook*. CRC Press, 1995.

[74] W.S. Janna. *Engineering heat transfer*. CRC press, 2018.

[75] M.I. Hasan et al. Influence of channel geometry on the performance of a counter flow microchannel heat exchanger. *International Journal of Thermal Sciences*, 48(8):1607–1618, 2009.

[76] I.-S. Oh, J.-S. Lee, and B.-R. Moon. Hybrid genetic algorithms for feature selection. *IEEE Transactions on pattern analysis and machine intelligence*, 26(11):1424–1437, 2004.

[77] W.F. Abd-El-Wahed, A. Mousa, and M.A. El-Shorbagy. Integrating particle swarm optimization with genetic algorithms for solving nonlinear optimization problems. *Journal of Computational and Applied Mathematics*, 235(5):1446–1453, 2011.

[78] Z. Cui, J. Zeng, and G. Sun. A fast particle swarm optimization. *International Journal of Innovative Computing, Information and Control*, 2(6):1365–1380, 2006.

[79] D.J. Lohan, E.M. Dede, and J.T. Allison. Topology optimization for heat conduction using generative design algorithms. *Structural and Multidisciplinary Optimization*, 55:1063–1077, 2017.

[80] K.M. Hamdia, X. Zhuang, and T. Rabczuk. An efficient optimization approach for designing machine learning models based on genetic algorithm. *Neural Computing and Applications*, 33(6):1923–1933, 2021.

12

Exploring the Method of Fundamental Solutions for Nonhomogeneous and Nonlinear Partial Differential Equations

Ankit Farkya and Anirudh Singh Rana

Department of Mathematics, Birla Institute of Technology and Science Pilani, Pilani Campus, India

Abstract

In the past decades, meshfree methods gained significant interest in finding the numerical solutions of the governing partial differential equations (PDEs) that arise in engineering and computational mathematics. The main advantages of meshfree methods are their applicability to complex geometries without the need for mesh generation and their use of a set of points scattered in the computational domain and boundary, which reduces computational costs. Among meshfree methods, the method of fundamental solution (MFS) is a popular numerical method with simple programming implementation and a truly meshless nature. Classical MFS is limited to homogeneous and linearized problems. This chapter presents the application of the MFS for the nonhomogeneous and nonlinear partial differential equations that arise in complex engineering systems. First, we present the MFS to the nonhomogeneous boundary value problems and then we apply it to nonlinear problems including electrohydrodynamic flow and thermal explosion equation. The results are compared with the different methods and literature to check the accuracy of the solution.

Keywords: Partial differential equations, extended MFS, numerical solutions

DOI: 10.1201/9788743806691-12

12.1 Introduction

The method of fundamental solutions is a meshless numerical technique used to solve various boundary value problems (BVPs) in applied mathematics and engineering. It is also referred to as the charge simulation method [3] and the superposition method [23]. The MFS was first proposed by Kupradze and Aleksidze [24], who referred to it as the method of functional equations. Later, Mathon, and Johnston [25] developed the first numerical algorithm for its implementation. The MFS is particularly effective for solving problems governed by linear and homogeneous PDEs. By exploiting the linearity of the governing equation, the solution can be constructed by superimposing multiple fundamental solutions to satisfy the prescribed boundary conditions. This approach eliminates the need for domain discretization, which is typically required in conventional numerical methods. The MFS is closely related to the boundary element method (BEM) and is often regarded as a nonsingular BEM, as the fundamental solutions are placed on a fictitious boundary instead of the actual boundary of the domain. Unlike BEM, the MFS does not require the explicit evaluation of integrals over domain boundaries. In addition, being meshless and free from integral calculations, the MFS offers high accuracy, rapid convergence, and simple implementation. Consequently, it has been employed across diverse fields, including acoustics [4, 22], elastostatics [5, 30], Stokes flow [36, 34, 33], and wave scattering [12, 22]. Readers may refer to comprehensive review articles on the MFS [9] for further details.

The MFS was applied extensively by the research community to solve linear and homogeneous problems during its early development. For nonhomogeneous problems, it must be combined with other methods, such as radial basis functions (RBFs), to approximate the particular solution. For the nonhomogeneous problem, the accuracy of the solution depends on the approximate particular solution. Over the past few decades, significant advancements have been made to extend the MFS to handle more complex problems. In the 1990s, Goldberg and Chen introduced an extension of the MFS for nonhomogeneous problems using radial basis functions. The RBFs are very popular in the meshless community for their simple implementation, high accuracy, and strong mathematical foundation. Radial basis functions are widely employed in neural networks [18], machine learning [6], and data processing [37]. The origins of RBFs can be traced back to Hardy, who introduced the first RBF, known as the multiquadrics (MQ), for surface fitting on irregular surfaces [17]. Later, Kansa [19, 20] extended the application of MQ beyond surface fitting by employing it in the RBF collocation method to solve various types of PDEs, leading to the development of what is now known as the Kansa method. In addition to MQ, several other radial basis functions have been developed, including thin-plate splines (TPS) [11], polyharmonic splines [27], Gaussian [14], and inverse multiquadrics (IMQ) [10]. Another approach presented in [2] for solving nonhomogeneous Helmholtz and Poisson equations involves the approximation of the nonhomogeneous term using the fundamental solutions of the Helmholtz operator. The coefficients obtained from the function approximation are rescaled to compute a particular solution, which reduces the computational cost and avoids the need for a closed-form analytical expression.

The dual reciprocity method (DRM) [29] and multiple reciprocity method (MRM) [28] are classical techniques for obtaining particular solutions to such problems. In DRM, the nonhomogeneous term is approximated by a linear combination of basis functions, which requires interior points. However, MRM does not require interior nodes and remains a boundary-type method. However, MRM is computationally expensive due to the complexity of the interpolation matrices needed for the recursive application of fundamental solutions. The MFS with these methods has been widely used to address various nonlinear problems [21, 32, 13, 1]. The MFS is combined with these methods using the Picard iteration method to address the nonlinear problems. The MFS has also gained significant attention for solving nonhomogeneous and nonlinear systems, such as the Navier–Stokes equations [7, 31] and advection-diffusion equations [35].

In the present contribution, we employ the MFS to electrohydrodynamic (EHD) flow and thermal explosion problems, both governed by nonlinear equations. The numerical solutions obtained by the MFS are combined with a fixed-point iteration scheme and the method of particular solutions. At each iteration, the linear nonhomogeneous equation is solved, which is further split into the general and particular solutions. Numerical convergence is observed, and results for different parameters are obtained and validated against numerical results from mesh-based methods and existing literature.

12.2 MFS for Nonhomogeneous Problems

Consider the nonhomogeneous boundary value problem:

$$\mathcal{L}w(\mathbf{x}) = f(\mathbf{x}), \quad \mathbf{x} \in \Omega, \tag{12.1}$$

$$\mathcal{B}w(\mathbf{x}) = h(\mathbf{x}), \quad \mathbf{x} \in \Gamma, \tag{12.2}$$

where Ω is a connected and bounded domain in $\mathbb{R}^n$, $\mathcal{L}$ is the linear differential operator, and $\mathcal{B}$ is the boundary operator. The complete solution w of eqn (12.1)–(12.2) is split into general solution w_{H} and particular solution w_{P} such that:

$$\mathcal{L}w_{\mathrm{P}}(\mathbf{x}) = f(\mathbf{x}), \quad \mathbf{x} \in \Omega, \tag{12.3}$$

and the general solution satisfies the governing equation and boundary conditions:

$$\mathcal{L}w_{\mathrm{H}}(\mathbf{x}) = 0, \qquad \mathbf{x} \in \Omega, \tag{12.4}$$

$$\mathcal{B}w_{\mathrm{H}}(\mathbf{x}) = h(\mathbf{x}) - \mathcal{B}w_{\mathrm{P}}(\mathbf{x}). \quad \mathbf{x} \in \Gamma. \tag{12.5}$$

The homogeneous problem is solved using the classical MFS once the particular solution is obtained. Various radial basis function approaches are available for computing particular solutions. Detailed information on radial basis functions and collocation methods can be found in Ref. [8]. To find the particular solution of the nonhomogeneous problems, first, the nonhomogeneous term f is approximated using the basis functions ϕ_{i}, where $1 \leq i \leq M$ for the chosen M points in Ω, such that:

$$f(\mathbf{x}) = \sum_{i=1}^{M} a_{\mathrm{i}}\phi_{\mathrm{i}}(\mathbf{x}), \quad \mathbf{x} \in \Omega, \tag{12.6}$$

once the unknown coefficients a_{i} are determined uniquely the particular solution w_{p} can be approximated as

$$w_{\mathrm{p}}(\mathbf{x}) = \sum_{i=1}^{M} a_{\mathrm{i}}\Phi_{\mathrm{i}}(\mathbf{x}), \tag{12.7}$$

such that:

$$\mathcal{L}\Phi_{\mathrm{i}}(\mathbf{x}) = \phi_{\mathrm{i}}(\mathbf{x}), \quad \mathbf{x} \in \Omega, \tag{12.8}$$

where Φ_{i} is the particular solution corresponding to the RBF ϕ_{i}. It is important to note that the method for finding the particular solution depends on the availability of a closed-form expression for Φ_{i} corresponding to the radial basis function ϕ_{i}. For radial basis functions such as the Gaussian $e^{-(r^2/c^2)}$ or the multiquadrics $\sqrt{r^2 + c^2}$ with $c > 0$, known as the shape parameter, the resulting coefficient matrix is invertible. However, for RBFs such as thin plate splines $r^2 \ln r$, the matrix may not be invertible. The particular solutions for the corresponding radial basis function, that is thin plate spline $\phi(r) = r^2 \ln r$ for the Helmholtz operator $\mathcal{L} = (\nabla^2 - \lambda)$ in $\mathbb{R}^2$ are as follows:

$$\Phi(r) = \begin{cases} -\frac{r^2 \ln r}{\lambda} - \frac{4}{\lambda^2}(1 + \ln r + K_0(\sqrt{\lambda} r), \ r \neq 0, \\ -\frac{4}{\lambda^2} + \frac{4\gamma}{\lambda^2} + \frac{4}{\lambda^2} \ln\left(\frac{\sqrt{\lambda}}{2}\right), \ r = 0. \end{cases} \tag{12.9}$$

Another approach suggested in [2] for the Poisson and nonhomogeneous Helmholtz equations, where the nonhomogeneous function is approximated by the fundamental solutions of the Helmholtz equation and some set of frequency and it can be easily combined with classical MFS. The Laplace operator is the special case of the Helmholtz operator $\mathcal{L} = \Delta - \lambda$ for $\lambda = 0$. Hence in this approach, the function $f(\mathbf{x})$ and the particular solution $w_{\mathrm{P}}(\mathbf{x})$ are approximated with a linear combination of fundamental solutions of Helmholtz equation $\phi_{\lambda_{\mathrm{r}}}(\mathbf{x}, \mathbf{x}_{\mathrm{s}})$ where $\mathbf{x}_{\mathrm{s}}$ is the source point (also referred to as the singularity point) and $\{\lambda_{\mathrm{r}} : r \in \mathbb{N}, 1 \leq r \leq q\}$ is the selected frequency set [2]:

$$f(\mathbf{x}) = \sum_{r=1}^{q} \sum_{j=1}^{n_{\mathrm{s}}} a_{\mathrm{rj}} \phi_{\lambda_{\mathrm{r}}}(\mathbf{x}, \mathbf{x}_{\mathrm{s}}^{j}), \tag{12.10}$$

where $\lambda_{\mathrm{r}} \neq \lambda_{\mathrm{j}}$ for $r \neq j$.

$$w_{\mathrm{P}}(\mathbf{x}) = \sum_{r=1}^{q} \sum_{j=1}^{n_{\mathrm{s}}} \frac{a_{\mathrm{rj}}}{\lambda_{\mathrm{r}} - \lambda} \phi_{\lambda_{\mathrm{r}}}(\mathbf{x}, \mathbf{x}_{\mathrm{s}}^{j}), \tag{12.11}$$

where $\lambda_{\mathrm{r}} \neq \lambda$ for indexing $r = 1, ..., q$ and the particular solution for Poisson's equation ($\lambda = 0$) is given by

$$w_{\mathrm{P}}(\mathbf{x}) = \sum_{r=1}^{q} \sum_{j=1}^{n_{\mathrm{s}}} \frac{a_{\mathrm{rj}}}{\lambda_{\mathrm{r}}} \phi_{\lambda_{\mathrm{r}}}(\mathbf{x}, \mathbf{x}_{\mathrm{s}}^{j}). \tag{12.12}$$

The obtained values of coefficients a_{rj} for approximating the $f(\mathbf{x})$ are rescaled for the particular solution with the factor $1/(\lambda_{\mathrm{r}} - \lambda)$ and $1/\lambda_{\mathrm{r}}$ for Helmholtz and Poisson's equations, respectively. The general solution of the homogeneous equation is obtained using the classical MFS by utilizing the fundamental solution of the governing equation:

$$w_{\mathrm{H}}(\mathbf{x}) = \sum_{j=1}^{n_{\mathrm{s}}} a_{\mathrm{r}} \phi(\mathbf{x}, \mathbf{x}_{\mathrm{s}}^{j}). \tag{12.13}$$

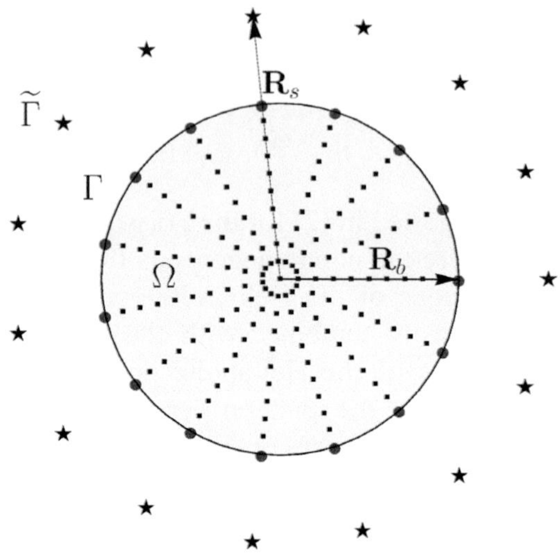

Figure 12.1 Schematic for the illustration of the extended MFS with boundary nodes (dots) on Γ, singularity points (stars) on the fictitious boundary $\widetilde{\Gamma}$, and some interior points in the domain Ω.

Once the value of a_r is obtained, the solution can be computed within the domain. Hence, for the nonlinear Helmholtz or Laplace operator, the above process can be readily extended by solving the linear nonhomogeneous equation at each iteration:

$$\mathcal{L}(w^{(n+1)}) = f(w^{(n)}), \qquad \text{for n} = 0,1,2,\ldots \tag{12.14}$$

The linear nonhomogeneous problem is solved by splitting it into finding the general and particular solutions. The particular solution is obtained using eqn (12.11) and (12.12) for the nonhomogeneous Helmholtz and Poisson equations, respectively. The general solution is obtained using eqn (12.13). Figure 12.1 depicts the schematic representation of the implementation of the MFS.

12.3 Application

1. *Electrohydrodynamic flow in a circular conduit*
   The EHD flow of a fluid in an "ion drag" configuration in a circular cylindrical conduit is governed by the following equation [26]:

$$\Delta w + H^2 \left(1 - \frac{w}{1 - \alpha w}\right) = 0, \tag{12.15}$$

$$w(x, y) = 0, \quad \text{on} \quad (x, y) \in \partial\Omega, \tag{12.16}$$

$$w'(x, y) = 0, \quad \text{on} \quad (x, y) = (0, 0), \tag{12.17}$$

where $w(x, y)$ is the velocity of the fluid, H is the Hartmann electric number, and α determines the strength of the nonlinearity:

$$H = \sqrt{\frac{j_0 a^2}{\mu K^2 E_0}}, \qquad \alpha = \frac{\partial p}{\partial z} \frac{K}{j_0} - 1. \tag{12.18}$$

The nonlinear parameter α can also be expressed as

$$\alpha = \frac{\frac{\partial p}{\partial z}}{j_0 / K} - 1. \tag{12.19}$$

The nonlinear parameter α characterizes the ratio of the pressure-induced force to the electric force exerted on the fluid. Here, j_0 is the uniform current density at the inlet, K is the mobility of ions, a is the radius of the conduit, and E_0 is the applied electrical field. Further technical details are available in [26, 13].

2. *Thermal explosion*
The modified form of the Lane–Emden equation of the second kind often used to model the thermal explosion in an infinite cylinder and a sphere. Consider the steady heat balance equation:

$$k\nabla^2 T + \rho Q A \exp\left(-\frac{E}{RT}\right) = 0, \tag{12.20}$$

where T is the temperature, Q is the heat of reaction, k is the thermal conductivity, ρ is the density, E is the energy of activation , R is the universal gas constant, and A is the frequency factor. Introducing the dimensionless variables as

$$\theta = \frac{E}{RT_0^2}(T - T_0), \tag{12.21}$$

where T_0 is the ambient temperature.

$$\nabla^2 \theta + \delta \exp\left(-\frac{\theta}{1 + \epsilon\theta}\right) = 0. \tag{12.22}$$

Define:

$$\delta = \frac{\rho Q A E}{k R T_0^2} \exp\left(-\frac{E}{R T_0^2}\right), \quad \epsilon = \frac{R T_0}{E}, \tag{12.23}$$

where δ is the Frank-Kamenetskii parameter and ϵ is the activation energy parameter. Boundary conditions for the above problem in a cylindrical geometry are:

$$\theta(x,y) = 0, \quad \text{on} \quad (x,y) \in \partial\Omega, \tag{12.24}$$

$$\theta'(x,y) = 0, \quad \text{on} \quad (x,y) = (0,0). \tag{12.25}$$

Both of the aforementioned problems are addressed by the following algorithm. The Algorithm for solving the nonlinear partial differential equation using extended MFS together with fixed-point iteration is as follows:

Algorithm

Algorithm 1:

Input: $R_b, R_s, n_b, n_s, n_l, \lambda_r, n, \epsilon, \delta, \alpha, H^2$
Output: w_P, w_H, w
Initial guess: $w^{(0)} = 0$

$f(\mathbf{x}) \leftarrow \sum_{r=1}^{q} \sum_{j=1}^{n_s} a_{rj} \phi_{\lambda_r}(\mathbf{x}, \mathbf{x}_s^j)$ ▷ Find a_{rj}

$w_P(\mathbf{x}) \leftarrow \sum_{r=1}^{q} \sum_{j=1}^{n_s} \frac{a_{rj}}{\lambda_r - \lambda} \phi_{\lambda_r}(\mathbf{x}, \mathbf{x}_s^j)$

$w_H(\mathbf{x}) \leftarrow \sum_{j=1}^{n_s} c_j \phi_{\lambda}(\mathbf{x}, \mathbf{x}_s^j)$ ▷ Find c_j

$w \leftarrow w_H + w_P$
for $i = 1$ **to** n **do**
  Update $f(\mathbf{x})$
  Compute $w_P(\mathbf{x}), w_H(\mathbf{x})$
  $w \leftarrow w_H + w_P$
end
Error $\leftarrow \|w^{(n+1)} - w^{(n)}\|_\infty$
if *Error* $\leq$ *tol* **then**
  Stop
end
else
  Increase n
end

12.4 Numerical Results

Figure 12.2 illustrates the effect of the activation energy parameter $\epsilon = 0, 1$, and 5 for a fixed value of $\delta = 1$ in the thermal explosion problem, which is described by the BVP (12.22)–(12.25). As values of ϵ increase, the solution decreases [13].

Figure 12.3 depicts the influence of different values of $\delta = 0.5, 1$, and 3 for a fixed value of $\epsilon = 1$ in the thermal explosion problem. As values of δ increase, the solution increases. The numerical solutions obtained by the MFS are compared with the finite element-based solver in Mathematica®. For MFS, parameters are taken as $n_{\mathrm{b}} = n_{\mathrm{s}} = 30$, $R_{\mathrm{b}} = 1$, and $R_{\mathrm{s}} = 3R_{\mathrm{b}}$ and interior points $n_{\mathrm{I}} = 361$. The chosen frequency set is $r^2 : 2 \leq r \leq 21$, ensuring $\lambda_{\mathrm{r}} \neq \lambda$. The MFS solutions are presented by solid lines and the FEM solutions are presented by dots.

Figures 12.4 and 12.5 show the iterative convergence using the extended MFS in conjunction with the fixed point iteration scheme by depicting the maximum absolute error between each iteration. Figure 12.4 shows the iterative convergence for the fixed value of $\epsilon = 1$ and varying the values of δ, and Figure 12.5 depicts the same for different values of $\epsilon = 0.5, 2$, and 5. The corresponding L^2 error in the MFS and FEM for the nonlinear thermal explosion problem in a cylinder is presented in Table 12.1.

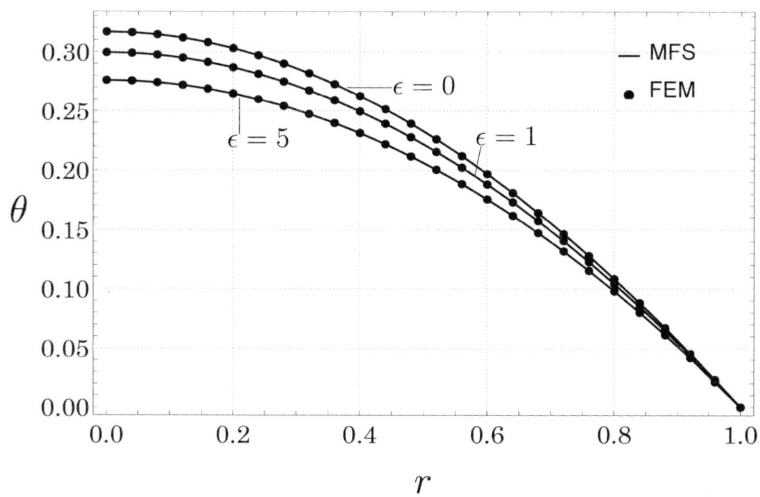

Figure 12.2 This figure depicts the solution θ in the radial direction for the boundary value problem governing the thermal explosion in an infinite cylinder eqn (12.22)–(12.25), with varying values of ϵ and a fixed value of $\delta = 1$.

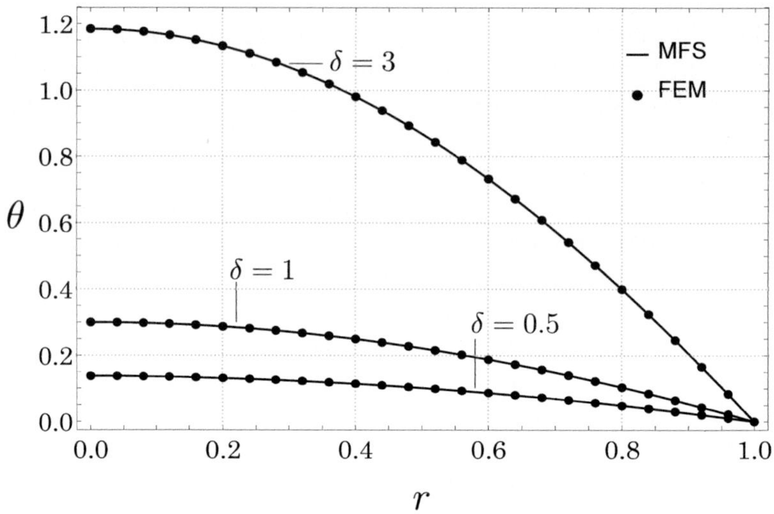

Figure 12.3 This figure illustrates the solution θ for the boundary value problem governing the thermal explosion in an infinite cylinder eqn (12.22)–(12.25), with varying the values of δ and a fixed value of $\epsilon = 1$.

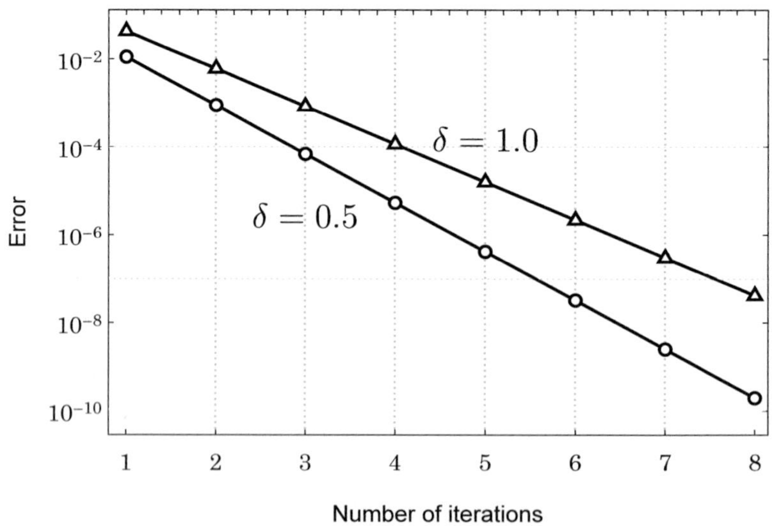

Figure 12.4 The figure shows the log plot of error at each iteration for the nonlinear problem of thermal explosion in a cylinder having variations in δ with a fixed value of $\epsilon = 1$.

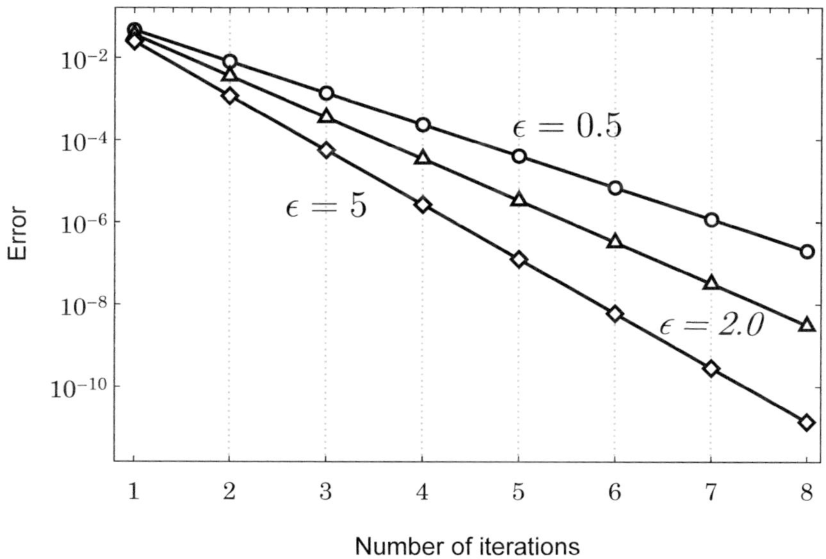

Figure 12.5 The figure presents the log plot of error at each iteration for nonlinear problem thermal explosion equation in a cylinder by varying ϵ with $\delta = 1$.

Table 12.1 This table shows the L^2 error in the numerical solution using the MFS and the FEM for the BVP (12.22)–(12.25) for three different values of $\epsilon = 0, 1$ and 5, with fixed values of $\delta = 0.5$ and 1.

ϵ	$\boldsymbol{\delta = 1}$	$\boldsymbol{\delta = 0.5}$
0	2.73843×10^{-5}	7.58790×10^{-6}
1	2.43123×10^{-5}	7.33276×10^{-6}
5	1.84926×10^{-5}	6.62534×10^{-6}

Figures 12.6 and 12.7 present the numerical solution of the radial velocity w for the EHD flow in a circular conduit, governed by (12.15)–(12.17). The solution is obtained using the extended MFS combined with the fixed-point iteration scheme. As the parameter α increases, the velocity decreases.

Figures 12.8 and 12.9 illustrate the velocity profile for the boundary value problem (12.15)–(12.17). Numerical results using the MFS are compared with the results provided in [16] for different values of Hartmann number with the fixed values of $\alpha = 1$ and $\alpha = 0.5$. The relative percentage error between the extended MFS with the least square method (LSM) given in the [16]

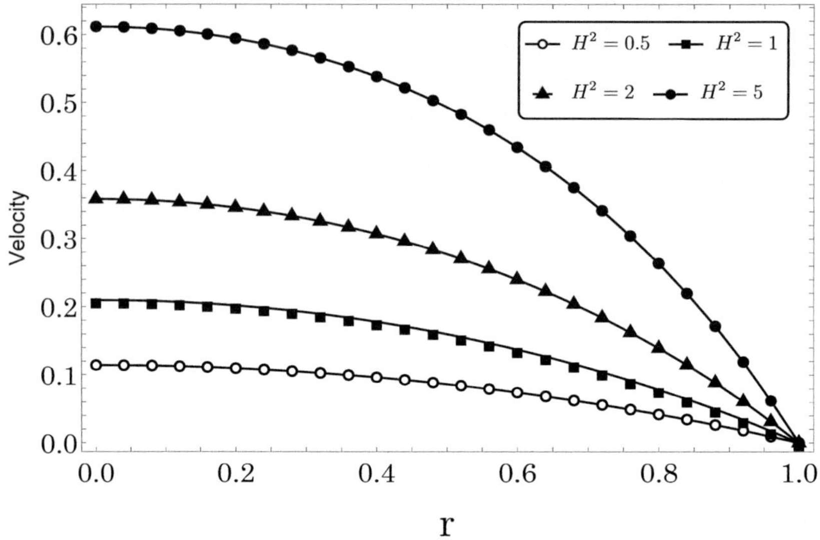

Figure 12.6 This figure displays the numerical solution w for the problem (12.15)–(12.17) in the r-direction by varying the values of H^2 with $\alpha = 0.1$.

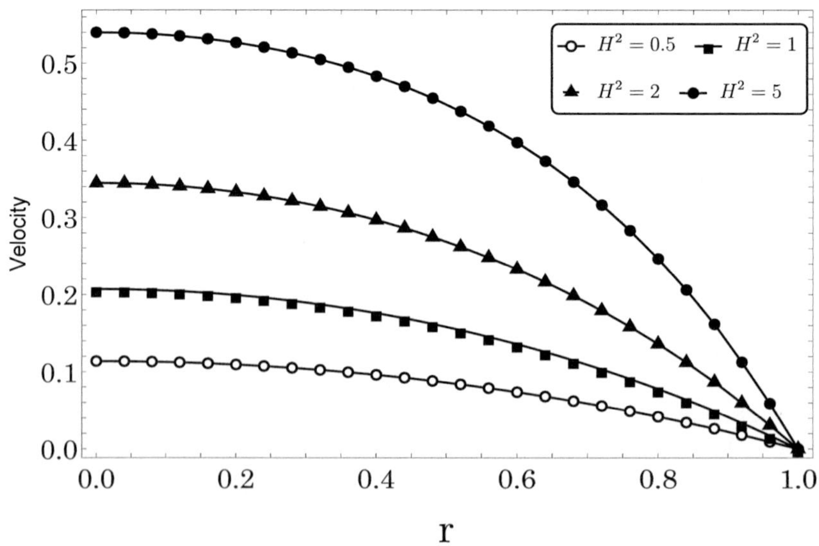

Figure 12.7 This figure shows the solution w for the problem (12.15)–(12.17) in the r-direction by varying the values of H^2 with $\alpha = 0.5$.

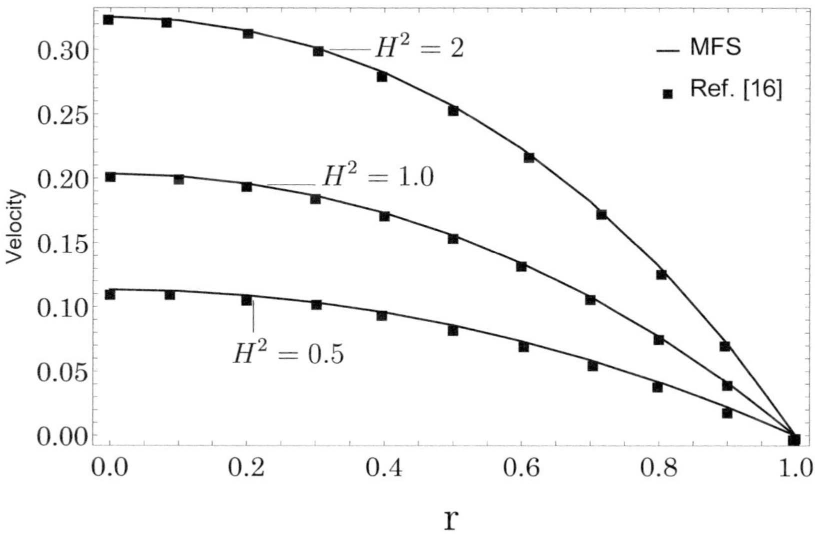

Figure 12.8 This figure compares the solution w obtained using the MFS with the solutions presented in Ref. [16] for the EHD problem by varying H^2 with a fixed value of $\alpha = 1$.

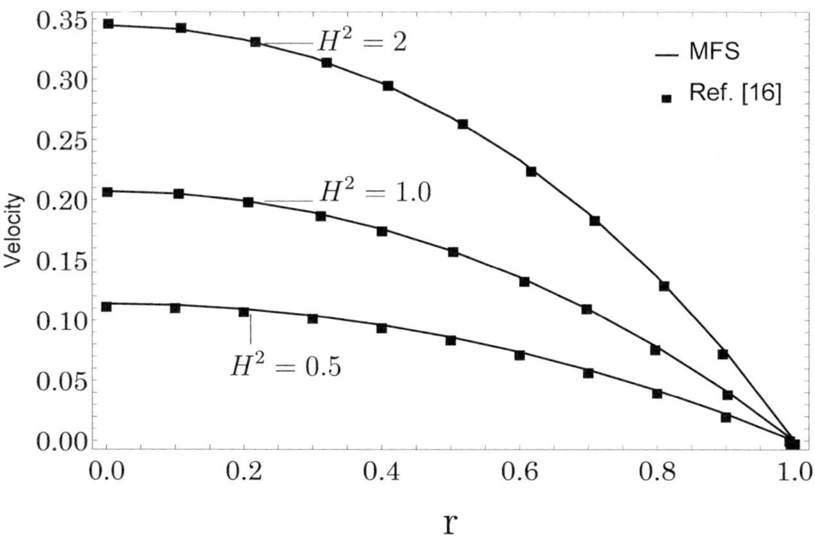

Figure 12.9 This figure compares the solution w obtained using the MFS with the solutions presented in Ref. [16] for the EHD problem by varying H^2 with a fixed value of $\alpha = 0.5$.

are detailed in Table 12.2. The iterative convergence for the EHD problem using the extended MFS is presented in Figure 12.10. For small nonlinearity solution converges in a fewer iteration.

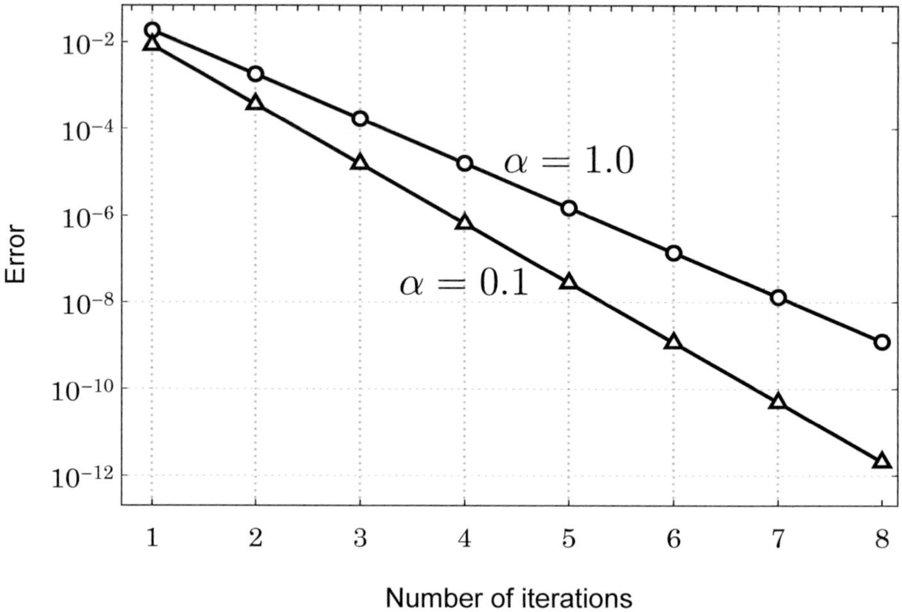

Figure 12.10 This figure demonstrates the log plot of error at each iteration for EHD flow in a circular conduit for different values of nonlinear parameter α with $H^2 = 2$.

Table 12.2 This table presents the relative % error in the velocity $w(r)$ for the boundary value problem governed by (12.15)–(12.17) with $H^2 = \alpha = 0.5$ using the MFS and solution provided in Ref. [16].

r	Ref. [16]	MFS	Relative % error
0.0	0.11374658	0.11376511	1.62906×10^{-2}
0.1	0.11264675	0.11267283	2.3152×10^{-2}
0.2	0.10934293	0.10938835	4.1539×10^{-2}
0.3	0.10382144	0.10388956	6.56127×10^{-2}
0.4	0.09605922	0.096142583	8.67829×10^{-2}
0.5	0.08602397	0.086105701	9.50096×10^{-2}
0.6	0.07367438	0.073734103	8.10635×10^{-2}
0.7	0.05896029	0.058983946	4.01219×10^{-2}
0.8	0.04182281	0.041813351	2.26168×10^{-2}
0.9	0.02219461	0.022176004	6.50158×10^{-2}
1.0	0.0	8.1508131×10^{-9}	–

12.5 Conclusion

In this chapter, we present the application of the MFS for nonhomogeneous and nonlinear partial differential equations. The MFS is combined with a fixed-point iteration scheme, where a linear nonhomogeneous boundary value problem is solved at each iteration, and convergence is checked. For small nonlinearities in both problems, the solution converges within fewer iterations. Numerical results are obtained using the MFS for different parameters and are further compared with known solutions from the literature and the other numerical methods.

References

[1] Carlos J.S. Alves and Ana L. Silvestre. On the application of the method of fundamental solutions to nonlinear partial differential equations. *Eng. Anal. Bound. Elem.*, 92:267–274, 2018.

[2] CJS Alves and CS Chen. A new method of fundamental solutions applied to nonhomogeneous elliptic problems. *Adv. Comput. Math.*, 23:125–142, 2005.

[3] K. Amano. A charge simulation method for the numerical conformal mapping of interior, exterior and doubly-connected domains. *J. Comput. Appl. Math.*, 53(3):353–370, 1994.

[4] Julieta António, Antonio Tadeu, and Luís Godinho. A three-dimensional acoustics model using the method of fundamental solutions. *Eng. Anal. Bound. Elem.*, 32(6):525–531, 2008.

[5] Omar Askour, Abdeljalil Tri, Bouazza Braikat, Hamid Zahrouni, and Michel Potier-Ferry. Method of fundamental solutions and high order algorithm to solve nonlinear elastic problems. *Eng. Anal. Bound. Elem.*, 89:25–35, 2018.

[6] Christopher JC Burges. A tutorial on support vector machines for pattern recognition. *Data Min. Knowl. Discov.*, 2(2):121–167, 1998.

[7] C.A. Bustamante, H. Power, and W.F. Florez. A global meshless collocation particular solution method for solving the two-dimensional navier–stokes system of equations. *Comput. Math. Appl.*, 65(12):1939–1955, 2013.

[8] Wen Chen, Zhuo-Jia Fu, and Ching-Shyang Chen. *Recent advances in radial basis function collocation methods*, volume 768. Springer, 2014.

[9] Alexander H.D. Cheng and Yongxing Hong. An overview of the method of fundamental solutions–solvability, uniqueness, convergence, and stability. *Eng. Anal. Bound. Elem.*, 120:118–152, 2020.

[10] Maggie E Chenoweth and Scott A Sarra. A numerical study of generalized multiquadric radial basis function interpolation. *SIAM Undergraduate Research Online*, 2(2):58–70, 2009.

[11] Jean Duchon. Splines minimizing rotation-invariant semi-norms in sobolev spaces. In *Constructive theory of functions of several variables: Proceedings of a conference held at Oberwolfach April 25–May 1, 1976*, pages 85–100. Springer, 1977.

[12] Graeme Fairweather, Andreas Karageorghis, and Paul A Martin. The method of fundamental solutions for scattering and radiation problems. *Eng. Anal. Bound. Elem.*, 27(7):759–769, 2003.

[13] Ankit Farkya and Anirudh Singh Rana. A novel meshfree method for nonlinear equations in flow through porous media and electrohydrodynamic flows. *J. Comput. Theor. Transp.*, 53(5):335–348, 2024.

[14] Bengt Fornberg, Elisabeth Larsson, and Natasha Flyer. Stable computations with gaussian radial basis functions. *SIAM J. Sci. Comput.*, 33(2):869–892, 2011.

[15] David Al'bertovich Frank-Kamenetskii. *Diffusion and heat transfer in chemical kinetics*. Plenum Press, 1969.

[16] S.E. Ghasemi, M. Hatami, GH.R. Mehdizadeh Ahangar, and D.D. Ganji. Electrohydrodynamic flow analysis in a circular cylindrical conduit using least square method. *J. Electrost.*, 72(1):47–52, 2014.

[17] Rolland L. Hardy. Multiquadric equations of topography and other irregular surfaces. *J. Geophys. Res.*, 76:1905–1915, 1971.

[18] Young-Sup Hwang and Sung-Yang Bang. An efficient method to construct a radial basis function neural network classifier. *Neural Netw.*, 10(8):1495–1503, 1997.

[19] E.J. Kansa. Multiquadric–a scattered data approximation scheme with applications to computational fluid-dynamics–i surface approximations and partial derivative estimates. *Comput. Math. Appl.*, 19:127–145, 1990.

[20] E.J. Kansa. Multiquadrics–a scattered data approximation scheme with applications to computational fluid-dynamics–ii solutions to parabolic, hyperbolic and elliptic partial differential equations. *Comput. Math. Appl.*, 19:147–161, 1990.

[21] Andreas Karageorghis and G Fairweather. The Method of fundamental solutions for the solution of nonlinear plane potential problems. *IMA J. Numer. Anal.*, 9:231–242, 04 1989.

[22] Prasanna S Kondapalli, David J Shippy, and Graeme Fairweather. Analysis of acoustic scattering in fluids and solids by the method of fundamental solutions. *J. Acoust. Soc. Am.*, 91(4):1844–1854, 1992.

[23] Gary H. Koopmann, Limin Song, and John B. Fahnline. A method for computing acoustic fields based on the principle of wave superposition. *J. Acoust. Soc. Am.*, 86(6):2433–2438, 12 1989.

[24] V.D. Kupradze and M.A. Aleksidze. The method of functional equations for the approximate solution of certain boundary value problems. *USSR Comput. Math. Math. Phys.*, 4:82–126, 1964.

[25] Rudolf Mathon and R. L. Johnston. The approximate solution of elliptic boundary-value problems by fundamental solutions. *SIAM J. Numer. Anal.*, 14(4):638–650, 1977.

[26] S. McKee, R. Watson, J. A. Cuminato, J. Caldwell, and M. S. Chen. Calculation of electrohydrodynamic flow in a circular cylindrical conduit. *ZAMM - Journal of Applied Mathematics and Mechanics*, 77(6):457–465, 1997.

[27] AS Muleshkov, MA Golberg, and CS Chen. Particular solutions of helmholtz-type operators using higher order polyhrmonic splines. *Comput. Mech.*, 23(5):411–419, 1999.

[28] Andrzej J Nowak and Paul W Partridge. Comparison of the dual reciprocity and the multiple reciprocity methods. *Eng. Anal. Bound. Elem.*, 10(2):155–160, 1992.

[29] Paul William Partridge, Carlos Alberto Brebbia, et al. *Dual reciprocity boundary element method*. Springer Science & Business Media, 2012.

[30] D Redekop. Fundamental solutions for the collation method in planar elastostatics. *Appl. Math. Model.*, 6(5):390–393, 1982.

[31] Shantanu Shahane, Anand Radhakrishnan, and Surya Pratap Vanka. A high-order accurate meshless method for solution of incompressible fluid flow problems. *J. Comput. Phys.*, 445:110623, 2021.

[32] A. Tri, H. Zahrouni, and M. Potier-Ferry. Perturbation technique and method of fundamental solution to solve nonlinear poisson problems. *Eng. Anal. Bound. Elem.*, 35:273–278, 2011.

[33] Chia-Cheng Tsai, DL Young, CM Fan, and CW Chen. Mfs with time-dependent fundamental solutions for unsteady stokes equations. *Eng. Anal. Bound. Elem.*, 30(10):897–908, 2006.

[34] D.L. Young, C.W. Chen, C.M. Fan, K. Murugesan, and C.C. Tsai. The method of fundamental solutions for stokes flow in a rectangular cavity with cylinders. *Eur. J. Mech. B/Fluids*, 24(6):703–716, 2005.

[35] DL Young, CM Fan, CC Tsai, CW Chen, and K Murugesan. Eulerian-lagrangian method of fundamental solutions for multi-dimensional advection-diffusion equation. In *International Mathematical Forum*, 2006.

[36] DL Young, SJ Jane, CM Fan, Krishnan Murugesan, and CC Tsai. The method of fundamental solutions for 2d and 3d stokes problems. *J. Comput. Phys.*, 211(1):1–8, 2006.

[37] Barbara Zitova and Jan Flusser. Image registration methods: a survey. *Image Vis. Comput.*, 21(11):977–1000, 2003.

13

A Fractional Order Taylor Wavelets Approach for the Numerical Simulation of Fractional Variational Problems

Monika Anand[1] and Vineet Kishore Srivastava[2]

[1]Department of Sciences, Quantum University, Roorkee–247667, Uttarakhand, India
[2]Department of Mathematics, School of Life and Applied Sciences, Uttaranchal University, India
E-mail: monikaanand895@gmail.com; vk_srivastava78@yahoo.co.in

Abstract

In this study, the fractional order Taylor wavelets approach is implemented to solve the fractional variational problems. The proposed approach employs the Caputo derivative together with the Riemann–Liouville integral in its fractional form. A key feature of this work is the use of Taylor wavelets of fractional order, which offers high accuracy due to their localized basis functions, making them well-suited for approximating solutions to fractional variational problems. Additionally, the Lagrange multipliers approach is utilized to merge the constraints of these problems with the performance index, after which the model is transformed into a set of algebraic equations. The proposed approach is applied to solve some illustrative examples, and then the results are presented through tables and graphs. The achieved outcomes demonstrated its efficiency in solving fractional variational problems, showing excellent agreement with the exact solutions. The present work offers improved accuracy and effectiveness when compared to other existing approaches reported in the literature.

Keywords: Fractional calculus, fractional order Taylor wavelets, fractional variational problems, Lagrange multipliers, convergence analysis

DOI: 10.1201/9788743806691-13

13.1 Introduction

Fractional calculus is considered a crucial field of mathematics that generalizes the concept of differentiation and integration to non-integer orders. It has been increasingly applied to real-world processes in recent years [1, 2]. Fractional calculus involves several types of fractional derivatives and integrals. Therefore, selecting the appropriate fractional operator for a specific problem is one of the primary challenges. Nowadays, numerous types of fractional operators, both with and without singular kernels, are currently available. Some of them are Caputo, Hilfer, Riemann-Liouville, Caputo–Fabrizio, Atangana–Baleanu, Grunwald–Letnikov, etc. The numerical computation of fractional integrals and derivatives is essential for understanding fractional calculus and solving fractional differential equations (FDEs) [3]. The study of FDEs has expanded considerably in recent years because of their wide applications in science and engineering. Their non-locality and non-commutativity make them particularly effective in modeling memory effects and hereditary characteristics of various physical phenomena [4].

The incorporation of fractional derivatives has led to the emergence of a variety of innovative mathematical models. Recently, the calculus of variations, a classical area of optimization theory, has gained considerable attention due to its growing importance in real-world applications. The classical calculus of variations is extended by incorporating fractional order derivatives in the objective function. This has resulted in the development of the fractional calculus of variations as a distinct research area [5]. Fractional variational problems (FVPs) are a specialized type of variational problem where the objective functional or performance index involves one or more fractional derivative terms. These problems have a crucial role in numerous fields, including physics, engineering and other areas where energy principles are utilized.

Besides this, wavelet theory is a recent advancement that is rapidly gaining interest in various scientific and engineering disciplines such as signal and image processing, time-frequency analysis, edge detection, and the development of fast algorithms [6]. Wavelets offer various useful properties like orthogonality, compact support, and the ability to represent functions at different resolution levels [7]. The wavelet-based numerical approaches have recently emerged as a promising research area. Consequently, the wavelet technique serves as an efficient method for solving FVPs.

Over the past few years, many numerical approaches have been utilized for solving FVPs. These methods are categorized into two main types, that is indirect and direct approaches. In the indirect approach, the fractional Euler–Lagrange equation is usually used to derive and solve the necessary conditions. On the other hand, the direct approach converts the FVPs into an optimization problem without explicitly solving the Euler–Lagrange equations. These methods for solving FVPs have been widely explored in the literature. For instance, Zaky et al. [3] proposed a scheme using Jacobi functions for the solution of FVPs. In [4], Wang and Xiao studied a scheme based on the discrete Euler–Lagrange equation and variational integrators of fractional order for solving a class of FVPs. Dehestani et al. [8] discussed an approach based on modified Bessel wavelets for solving FVPs. In [9], Khader and Hendy employed fractional Chebyshev finite difference and the Rayleigh–Ritz method to solve FVPs. Lofti and Yousefi [10] applied a special type of polynomial basis function for the solution of FVPs. In [11], Aggarwal presented a fractional finite element formulation and Euler–Lagrange equations for solving FVPs that satisfies the transversality/natural boundary conditions. In [5], Singh et al. suggested an approach for the solution of non-linear FVPs by applying the Ritz method using various orthogonal polynomials. Ordokhani and Rahimkhani [12] examined the FVPs with the help of Muntz–Legendre polynomials and the Rayleigh–Ritz method. Pooseh et al. [13] solved FVPs by employing a finite difference approach by discretizing the objective function with Grunwald–Letnikov approximations. In [14], Ezz-Eldien et al. investigated the FVPs by utilizing the shifted Legendre orthonormal polynomials and operational matrix of integration. Almeida and Morgado [15] employed a variational approach to solve FVPs by incorporating the tempered Caputo fractional derivative. In [16], Mohammed solved a class of FVPs by utilizing a direct method based on an operational matrix of generalized hat basis functions. Khader [17] investigated the FVPs by using the properties of the Rayleigh–Ritz method and the chain rule. In [18], Keshavarz et al. proposed a numerical technique for the solution of linear and non-linear problems in the calculus of variations by utilizing Bernoulli wavelets.

Motivated by the existing literature, this study seeks to propose an effective approach by employing fractional order Taylor wavelets (FOTWs) to approximate the solutions of FVPs. Firstly, the unknown function is represented by using the FOTWs, and then Lagrange multipliers are used to integrate the performance index with constraint equations. The constrained extrema approach is utilized to transform the considered model into a set of algebraic equations with unknown coefficients. By applying the standard iterative approach, the unknown coefficients are determined and further used to find the solution of FVPs. In this study, we focus on minimizing FVPs of the following form:

$$Min\, J\left[x(\omega)\right] = \int_a^b F\left(\omega, \mu(\omega), x(\omega), D_\omega^\alpha x(\omega)\right) d\omega, \quad 0 < \alpha \leq 1, \tag{13.1}$$

with the initial boundary conditions

$$x(0) = a_0,\; x(b) = a_1, \tag{13.2}$$

where $a_0, a_1 \in R, \mu(\omega)$ is a known function and D_ω^α denotes the Caputo fractional derivative of $x(\omega)$.

The organization of the rest paper is structured as follows. In Section 13.2, essential definitions of fractional calculus are discussed. Section 13.3 provides the definition of Taylor wavelets and FOTWs. Section 13.4 presents the idea of function approximation via FOTWs. In Section 13.5, a numerical method is suggested for the solution of the considered model. Section 13.6 offers a detailed analysis of error and convergence for the suggested method. In Section 13.7, some illustrative examples are solved by utilizing the proposed approach. Finally, Section 13.8 concludes the paper with the main findings.

13.2 Preliminaries

Definition 13.2.1: The Caputo fractional derivative having order $\alpha > 0$ is stated as [1]:

$$D_\omega^\alpha h(\omega) = \begin{cases} \frac{1}{\lceil(\eta-\alpha)} \int_0^\omega \frac{h^{(\eta)}(\tau)}{(\omega-\tau)^{\alpha+1-\eta}} d\tau, & \eta - 1 < \alpha \leq \eta, \eta \in N \\ h^{(\eta)}(\omega), & \alpha = \eta \end{cases}. \tag{13.3}$$

Definition 13.2.2: The Riemann–Liouville fractional integral having order $\alpha \geq 0$ is stated as [1]:

$$I_\omega^\alpha h(\omega) = \begin{cases} \frac{1}{\lceil(\alpha)} \int_0^\omega (\omega-\tau)^{\alpha-1} h(\tau)\, d\tau = \frac{1}{\lceil(\alpha)} \omega^{\alpha-1} * h(\omega), & \alpha > 0 \\ h(\omega), & \alpha = 0 \end{cases}, \tag{13.4}$$

where $\omega^{\alpha-1} * h(\omega)$ is the convolution product of $\omega^{\alpha-1}$ and $h(\omega)$ The operators defined in eqn (13.3) and (13.4) fulfils the following properties [19]:

(i) $I_\omega^\alpha\left(\phi x(\omega) + \sigma y(\omega)\right) = \phi I_\omega^\alpha x(\omega) + \sigma I_\omega^\alpha y(\omega)$, where ϕ & σ are constants.

(ii)

$$D_{\omega}^{\alpha}\left(I_{\omega}^{\alpha}x(\omega)\right) = x(\omega). \tag{13.5}$$

(iii)

$$I_{\omega}^{\alpha}\left(D_{\omega}^{\alpha}x(\omega)\right) = x(\omega) - \sum_{r=0}^{\lceil\alpha\rceil-1} \frac{x^{(r)}(0)\omega^{r}}{\lceil(r+1)}, \tag{13.6}$$

where $\lceil\alpha\rceil$ denotes the ceiling function of α.

(iv)

$$D_{\omega}^{\alpha}x(\omega) = I_{\omega}^{n-\alpha}D_{\omega}^{\alpha}x(\omega). \tag{13.7}$$

Further, the following concepts are utilized in this study.

- The Sobolev norm of a function of order for [0,1] is stated as [20]:

$$\|x(\omega)\|_{H^{\vartheta}[0,1]} = \left(\sum_{\gamma=0}^{\vartheta}\int_{0}^{1}\left|x^{(\gamma)}(\omega)\right|^{2}d\omega\right)^{\frac{1}{2}}$$
$$= \left(\sum_{\gamma=0}^{\vartheta}\left\|x^{(\gamma)}(\omega)\right\|_{2}^{2}\right)^{\frac{1}{2}}, \tag{13.8}$$

  where$x^{(\gamma)}(\omega)$ is the differentiation of $x(\omega)$ with order (γ).
- If $g_1 \in L^1(R)$ and $g_2 \in L^p(R)$ for some $p \in [1,\infty)$, then we have the following property as [21]:

$$\|g_1 * g_2\|_{\mathrm{p}} = \|g_1\|_1\,\|g_2\|_{\mathrm{p}}. \tag{13.9}$$

13.3 Fractional Order Taylor Wavelets

In this section, the definitions of Taylor wavelets and FOTWs are recalled.

Taylor wavelets: The Taylor wavelets $\psi_{\mathrm{n,m}}(\omega) = \psi(k,n,m,\omega)$ have four arguments, and it is described as [22]:

$$\psi_{\mathrm{n,m}}(\omega) = \begin{cases} 2^{\frac{k-1}{2}}\hat{T}_{\mathrm{m}}\left(2^{k-1}\omega - n + 1\right), & \frac{n-1}{2^{k-1}} \le \omega \le \frac{n}{2^{k-1}} \\ 0, & otherwise \end{cases}, \tag{13.10}$$

where $k \in Z^{+}, m \in Z_{0}^{+}, \omega$ is a normalized time of Taylor polynomial for $[0,1)$ and $\hat{T}_{\mathrm{m}}(\omega)$ is normalized Taylor polynomial of degree m which is given as

$$\hat{T}_{\mathrm{m}}(\omega) = \sqrt{2m+1}\,\omega^{m}. \tag{13.11}$$

Fractional order Taylor wavelets: The FOTWs $\psi^{\theta}_{\mathrm{n,m}}(\omega)$ are described by using eqn (13.10) as [23]

$$\psi^{\theta}_{\mathrm{n,m}}(\omega) = \begin{cases} 2^{\frac{k-1}{2}} \hat{T}_{\mathrm{m}}\left(2^{k-1}\omega^{\theta} - n + 1\right), & \left(\frac{n-1}{2^{k-1}}\right)^{\frac{1}{\theta}} \leq \omega \leq \left(\frac{n}{2^{k-1}}\right)^{\frac{1}{\theta}} \\ 0, & otherwise \end{cases}, \tag{13.12}$$

where $\hat{T}_{\mathrm{m}}(\omega)$ is defined in eqn (13.11).

13.4 Function Approximation via FOTWs

In this section, function approximation is provided by utilizing FOTWs. It simplifies the complex functions, enabling easier analysis and interpretation of data.

Now, any function $x(\omega)$ in $L_2[0, 1]$ is expressed using FOTWs as

$$x(\omega) \simeq \sum_{n=1}^{\infty} \sum_{m=0}^{\infty} c_{\mathrm{n,m}} \psi^{\theta}_{\mathrm{n,m}}(\omega), \tag{13.13}$$

where $c_{\mathrm{n,m}} = \left\langle x(\omega), \psi^{\theta}_{\mathrm{n,m}}(\omega) \right\rangle$ and $\langle .,. \rangle$ shows the inner product under $L_2[0, 1]$. The truncated series of eqn (13.13) is provided as

$$\begin{aligned} x(\omega) &\simeq \sum_{n=1}^{2^{k-1}} \sum_{m=0}^{M-1} c_{\mathrm{n,m}} \psi^{\theta}_{\mathrm{n,m}}(\omega) \\ &= C^T \Psi^{\theta}_{\hat{m}}(\omega), \end{aligned} \tag{13.14}$$

where C and $\Psi^{\theta}_{\hat{m}}(\omega)$ are $(2^{k-1}M \times 1)$ order matrices provided as

$$\begin{aligned} C = [&c_{1,0}, c_{1,1}, ..., c_{1,M-1}; c_{2,0}, c_{2,1}, ..., c_{2,M-1}; ...; \\ &c_{2^{k-1},0}, c_{2^{k-1},1} ..., c_{2^{k-1},M-1}]^T, \end{aligned} \tag{13.15}$$

and

$$\begin{aligned} \Psi^{\theta}_{\hat{m}}(\omega) = \Big[&\psi^{\theta}_{1,0}(\omega), ..., \psi^{\theta}_{1,M-1}(\omega); \psi^{\theta}_{2,0}(\omega), ..., \psi^{\theta}_{2,M-1}(\omega); ...; \\ &\psi^{\theta}_{2^{k-1},0}(\omega), ..., \psi^{\theta}_{2^{k-1},M-1}(\omega)\Big]^T. \end{aligned} \tag{13.16}$$

In numerical computation, $\hat{m} = 2^{k-1}M$ denotes the number of FOTWs basis.

13.5 Numerical Method for FVPs

This section provides a stepwise solution of the considered model defined in eqn (13.1) and (13.2) by FOTWs with function approximation and Lagrange multipliers.

Consider the FVPs in eqn (13.1) and approximate $D^{\alpha}_{\omega}x(\omega)$ by using eqn (13.14) as

$$D^{\alpha}_{\omega}x(\omega) \simeq \sum_{n=1}^{2^{k-1}} \sum_{m=0}^{M-1} c_{\mathrm{n,m}} \psi^{\theta}_{\mathrm{n,m}}(\omega) = C^T \Psi^{\theta}_{\hat{m}}(\omega), \tag{13.17}$$

where C and $\Psi^{\theta}_{\hat{m}}(\omega)$ are defined in eqn (13.15) and (13.16), respectively.

Applying the operator I^{α}_{ω} of eqn (13.4) on eqn (13.17) and utilizing eqn (13.2) and (13.6), we obtain:

$$\begin{aligned} I^{\alpha}_{\omega}\left(D^{\alpha}_{\omega}x(\omega)\right) &= I^{\alpha}_{\omega}\left(C^T \Psi^{\theta}_{\hat{m}}(\omega)\right) \\ x(\omega) - x(0) &= C^T \left(I^{\alpha}_{\omega} \Psi^{\theta}_{\hat{m}}(\omega)\right) \\ x(\omega) &= C^T \aleph(\alpha, \theta, \omega) + a_0, \end{aligned} \tag{13.18}$$

where $\aleph(\alpha, \theta, \omega) = I^{\alpha}_{\omega} \Psi^{\theta}_{\hat{m}}(\omega)$.

The approximation form of eqn (13.1) using eqn (13.17) and (13.18) is expressed as

$$J\left[x(\omega)\right] = \int_a^b F\left(\omega, \mu(\omega), C^T \aleph(\alpha, \theta, \omega) + a_0, C^T \Psi^{\theta}_{\hat{m}}(\omega)\right) d\omega. \tag{13.19}$$

Now, using eqn (13.18), the boundary condition given in eqn (13.2) is provided as

$$x(b) = C^T \aleph(\alpha, \theta, b) + a_0. \tag{13.20}$$

By using an unknown Lagrange multiplier (λ), eqn (13.19) is merged with eqn (13.20) as

$$J^*\left[x(\omega)\right] = J\left[x(\omega)\right] + \lambda\left(C^T \aleph(\alpha, \theta, b) + a_0 - a_1\right).$$

The extremum conditions for FVPs defined in eqn (13.1) and (13.2) is expressed as

$$\frac{\partial J^*}{\partial C} = 0 \text{ and } \frac{\partial J^*}{\partial \lambda} = 0. \tag{13.21}$$

The considered model is transformed into a set of algebraic equations given in eqn (13.21). These equations are solved for C and λ using a standard iterative method. Once C is obtained, it is used to compute the approximate solution of $x(\omega)$ by using eqn (13.18).

13.6 Error and Convergence

This section presents a detailed discussion of the error and *convergence analysis* of the suggested approach.

- The absolute errors (AEs) between the analytical solution $x(\omega)$ and the wavelet approximate solution $x_{\hat{m}}(\omega)$ are evaluated as

$$\text{AEs} = |x(\omega) - x_{\hat{m}}(\omega)|\,.$$

- The maximum absolute errors (MAEs) associated with the considered model are obtained as

$$\text{Maximum absolute errors, L}_\infty = \max_{\omega\in[0,1]} \text{AEs}.$$

- The L_2 norm error is evaluated as

$$L_2 = \|x(\omega) - x_{\hat{m}}(\omega)\|_2 = \left(\int_0^1 |\text{AEs}|^2\, d\omega\right)^{\frac{1}{2}}.$$

Lemma 13.6.1: Let $x(\omega) \in H^{\vartheta}[0,1)$ for $\vartheta \geq 0$ and $x_{\hat{m}}(\omega) = C^T\Psi^{\theta}_{\hat{m}}(\omega)$ is the best approximation to $x(\omega)$ from $Y = \text{span}\left\{\Psi^{\theta}_{\hat{m}}(\omega) : \hat{m} = 2^{k-1}M\right\}$, then the error bound of the approximate solution $x_{\hat{m}}(\omega)$ is given as

$$\|x(\omega) - x_{\hat{m}}(\omega)\|_2 \leq c(M-1)^{-\vartheta}(2^{k-1})^{-\vartheta}\left\|x^{(\vartheta)}\right\|_2,$$

and for $1 \leq u \leq \vartheta$,

$$\|x(\omega) - x_{\hat{m}}(\omega)\|_{H^{\vartheta}[0,1)} \leq c(M-1)^{2u-\vartheta\frac{1}{2}}(2^{k-1})^{u-\vartheta}\left\|x^{(\vartheta)}\right\|_2, \tag{13.22}$$

where c depends on ϑ.

Theorem 13.6.1: Let $x(\omega) \in H^{\vartheta}[0,1)$ for $\vartheta \geq 0$, $M \geq \vartheta - 1, \alpha \in (0,1]$ and $u \geq 1$, then the error bound associated with the derivative of fractional form is given by

$$\|D^{\alpha}_{\omega}x(\omega) - D^{\alpha}_{\omega}x_{\hat{m}}(\omega)\|_2 \leq \frac{c}{\Gamma(2-\alpha)}(M-1)^{2u-\vartheta\frac{1}{2}}(2^{k-1})^{u-\vartheta}\left\|x^{(\mu)}\right\|_2. \tag{13.23}$$

Proof: By using eqn (13.4) and (13.7), we obtain:

$$\|D_\omega^\alpha x(\omega) - D_\omega^\alpha x_{\hat{m}}(\omega)\|_2^2 = \left\|I^{1-\alpha}(Dx(\omega) - Dx_{\hat{m}}(\omega))\right\|_2^2$$
$$= \left\|\frac{\omega^{-\alpha}}{\Gamma(1-\alpha)} * (Dx(\omega) - Dx_{\hat{m}}(\omega))\right\|_2^2. \quad (13.24)$$

Now, by using eqn (13.9) in eqn (13.24), we get:

$$\|D_\omega^\alpha x(\omega) - D_\omega^\alpha x_{\hat{m}}(\omega)\|_2^2 \leq \left\|\frac{\omega^{-\alpha}}{\Gamma(1-\alpha)}\right\|_1^2 \|(Dx(\omega) - Dx_{\hat{m}}(\omega))\|_2^2$$
$$\leq \left(\frac{1}{\Gamma(2-\alpha)}\right)^2 \|(Dx(\omega) - Dx_{\hat{m}}(\omega))\|_2^2. \quad (13.25)$$

By using eqn (13.8), we have:

$$\|D_\omega^\alpha x(\omega) - D_\omega^\alpha x_{\hat{m}}(\omega)\|_2^2 \leq \left(\frac{1}{\Gamma(2-\alpha)}\right)^2 \|x(\omega) - x_{\hat{m}}(\omega))\|_{H^\vartheta(0,1)}^2. \quad (13.26)$$

Taking the square root of eqn (13.26) and using eqn (13.22), we get:

$$\|D_\omega^\alpha x(\omega) - D_\omega^\alpha x_{\hat{m}}(\omega)\|_2 \leq \frac{c}{\Gamma(2-\alpha)}(M-1)^{2u-\vartheta\frac{1}{2}}(2^{k-1})^{u-\vartheta}\left\|x^{(\vartheta)}\right\|_2.$$

Theorem 13.6.2: Let $x(\omega) \in H^\vartheta(0,1)$ for $\vartheta \geq 0, M \geq \vartheta - 1, \alpha \in (0,1]$ and F in eqn (13.1) satisfy the Lipschitz condition. Then the error bound E_J of the suggested method is given by

$$\|E_J\|_2 \leq \left(1 + \frac{1}{\Gamma(2-\alpha)}\right)\rho c(M-1)^{2u-\vartheta\frac{1}{2}}(2^{k-1})^{u-\vartheta}\left\|x^{(\vartheta)}\right\|_2,$$

where ρ is the Lipschitz constant.

Proof: Based on the problem given in eqn (13.1), we have:

$$\|E_J\|_2 = \|J[x(\omega)] - J[x_{\hat{m}}(\omega)]\|_2.$$

From eqn (13.1), we obtain:

$$\|E_J\|_2 = \left\|\int_a^b F(\omega, \mu(\omega), x(\omega), D_\omega^\alpha x(\omega))\, d\omega\right.$$
$$\left. - \int_a^b F(\omega, \mu(\omega), x_{\hat{m}}(\omega), D_\omega^\alpha x_{\hat{m}}(\omega))\, d\omega\right\|_2.$$

As F satisfies the Lipschitz condition, therefore we get:

$$\|E_J\|_2 \leq \rho \int_a^b \|x(\omega) - x_{\hat{m}}(\omega)\|_2 \, d\omega + \rho \int_a^b \|D_\omega^\alpha x(\omega) - D_\omega^\alpha x_{\hat{m}}(\omega)\|_2 \, d\omega. \tag{13.27}$$

From eqn (13.27), we have:

$$\|E_J\|_2 \leq \rho \|x(\omega) - x_{\hat{m}}(\omega)\|_2 + \rho \|D_\omega^\alpha x(\omega) - D_\omega^\alpha x_{\hat{m}}(\omega)\|_2 . \tag{13.28}$$

By using eqn (13.22) and (13.23), we obtain:

$$\begin{aligned}
\|E_J\|_2 &\leq \rho c (M-1)^{2u-\vartheta \frac{1}{2}} (2^{k-1})^{u-\vartheta} \left\| x^{(\beta)} \right\|_2 \\
&\quad + \rho \frac{c}{\Gamma(2-\alpha)} (M-1)^{2u-\vartheta \frac{1}{2}} (2^{k-1})^{u-\vartheta} \left\| x^{(\vartheta)} \right\|_2 \\
&\leq \left(1 + \frac{1}{\Gamma(2-\alpha)}\right) \rho c (M-1)^{2u-\vartheta \frac{1}{2}} (2^{k-1})^{u-\vartheta} \left\| x^{(\vartheta)} \right\|_2 .
\end{aligned}$$

Thus, this completes the proof.

13.7 Numerical Examples

This section demonstrates the effectiveness of the proposed approach through graphical and tabular representations. To validate the method, several test problems are solved and the achieved outcomes are compared with existing approaches using tables to highlight the accuracy and reliability of the suggested method. All calculations are performed using Mathematica software.

Example 13.7.1: Consider the following FVPs as [8, 14]

$$\text{MinJ} = \frac{1}{2} \int_0^1 \left(D_\omega^\alpha x(\omega) - \frac{\Gamma(1+\beta)}{\Gamma(1+\beta-\alpha)} \omega^{\beta-\alpha} \right)^2 d\omega,$$

with the initial boundary conditions:

$$x(0) = 0, x(1) = 1.$$

The analytical solution is given as $x(\omega) = \omega^\beta$.

The given problem is solved for various $\theta, \alpha, \beta,$ and $\hat{\text{m}}$ through the approach described in Section 13.6. In Table 13.1, the AEs are provided for $\alpha = 0.9$ and $\beta = 3$, which clearly demonstrate that the errors continuously decrease as the number of wavelet basis increases. Figure 13.1 depicts the AEs at different α and the behaviour of approximate solutions across distinct values of β. In Figure 13.2, the AEs are represented at different θ. Table 13.2 demonstrates MAEs for different wavelet basis and compares with the techniques based on fractional variational integrators (FVIs) [4] and shifted orthonormal Legendre polynomials (SOLPs) [14]. In Table 13.3, the approximate values of *J* are given for distinct α and compared with SOLPs [14]. The presented tables, as well as figures confirm that the suggested approach efficiently produces good results with fewer wavelet basis.

Table 13.1 AEs of Example 13.7.1 for $\alpha = 0.9, \theta = 1,$ and $\beta = 3$.

ω	$\hat{m} = 4$	$\hat{m} = 6$	$\hat{m} = 8$	$\hat{m} = 10$
0.1	5.3×10^{-5}	4.8×10^{-6}	8.9×10^{-7}	1.0×10^{-7}
0.2	6.2×10^{-5}	3.1×10^{-6}	1.0×10^{-6}	7.0×10^{-8}
0.3	9.9×10^{-5}	5.2×10^{-6}	3.0×10^{-7}	1.0×10^{-7}
0.4	5.1×10^{-5}	4.0×10^{-6}	8.2×10^{-7}	2.0×10^{-7}
0.5	2.7×10^{-5}	2.9×10^{-6}	4.2×10^{-7}	1.2×10^{-7}
0.6	7.7×10^{-5}	5.5×10^{-6}	5.7×10^{-7}	8.4×10^{-8}
0.7	6.2×10^{-5}	6.5×10^{-7}	5.7×10^{-7}	1.5×10^{-7}
0.8	8.6×10^{-6}	4.0×10^{-6}	3.0×10^{-7}	1.2×10^{-7}
0.9	7.2×10^{-5}	4.7×10^{-7}	4.6×10^{-7}	6.2×10^{-8}
L_2	6.1×10^{-5}	3.9×10^{-6}	6.0×10^{-7}	1.4×10^{-7}

Table 13.2 MAEs for Example 13.7.1 at $\alpha = 0.99$, $\theta = 1$, and $\beta = 1$.

Proposed method		**FVIs [4]**		**SOLPs [14]**	
$\hat{m}$	**MAEs**	$\hat{m}$	**MAEs**	$\hat{m}$	**MAEs**
5	1.5×10^{-4}	10	5.8×10^{-4}	5	2.0×10^{-4}
7	7.8×10^{-5}	20	3.9×10^{-4}	9	5.3×10^{-5}
9	4.7×10^{-5}	40	2.4×10^{-4}	13	2.6×10^{-5}
11	3.1×10^{-5}	–	–	–	–
12	2.6×10^{-5}	–	–	–	–

Table 13.3 Approximate values of J for Example 13.7.1 at $\theta = 1$ and $\beta = 3$.

$\hat{m}$	**Proposed method**			**SOLPs [14]**		
	$\alpha = 0.5$	$\alpha = 0.7$	$\alpha = 0.9$	$\alpha = 0.5$	$\alpha = 0.7$	$\alpha = 0.9$
4	7.8×10^{-7}	7.8×10^{-9}	1.9×10^{-7}	8.3×10^{-7}	8.4×10^{-7}	2.0×10^{-7}
6	2.6×10^{-9}	4.5×10^{-9}	1.8×10^{-9}	2.8×10^{-9}	4.7×10^{-9}	1.8×10^{-9}
8	6.5×10^{-11}	1.4×10^{-10}	7.8×10^{-11}	7.0×10^{-11}	1.5×10^{-10}	8.0×10^{-11}
10	4.0×10^{-12}	1.1×10^{-11}	7.2×10^{-12}	4.2×10^{-12}	1.1×10^{-11}	7.3×10^{-12}

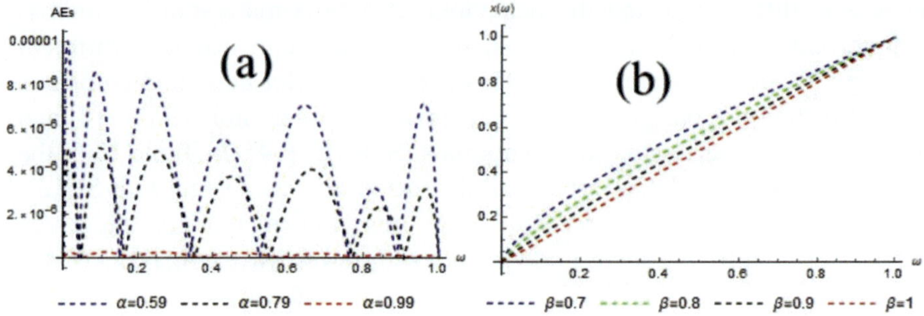

Figure 13.1 Graph of Example 13.7.1 at $\theta = 1$ for a) AEs at $\hat{m} = 6$ and $\beta = 3$ and b) approximate solutions of $x(\omega)$ at $\hat{m} = 7$ and $\alpha = 0.75$.

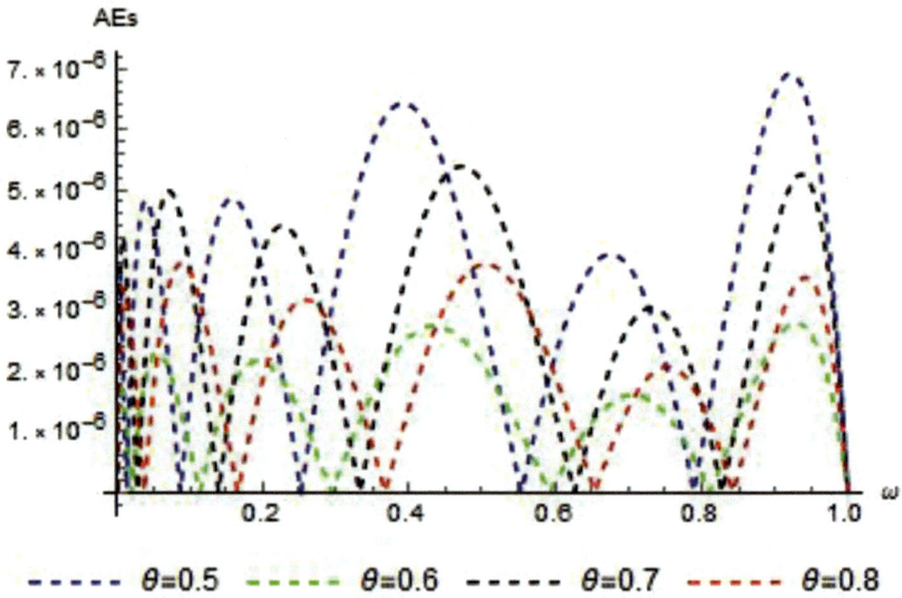

Figure 13.2 Graph of AEs of Example 13.7.1 for $\hat{m} = 6$ $\alpha = 0.7$, and $\beta = 3$

Example 13.7.2: Consider the following FVPs as [14]:

$$\text{MinJ} = \frac{1}{2}\int_0^1 \left(D_\omega^\alpha x(\omega) - \frac{\Gamma(3+2\alpha)}{\Gamma(3+\alpha)}\omega^{2+\alpha} - \Gamma(2+\alpha)\omega \right)^2 d\omega,$$

with the initial boundary conditions

$$x(0) = 1, x(1) = 3.$$

The analytical solution is given as $x(\omega) = \omega^{2\alpha+2} + \omega^{\alpha+1} + 1$.

This problem is solved for various values of θ, α and $\hat{m}$. Table 13.4 represents the AEs at $\alpha = 0.99$ for increasing value of $\hat{m}$. In Table 13.5, the MAEs are shown for different $\hat{m}$ and α, which are compared with the technique based on SOLPs [14]. Figure 13.3 demonstrates the graph of AEs for different wavelet basis and approximate solutions across distinct values of α. In Table 13.6, the approximate values of J are given for distinct α and compared with SOLPs [14]. The obtained results clearly indicate that

Table 13.4 AEs of Example 13.7.2 for ($\alpha = 0.99$ and $\theta = 1$).

ω	$\hat{m} = 5$	$\hat{m} = 7$	$\hat{m} = 9$	$\hat{m} = 11$
0.1	1.7×10^{-7}	6.1×10^{-8}	3.9×10^{-9}	1.1×10^{-9}
0.2	1.3×10^{-6}	1.6×10^{-8}	6.4×10^{-9}	1.1×10^{-9}
0.3	6.1×10^{-7}	6.0×10^{-8}	7.2×10^{-9}	5.7×10^{-10}
0.4	8.6×10^{-7}	2.4×10^{-8}	6.3×10^{-10}	3.9×10^{-10}
0.5	1.3×10^{-6}	5.7×10^{-8}	6.3×10^{-9}	1.0×10^{-9}
0.6	3.8×10^{-7}	1.2×10^{-8}	4.0×10^{-9}	9.4×10^{-10}
0.7	8.7×10^{-7}	5.2×10^{-8}	2.7×10^{-9}	2.9×10^{-10}
0.8	9.6×10^{-7}	1.3×10^{-8}	5.2×10^{-9}	1.7×10^{-10}
0.9	3.8×10^{-7}	2.8×10^{-8}	3.7×10^{-9}	1.3×10^{-10}
L_2	8.6×10^{-7}	4.0×10^{-8}	4.6×10^{-9}	8.2×10^{-10}

Table 13.5 MAEs for Example 13.7.2 at $\theta = 1$.

$\hat{m}$	Proposed method			SOLPs [14]		
	$\alpha = 0.3$	$\alpha = 0.6$	$\alpha = 0.9$	$\alpha = 0.3$	$\alpha = 0.6$	$\alpha = 0.9$
5	1.9×10^{-4}	7.8×10^{-5}	1.5×10^{-5}	3.9×10^{-3}	2.8×10^{-3}	7.7×10^{-4}
7	2.8×10^{-5}	7.1×10^{-6}	8.2×10^{-7}	1.7×10^{-3}	8.4×10^{-4}	1.2×10^{-4}
9	4.4×10^{-6}	1.3×10^{-6}	1.0×10^{-7}	9.0×10^{-4}	3.6×10^{-4}	4.4×10^{-5}
11	2.4×10^{-6}	3.4×10^{-7}	2.2×10^{-8}	5.3×10^{-4}	1.8×10^{-4}	2.0×10^{-5}

Table 13.6 Approximate values of J for Example 13.7.2 at $\theta = 1$.

$\hat{m}$	Proposed method			SOLPs [14]		
	$\alpha = 0.3$	$\alpha = 0.6$	$\alpha = 0.9$	$\alpha = 0.3$	$\alpha = 0.6$	$\alpha = 0.9$
5	3.5×10^{-8}	6.4×10^{-8}	1.3×10^{-8}	2.3×10^{-7}	8.9×10^{-7}	2.1×10^{-7}
7	5.7×10^{-10}	6.2×10^{-10}	7.1×10^{-11}	1.6×10^{-8}	2.7×10^{-8}	1.5×10^{-9}
9	2.9×10^{-11}	2.3×10^{-11}	1.8×10^{-12}	2.1×10^{-9}	2.7×10^{-9}	9.6×10^{-11}
11	2.9×10^{-12}	1.7×10^{-12}	1.0×10^{-13}	4.3×10^{-10}	4.5×10^{-10}	1.2×10^{-11}

increasing the number of wavelet basis results in improved accuracy with lower errors.

Example 13.7.3: Consider the following FVPs as [13, 15]

$$\text{MinJ} = \int_0^1 \left(D_\omega^\alpha x(\omega) - \frac{2\omega^{1+\alpha}}{\Gamma(2+\alpha)} \right)^2 d\omega,$$

with the initial boundary conditions

$$x(0) = 0, x(1) = 1.$$

The analytical solution for $\alpha = 0.5$ is given as $x(\omega) = \omega^2$.

This problem is solved for various values of $\theta, \alpha,$ and $\hat{\text{m}}$. In Table 13.7, the AEs for increasing value of $\hat{m}$ are presented at $\alpha = 0.5$. In Table 13.8, the approximate values of J are provided for distinct α. Figure 13.4 shows

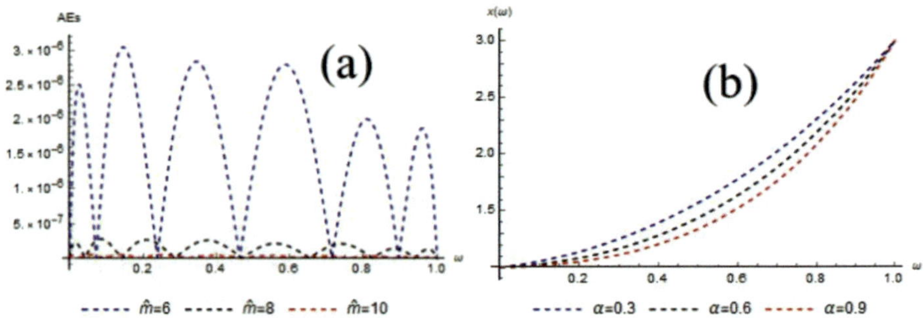

Figure 13.3 Graph of Example 13.7.2 at $\theta = 1$ for a) AEs at $\alpha = 0.9$ and b) approximate solutions of $x(\omega)$ at $\hat{m} = 6$.

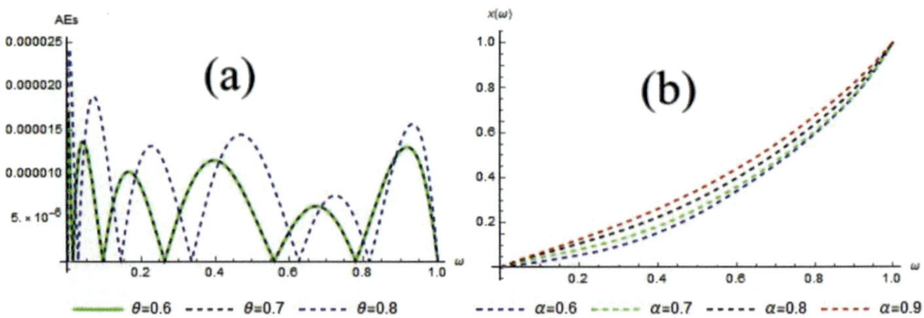

Figure 13.4 Graph of Example 13.7.3 at $\hat{m} = 6$ for a) AEs at $\alpha = 0.5$ and b) approximate solutions of $x(\omega)$ at $\theta = 1$.

Table 13.7 AEs of Example 13.7.3 for $\alpha = 0.5$ and $\theta = 1$.

ω	$\hat{m} = 5$	$\hat{m} = 7$	$\hat{m} = 9$	$\hat{m} = 11$
0.1	2.0×10^{-4}	4.9×10^{-5}	9.4×10^{-6}	7.7×10^{-6}
0.2	1.6×10^{-4}	5.3×10^{-5}	5.9×10^{-6}	5.2×10^{-6}
0.3	1.4×10^{-4}	1.4×10^{-5}	9.8×10^{-6}	5.4×10^{-6}
0.4	2.5×10^{-4}	3.5×10^{-5}	1.1×10^{-5}	2.2×10^{-6}
0.5	1.2×10^{-4}	6.1×10^{-6}	4.7×10^{-6}	3.2×10^{-7}
0.6	6.9×10^{-5}	4.0×10^{-5}	8.3×10^{-6}	4.1×10^{-6}
0.7	1.3×10^{-4}	2.5×10^{-5}	5.8×10^{-6}	2.4×10^{-6}
0.8	3.0×10^{-5}	1.8×10^{-5}	8.3×10^{-6}	3.0×10^{-6}
0.9	2.4×10^{-4}	1.5×10^{-5}	4.5×10^{-6}	7.1×10^{-7}
L_2	1.7×10^{-4}	3.4×10^{-5}	1.0×10^{-5}	3.8×10^{-6}

Table 13.8 Approximate value of J for Example 13.7.3 at $\theta = 1$.

$\hat{m}$	$\alpha = 0.5$	$\alpha = 0.6$	$\alpha = 0.7$	$\alpha = 0.8$	$\alpha = 0.9$
5	5.7×10^{-7}	0.0177606	0.0769947	0.1751621	0.3012257
7	3.4×10^{-8}	0.0171061	0.0760245	0.1746221	0.3011063
9	4.4×10^{-9}	0.0166918	0.0754586	0.1743387	0.3010505
11	8.7×10^{-10}	0.0164011	0.0750856	0.1741665	0.3010195

the graphical representation of AEs at different θ and approximate solutions across distinct values of α. In Table 13.9, the MAEs are shown for different wavelet basis at $\alpha = 0.5$, which are compared with the techniques based on discrete direct methods (DDMs) [13] and Euler–Lagrange equation (ELE) [15]. As illustrated by the tables, as well as figures, the proposed approach is more efficient and easier to implement.

Example 13.7.4: Consider the following FVPs as [14]

$$\text{MinJ} = \int_0^{\frac{\pi}{2}} \left(D_\omega^\alpha x(\omega) - \cos(\omega)\right)^2 d\omega,$$

with the initial boundary conditions

$$x(0) = 0, x\left(\frac{\pi}{2}\right) = 1.$$

The analytical solution for $\alpha = 1$ is given as $x(\omega) = \sin(\omega)$.

The given problem is solved for various values of $\theta, \alpha,$ and $\hat{m}$. Table 13.10 demonstrates the AEs for increasing value of $\hat{m}$ at $\alpha = 1$ & $\theta = 1$. This clearly shows that the proposed approach provides good results with a very less number of wavelet basis. Figure 13.5 represents the graph of AEs at different θ and approximate values of $x(\omega)$ across distinct values of α. In Table 13.11, the MAEs and approximate values of J are given for different

Table 13.9 MAEs for Example 13.7.3 at $\alpha = 0.5$ and $\theta = 1$.

Proposed method		DDMs [13]		ELE [15]	
$\hat{m}$	MAEs	$\hat{m}$	MAEs	$\hat{m}$	MAEs
5	3.9×10^{-4}	5	2.6×10^{-2}	5	3.5×10^{-2}
7	9.6×10^{-5}	10	1.5×10^{-2}	10	1.4×10^{-2}
9	3.4×10^{-5}	30	6.5×10^{-3}	30	2.9×10^{-3}
11	1.0×10^{-5}	–	–	–	–

Table 13.10 AEs of Example 13.7.4 for $\alpha = 1$ and $\theta = 1$.

ω	$\hat{m} = 3$	$\hat{m} = 5$	$\hat{m} = 7$	$\hat{m} = 9$
0.1	1.3×10^{-3}	6.2×10^{-6}	7.2×10^{-10}	4.6×10^{-11}
0.2	1.6×10^{-3}	1.0×10^{-6}	2.3×10^{-8}	1.0×10^{-10}
0.3	1.2×10^{-3}	8.0×10^{-6}	6.9×10^{-9}	8.5×10^{-11}
0.4	4.6×10^{-4}	9.3×10^{-6}	2.0×10^{-8}	1.6×10^{-10}
0.5	4.7×10^{-4}	4.9×10^{-6}	2.6×10^{-8}	2.0×10^{-11}
0.6	1.3×10^{-3}	2.2×10^{-6}	6.9×10^{-9}	1.7×10^{-10}
0.7	1.9×10^{-3}	8.6×10^{-6}	1.8×10^{-8}	1.8×10^{-10}
0.8	2.1×10^{-3}	1.0×10^{-5}	2.9×10^{-8}	1.0×10^{-11}
0.9	2.0×10^{-3}	7.9×10^{-6}	1.4×10^{-8}	2.1×10^{-10}
L_2	1.4×10^{-3}	6.9×10^{-6}	1.8×10^{-8}	1.3×10^{-10}

Table 13.11 MAEs of $x(\omega)$ and approximate values of J for Example 13.7.4 at $\alpha = 1$ and $\theta = 1$.

$\hat{m}$	Proposed method		SOLPs [14]	
	MAEs	J	MAEs	J
4	1.6×10^{-4}	1.0×10^{-6}	2.6×10^{-3}	6.7×10^{-6}
5	1.0×10^{-5}	6.8×10^{-9}	2.3×10^{-4}	1.0×10^{-7}
6	5.9×10^{-7}	2.9×10^{-11}	1.6×10^{-5}	2.9×10^{-10}
7	2.9×10^{-8}	9.3×10^{-14}	9.9×10^{-7}	2.4×10^{-12}
8	1.2×10^{-9}	2.2×10^{-16}	5.9×10^{-8}	4.4×10^{-15}

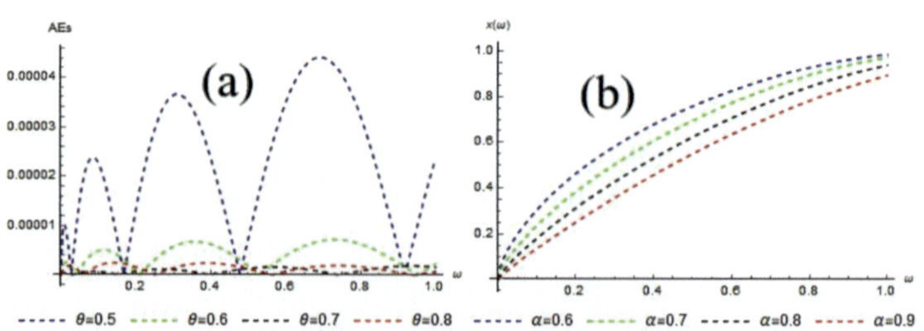

Figure 13.5 Graph of Example 13.7.4 at $\hat{m} = 6$ for a) AEs at $\alpha = 1$ and b) approximate solutions of $x(\omega)$ at $\theta = 1$.

wavelet basis and compared with the technique based on SOLPs [14]. The suggested method proves to be highly accurate as the approximate solutions closely align with the exact solution.

13.8 Conclusion

In this study, an efficient approach for solving the FVPs is described by utilizing FOTWs. The method employs FOTWs basis functions, which offer excellent accuracy, rapid convergence, and significant computational efficiency. By incorporating the Lagrange multiplier, the given constraints are integrated with the performance index. The constrained extrema approach is used to transform the model into algebraic equations, which are then solved to determine the unknown coefficients. Several numerical examples are illustrated to validate the reliability and effectiveness of the suggested technique. The achieved results demonstrate high precision, which are presented by tables and graphs. Additionally, a comparative analysis with existing techniques confirms the efficiency and fast convergence of the proposed approach. Notably, this approach requires only a limited number of wavelet basis functions to achieve satisfactory results. Overall, this study demonstrates the potential of FOTWs in solving the FVPs. For future work, this work can be extended to multi-dimensional FVPs and explore various fractional operators for improved accuracy.

Acknowledgements

I sincerely thank the editors for giving me the opportunity to contribute to this book chapter and for their valuable guidance throughout the process.

References

[1] Kenneth S. Miller and Bertram Ross. *An introduction to the fractional calculus and fractional differential equations*. Wiley, 1993.

[2] Paul L. Butzer and U. Westphal. An introduction to fractional calculus. In *Applications of fractional calculus in physics*, pages 1–85. World Scientific, 2000.

[3] Mahmoud A. Zaky, Eid H. Doha, and José Tenreiro Machado. A spectral framework for fractional variational problems based on fractional jacobi functions. *Applied Numerical Mathematics*, 132:51–72, 2018.

[4] Dongling Wang and Aiguo Xiao. Fractional variational integrators for fractional variational problems. *Communications in Nonlinear Science and Numerical Simulation*, 17(2):602–610, 2012.

[5] Harendra Singh, Ram K. Pandey, and Hari M. Srivastava. Solving non-linear fractional variational problems using jacobi polynomials. *Mathematics*, 7(3):224, 2019.

[6] Amara Graps. An introduction to wavelets. *IEEE Computational Science and Engineering*, 2(2):50–61, 1995.

[7] I. M. Dremin, O. V. Ivanov, and V. A. Nechitailo. Wavelets and their uses. *Physics-Uspekhi*, 44(5):447, 2001.

[8] H. Dehestani, Y. Ordokhani, and M. Razzaghi. Modified wavelet method for solving fractional variational problems. *Journal of Vibration and Control*, 27(5-6):582–596, 2021.

[9] M. M. Khader and A. S. Hendy. A numerical technique for solving fractional variational problems. *Mathematical Methods in the Applied Sciences*, 36(10):1281–1289, 2013.

[10] A. Lotfi and S. A. Yousefi. A numerical technique for solving a class of fractional variational problems. *Journal of Computational and Applied Mathematics*, 237(1):633–643, 2013.

[11] Om P. Agrawal. A general finite element formulation for fractional variational problems. *Journal of Mathematical Analysis and Applications*, 337(1):1–12, 2008.

[12] Y. Ordokhani and P. Rahimkhani. A numerical technique for solving fractional variational problems by müntzâĂŞlegendre polynomials. *Journal of Applied Mathematics and Computing*, 58:75–94, 2018.

[13] S. Pooseh, R. Almeida, and D. F. Torres. Discrete direct methods in the fractional calculus of variations. *Computers & Mathematics with Applications*, 66(5):668–676, 2013.

[14] S. S. Ezz-Eldien, R. M. Hafez, A. H. Bhrawy, D. Baleanu, and A. A. El-Kalaawy. New numerical approach for fractional variational problems using shifted legendre orthonormal polynomials. *Journal of Optimization Theory and Applications*, 174:295–320, 2017.

[15] R. Almeida and M. L. Morgado. Analysis and numerical approximation of tempered fractional calculus of variations problems. *Journal of Computational and Applied Mathematics*, 361:1–12, 2019.

[16] Omar H. Mohammed. A direct method for solving fractional order variational problems by hat basis functions. *Ain Shams Engineering Journal*, 9(4):1513–1518, 2018.

[17] M. M. Khader. An efficient approximate method for solving fractional variational problems. *Applied Mathematical Modelling*, 39(5-6):1643–1649, 2015.

[18] E. Keshavarz, Y. Ordokhani, and M. Razzaghi. The bernoulli wavelets operational matrix of integration and its applications for the solution of linear and nonlinear problems in calculus of variations. *Applied Mathematics and Computation*, 351:83–98, 2019.

[19] N. Kumar and M. Mehra. Legendre wavelet collocation method for fractional optimal control problems with fractional bolza cost. *Numerical Methods for Partial Differential Equations*, 37(2):1693–1724, 2021.

[20] Claudio Canuto, M. Yousuff Hussaini, Alfio Quarteroni, and Thomas A. Zang. *Spectral methods: fundamentals in single domains*. Springer Science & Business Media, 2007.

[21] Richard F. Bass. *Real analysis for graduate students*. CreateSpace Independent Publishing, 2013.

[22] P. Vichitkunakorn, Tuan N. Vo, and M. Razzaghi. A numerical method for fractional pantograph differential equations based on taylor wavelets. *Transactions of the Institute of Measurement and Control*, 42(7):1334–1344, 2020.

[23] Tuan N. Vo, M. Razzaghi, and Pham T. Toan. Fractional-order generalized taylor wavelet method for systems of nonlinear fractional differential equations with application to human respiratory syncytial virus infection. *Soft Computing*, 26:165–173, 2022.

14

New Fourth-order Efficient Numerical Solutions of the Klein–Gordon Equation

Mukesh Gupta[1,2], Brajesh Kumar Singh[2], Mukesh Kumar Awasthi[2], and Mohammad Tamsir[3]

[1]Department of Mathematics, Govt. Degree College, India
[2]Department of Mathematics, Babasaheb Bhimrao Ambedkar University, India
[3]Department of Mathematics, College of Science, Jazan University, Saudi Arabia
E-mail: mukeshgupta606@gmail.com; bksingh0584@gmail.com; mukeshiitr.kumar@gmail.com; tamsiriitm@gmail.com

Abstract

In this chapter, Klein–Gordon (K–G) equation with cubic nonlinearity arises in quantum field theory (QFT), nonlinear optics, condensed matter physics, relativistic field theory, and many other fields have been studied by using a fourth-order numerical scheme. At the first, K–G equation is transformed into a coupled system of differential equations, and then, we discretize the time derivative using a finite difference approximation and the spatial derivatives using Crank–Nicolson. After that fourth-order cubic B-spline approximation is utilized in the obtained coupled system of differential equations to convert it into a linear system of simultaneous equations. We implemented the proposed scheme on the K–G equation to assess the efficiency and accuracy of the scheme and found good results in terms of error norms like L_∞ and L_2. Fourth-order spatial convergence of the technique is also shown numerically in the considered test problems. The proposed technique is found to be economical, powerful, easy, and efficient for the study of various kinds of linear and nonlinear physical models as compared to the existing techniques.

DOI: 10.1201/9788743806691-14

Keywords: Collocation technique, Klein–Gordon equation, cubic B-splines, Gauss elimination

14.1 Introduction

In recent years, it has become increasingly evident that a vast range of real-world phenomena can be modeled using ordinary and partial differential equations (ODEs and PDEs). The accurate solution of these equations is essential for understanding the underlying physical processes. One such important equation is the nonlinear Klein–Gordon (K–G) equation, which appears in various scientific and engineering fields, including the study of dislocation flux in crystals and Josephson junctions [1]. The general form of the K–G equation is given as

$$\frac{\partial^2 \phi}{\partial t^2} + \alpha \frac{\partial^2 \phi}{\partial \xi^2} + P'(\phi) = 0. \tag{14.1}$$

Originally proposed in 1926 by O. Klein and W. Gordon to describe the behavior of relativistic electrons, the K–G equation was later investigated by Schrödinger in the context of quantum wave theory. Over time, the equation has found applications across multiple disciplines such as relativistic mechanics, dispersive wave phenomena, condensed matter physics, quantum mechanics, plasma physics, solid-state physics, nonlinear optics, and quantum field theory (QFT). Notably, the K–G equation also admits soliton-type solutions, which are of particular interest in nonlinear analysis [2, 3, 4, 5].

Due to its broad applicability, the K–G equation has attracted significant attention from researchers who have developed numerous analytical and numerical approaches for its solution. Various numerical methods have been proposed, including the radial basis function method [6], Adomian decomposition method [7], exponential-type potential techniques [8], the novel (G'/G)-expansion method [9], and B-spline-based collocation schemes [10]. Other notable techniques include the auxiliary equation approach [11], exponential cubic B-spline collocation method [12], tension spline method [13], Sobolev gradient-based techniques [14], spectral methods involving Legendre wavelets [15], the homotopy-perturbation method [16], polynomial wavelets [17], differential transform method [18], and clique polynomial methods [19].

In addition to its classical form, the K–G equation has been extended to fractional time and space domains, which offer better modeling for certain physical processes. These fractional variants have also been studied using a variety of numerical schemes, including the natural homotopy perturbation method [20], Sinc-Chebyshev collocation method [21], and Haar wavelet techniques [22].

The primary objective of the present chapter is to explore the numerical solution of the nonlinear K–G equation using a fourth-order accurate collocation scheme based on cubic B-spline basis functions. This scheme has previously shown promising results when applied to parabolic PDEs such as the Burgers, modified Burgers, Burgers-Fisher, and Fisher equations [23, 24, 25]. Moreover, modified cubic B-splines have been employed effectively for constructing fourth-order accurate spatial derivatives in differential quadrature and collocation frameworks, particularly in solving equations like the Burgers equation, sine-Gordon equation, Fisher equation, and the K–G equation itself [26, 27, 28, 29, 30, 31]. However, to the best of the authors' knowledge, the application of this high-accuracy collocation scheme to hyperbolic PDEs such as the K–G equation remains largely unexplored.

The remainder of this chapter is organized as follows: Section 14.2 outlines the procedure for developing fourth-order spatial approximations and the associated differentiation strategy [23]. Section 14.3 presents the implementation details of the scheme and construction of the initial vector. Section 14.4 contains numerical experiments and analysis to evaluate the performance of the proposed method on the K–G equation. Finally, the conclusions drawn from this study are summarized in Section 14.5.

14.2 Cubic B-spline Fourth-Order Collocation Scheme

Let us consider the domain of the problem as $\Omega_{\mathrm{LR}} = [L, R]$ and its partition as $\chi = \{L = \xi_0 < \xi_1 < \xi_2 < \ldots < \xi_{\mathrm{n}-1} < \xi_{\mathrm{n}} = \mathrm{R}\}$ such that $\xi_\ell - \xi_{\ell-1} = h = \frac{R-L}{n}$. The cubic B-spline functions over the considered partition are piecewise continuous and smooth functions having minimal support for a given degree. Cubic B-splines $\mathcal{B}_\kappa = \mathcal{B}_\kappa(\xi)$ at the point ξ is defined as

$$\mathcal{B}_\kappa(\xi) = \frac{1}{h^3}\begin{cases} \sigma_{\kappa-2}^3, & \xi \in [\xi_{\kappa-2}, \xi_{\kappa-1}) \\ \sigma_{\kappa-2}^3 - 4\sigma_{\kappa-1}^3, & \xi \in [\xi_{\kappa-1}, \xi_\kappa) \\ 4\sigma_{\kappa+1}^3 - \sigma_{\kappa+2}^3, & \xi \in [\xi_\kappa, \xi_{\kappa+1}) \\ -\sigma_{\kappa+2}^3, & \xi \in [\xi_{\kappa+1}, \xi_{\kappa+2}) \\ 0, & \text{else} \end{cases}, \tag{14.2}$$

where $\sigma_\kappa = (\xi - \xi_\kappa)$. The values at grid points ξ_ℓ of $\mathcal{B}_\kappa$ and its first and second derivatives , denoted by $\mathcal{B}_{\kappa\ell} := \mathcal{B}_\kappa(\xi_\ell)$, $\mathcal{B}'_{\kappa\ell} := \mathcal{B}'_\kappa(\xi_\ell)$ and $\mathcal{B}''_{\kappa\ell} := \mathcal{B}''_\kappa(\xi_\ell)$, respectively, read as

$$\mathcal{B}_{\kappa\ell} = \begin{cases} 4, & \text{if } \kappa - \ell = 0 \\ 1, & \text{if } \kappa - \ell = \pm 1 \\ 0, & \text{else} \end{cases};$$
$$\mathcal{B}'_{\kappa\ell} = \begin{cases} \frac{3}{h}, & \text{if } \kappa - \ell = 1 \\ -\frac{3}{h}, & \text{if } \kappa - \ell = -1 \\ 0, & \text{else} \end{cases}; \tag{14.3}$$
$$\mathcal{B}''_{\kappa\ell} = \begin{cases} -\frac{12}{h^2}, & \text{if } \kappa - \ell = 0 \\ \frac{6}{h^2}, & \text{if } \kappa - \ell = \pm 1 \\ 0 & \text{else.} \end{cases}.$$

The set $\{\mathcal{B}_\ell : \ell \in \Delta_{n+1} \bigcup \{-1\}$, where $\Delta_n = \{0, 1, 2, \ldots, n\}\}$ forms a base for the functions defined over the domain Ω_{LR}. In the collocation scheme, $\phi(\xi, t)$, the function under study is approximated by $\Theta(\xi, t)$ the interpolating function. For this approximation, the base functions are considered as cubic B-spline functions and express this as

$$\Theta(\xi, t) = \sum_{j=-1}^{n+1} \omega_j \, \mathcal{B}_j(\xi). \tag{14.4}$$

Here, $\omega_j = \omega_j(t)$' are unknowns, which will be determined by combining the boundary conditions (BCs) and initial conditions (ICs) of the problem with the collocation approach. Moreover, boundary conditions and interpolation conditions are satisfied for the approximation $\Theta(\xi, t)$:

i) $\Theta(\xi_i, t) \equiv \phi_i(t) = \omega_{i-1} + 4\omega_i + \omega_{i+1}, \forall i \in \Delta_n,$
ii) $\Theta''(\xi_i, t) = \phi_i''(t) - \frac{1}{12} h^2 \phi^{(4)}(\xi_i, t)$, for $i = 0, n$,

where $\phi_i(t) = \phi(\xi_i, t)$, $\phi_i''(t) = \phi''(\xi_i, t)$ and $\Delta_n = \{0, 1, 2, \ldots, n\}$. If $\phi(\xi, t)$ is a sufficiently smooth function and unique cubic B-spline interpolant of the same is $\Theta(\xi, t)$ which meets the above end conditions, then at node ξ_j [23]:

$$\begin{aligned} &\Theta'(\xi_j, t) = \phi'(\xi_j, t) + O(h^4), \text{ for all } j \in \Delta_n, \text{and,} \\ &\Theta''(\xi_j, t) = \phi''(\xi_j, t) - \frac{1}{12} h^2 \phi^{(4)}(\xi_j, t) + O(h^4), \text{ for all } j \in \Delta_n. \end{aligned} \tag{14.5}$$

On utilizing Taylor's series expansion formula along with finite difference formula, the approximations of $\phi^{(4)}(\xi_j, t)$ at node ξ_j can be found as a linear sum of $\Theta''(\xi_j, t)$ as in eqn (14.6):

$$\begin{aligned}
\phi^{(4)}(\xi_0, t) &= \frac{2\Theta''(\xi_0, t) - 5\Theta''(\xi_1, t) + 4\Theta''(\xi_2, t) - \Theta''(\xi_3, t)}{h^2} + O(h^2), \\
\phi^{(4)}(\xi_j, t) &= \frac{\Theta''(\xi_{j-1}, t) - 2\Theta''(\xi_j, t) + \Theta''(\xi_{j+1}, t)}{h^2} + O(h^2), \\
&\qquad j \in \Delta_{n-1} \setminus \{0\}, \\
\phi^{(4)}(\xi_n, t) &= \frac{2\Theta''(\xi_n, t) - 5\Theta''(\xi_{n-1}, t) + 4\Theta''(\xi_{n-2}, t) - \Theta''(\xi_{n-3}, t)}{h^2} \\
&\qquad + O(h^2).
\end{aligned} \tag{14.6}$$

Now utilizing eqn (14.6) in eqn (14.5), the fourth-order approximation of ϕ'' at various grids is obtained as

$$\begin{aligned}
\phi'(\xi_j, t) &= \Theta'(\xi_j, t) + O(h^4), \qquad j \in \Delta_n, \\
\phi''(\xi_0, t) &= \frac{14\Theta''(\xi_0, t) - 5\Theta''(\xi_1, t) + 4\Theta''(\xi_2, t) - \Theta''(\xi_3, t)}{12} + O(h^4), \\
\phi''(\xi_j, t) &= \frac{\Theta''(\xi_{j-1}, t) + 10\Theta''(\xi_j, t) + \Theta''(\xi_{j+1}, t)}{12} + O(h^4), \quad j \in \Delta_{n-1} \setminus \{0\}, \\
\phi''(\xi_n, t) &= \frac{14\Theta''(\xi_n, t) - 5\Theta''(\xi_{n-1}, t) + 4\Theta''(\xi_{n-2}, t) - \Theta''(\xi_{n-3}, t)}{12} + O(h^4).
\end{aligned} \tag{14.7}$$

Eqn (14.4) and (14.7) yields the following:

$$\begin{aligned}
\phi''(\xi_0, t) =& \frac{1}{12}\Bigg[14 \sum_{j=-1}^{n+1} \omega_j \mathcal{B}_j''(\xi_0) - 5 \sum_{j=-1}^{n+1} \omega_j \mathcal{B}_j''(\xi_1) + 4 \sum_{j=-1}^{n+1} \omega_j \mathcal{B}_j''(\xi_2) \\
& - \sum_{j=-1}^{n+1} \omega_j \mathcal{B}_j''(\xi_3)\Bigg].
\end{aligned} \tag{14.8}$$

$$\begin{aligned}
\phi''(\xi_j, t) =& \frac{1}{12}\Bigg[\sum_{j=-1}^{n+1} \omega_j \mathcal{B}_j''(\xi_{j-1}) + 10 \sum_{j=-1}^{n+1} \omega_j \mathcal{B}_j''(\xi_j) \\
& + \sum_{j=-1}^{n+1} \omega_j \mathcal{B}_{j+1}''(\xi_{j+1})\Bigg], \qquad j \in \Delta_{n-1} \setminus \{0\},
\end{aligned} \tag{14.9}$$

$$\phi''(\xi_n, t) = \frac{1}{12}\Bigg[14 \sum_{j=-1}^{n+1} \omega_j \mathcal{B}_j''(\xi_n) - 5 \sum_{j=-1}^{n+1} \omega_j \mathcal{B}_j''(\xi_{n-1})$$

$$+4\sum_{j=-1}^{n+1}\omega_j\mathcal{B}_j''(\xi_{n-2})-\sum_{j=-1}^{n+1}\omega_j\mathcal{B}_j''(\xi_{n-3})\Bigg]. \tag{14.10}$$

Simplifying eqn (14.8)–(14.10) by utilizing the values of $\mathcal{B}_j''$ from eqn (14.3), we get the following fourth order approximation of ϕ'' at various grids:

$$\phi''(\xi_0,t)=\frac{14\omega_{-1}-33\omega_0+28\omega_1-14\omega_2+6\omega_3-\omega_4}{2h^2}, \tag{14.11}$$

$$\phi''(\xi_i,t)=\frac{\omega_{i-2}+8\omega_{i-1}-18\omega_i+8\omega_{i+1}+\omega_{i+2}}{2h^2}, i\in\Delta_{n-1}\setminus\{0\}, \tag{14.12}$$

$$\phi''(\xi_n,t)=\frac{14\omega_{n+1}-33\omega_n+28\omega_{n-1}-14\omega_{n-2}+6\omega_{n-3}-\omega_{n-4}}{2h^2}. \tag{14.13}$$

The next section will describe the utilization of spatial fourth-order approximations to study the K–G equation.

14.3 Application of Scheme on K–G Equation

This section deals with the application of fourth-order collocation scheme on the K–G eqn (14.1). In order to implement the 4th order technique and to solve the K–G eqn (14.1) numerically, we reduce the order of derivative in time by considering $\phi_t=\psi$.

For the particular case $P'(\phi)=\beta\phi+\gamma\phi^3$, an initial boundary value problem (IBVP) is setup for the K–G eqn (14.1) as

$$\left.\begin{aligned}\frac{\partial\psi}{\partial t}&=-\alpha\frac{\partial^2\phi}{\partial\xi^2}-\beta\phi-\gamma\phi^3\\ \frac{\partial\phi}{\partial t}&=\psi\end{aligned}\right\}, \tag{14.14}$$

with initial conditions (ICs):

$$\left.\begin{aligned}\phi(\xi,0)&=f_1(\xi) &\quad \xi\in[L,R]\\ \psi(\xi,0)&=f_2(\xi) &\quad \xi\in[L,R]\end{aligned}\right\}, \tag{14.15}$$

and boundary conditions (BCs) as

$$\phi_\xi(L,t)=\phi_\xi(R,t)=0\ ;\ \psi_\xi(L,t)=\psi_\xi(R,t)=0.$$

Now the Crank–Nicolson method and time discretization of equation (14.14) by a forward finite difference scheme leads to eqn (14.16):

$$\left.\begin{aligned} \frac{\psi^{(k+1)}-\psi^{(k)}}{\Delta t}+\frac{\alpha}{2}\left[(\phi_{\xi\xi})^{(k+1)}+(\phi_{\xi\xi})^{(k)}\right]+\frac{\beta}{2}\left[\phi^{(k+1)}+\phi^{(k)}\right]+\\ \frac{\gamma}{2}\left[(\phi^3)^{(k+1)}+(\phi^3)^{(k)}\right]=0,\\ \frac{\phi^{(k+1)}-\phi^{(k)}}{\Delta t}-\frac{1}{2}\left[\psi^{(k+1)}+\psi^{(k)}\right]=0. \end{aligned}\right\} \quad (14.16)$$

On linearizing the nonlinear term ϕ^3 by the relation:

$$\left(\phi^3\right)^{(k+1)}=3\phi^{(k+1)}\left(\phi^2\right)^{(k)}-2\left(\phi^3\right)^{(k)}.$$

Thereafter, $(k+1)$th time level terms and kth time level terms are separated, the above eqn (14.16) reduces to:

$$\left.\begin{aligned} \phi^{(k+1)}\left[3\gamma\left(\phi^2\right)^{(k)}+\beta\right]+\alpha\phi_{\xi\xi}^{(k+1)}+\frac{2}{\Delta t}\psi^{(k+1)}=\\ \phi^{(k)}\left[\gamma\left(\phi^2\right)^{(k)}-\beta\right]-\alpha\phi_{\xi\xi}^{(k)}+\frac{2}{\Delta t}\psi^{(k)},\\ \frac{2}{\Delta t}\phi^{(k+1)}-\psi^{(k+1)}=\frac{2}{\Delta t}\phi^{(k)}+\psi^{(k)}. \end{aligned}\right\} \quad (14.17)$$

Taking $(\phi_i)^{(k)}=\mathrm{K}_i$ and approximating ϕ and ψ in eqn (14.17) at each nodal point by using 4th order B-spline approximation.

At ξ_0 first eqn of (14.17) becomes:

$$\begin{aligned}&\left(\omega_{-1}^{(k+1)}+4\omega_0^{(k+1)}+\omega_1^{(k+1)}\right)\left(3\gamma\mathrm{K}_0^2+\beta\right)+\frac{\alpha}{2h^2}\left(14\omega_{-1}^{(k+1)}-33\omega_0^{(k+1)}\right.\\ &\left.+28\omega_1^{(k+1)}-14\omega_2^{(k+1)}+6\omega_3^{(k+1)}-\omega_4^{k+1}\right)+\frac{2}{\Delta t}\left(\eta_{-1}^{(k+1)}+4\eta_0^{(k+1)}\right.\\ &\left.+\eta_1^{(k+1)}\right)=\left(\omega_{-1}^{(k)}+4\omega_0^{(k)}+\omega_1^{(k)}\right)\left(\gamma\mathrm{K}_0^2-\beta\right)-\frac{\alpha}{2h^2}\left(14\omega_{-1}^{(k)}-33\omega_0^{(k)}\right.\\ &\left.+28\omega_1^{(k)}-14\omega_2^{(k)}+6\omega_3^{(k)}-\omega_4^{k}\right)+\frac{2}{\Delta t}\left(\eta_{-1}^{(k)}+4\eta_0^{(k)}+\eta_1^{(k)}\right). \quad (14.18)\end{aligned}$$

For nodal point $\xi_j,\ 1<j<n$ first eqn of (14.17) is:

$$\begin{aligned}&\left(\omega_{j-1}^{(k+1)}+4\omega_j^{(k+1)}+\omega_{j+1}^{(k+1)}\right)\left(3\gamma\mathrm{K}_j^2+\beta\right)+\frac{\alpha}{2h^2}\left(\omega_{j-2}^{(k+1)}+8\omega_{j-1}^{(k+1)}-\right.\\ &\left.18\omega_j^{(k+1)}+8\omega_{j+1}^{(k+1)}+\omega_{j+2}^{(k+1)}\right)+\frac{2}{\Delta t}\left(\eta_{j-1}^{(k+1)}+4\eta_j^{(k+1)}+\eta_{j+1}^{(k+1)}\right)\\ &=\left(\omega_{j-1}^{(k)}+4\omega_j^{(k)}+\omega_{j+1}^{(k)}\right)\left(\gamma\mathrm{K}_j^2-\beta\right)-\frac{\alpha}{2h^2}\left(\omega_{j-2}^{(k)}+8\omega_{j-1}^{(k)}-18\omega_j^{(k)}\right.\end{aligned}$$

$$+8\omega_{j+1}^{(k)} + \omega_{j+2}^{(k)}\Big) + \frac{2}{\Delta t}\left(\eta_{j-1}^{(k)} + 4\eta_j^{(k)} + \eta_{j+1}^{(k)}\right) \tag{14.19}$$

Now, at last nodal point ξ_n, first eqn of (14.17) is:

$$\begin{aligned}
&\left(\omega_{n-1}^{(k+1)} + 4\omega_n^{(k+1)} + \omega_{n+1}^{(k+1)}\right)\left(3\gamma K_n^2 + \beta\right) + \frac{\alpha}{2h^2}\Big(14\omega_{n+1}^{(k+1)} - 33\omega_n^{(k+1)} \\
&+28\omega_{n-1}^{(k+1)} - 14\omega_{n-2}^{(k+1)} + 6\omega_{n-3}^{(k+1)} - \omega_{n-4}^{(k+1)}\Big) + \frac{2}{\Delta t}\Big(\eta_{n-1}^{(k+1)} + 4\eta_n^{(k+1)} \\
&+\eta_{n+1}^{(k+1)}\Big) = \left(\omega_{n-1}^{(k)} + 4\omega_n^{(k)} + \omega_{n+1}^{(k)}\right)\left(\gamma K_n^2 - \beta\right) - \frac{\alpha}{2h^2}\Big(14\omega_{n+1}^{(k)} - 33\omega_n^{(k)} \\
&+28\omega_{n-1}^{(k)} - 14\omega_{n-2}^{(k)} + 6\omega_{n-3}^{(k)} - \omega_{n-4}^{(k+1)}\Big) + \frac{2}{\Delta t}\left(\eta_{n-1}^{(k)} + 4\eta_n^{(k)} + \eta_{n+1}^{(k)}\right)
\end{aligned} \tag{14.20}$$

Also, the second eqn of (14.17) for nodal point $\xi_j,\ 0 \le j \le n$ is:

$$\begin{aligned}
&\frac{2}{\Delta t}\left(\omega_{j-1}^{(k+1)} + 4\omega_j^{(k+1)} + \omega_{j+1}^{(k+1)}\right) - \left(\eta_{j-1}^{(k+1)} + 4\eta_j^{(k+1)} + \eta_{j+1}^{(k+1)}\right) \\
&= \frac{2}{\Delta t}\left(\omega_{j-1}^{(k)} + 4\omega_j^{(k)} + \omega_{j+1}^{(k)}\right) + \left(\eta_{j-1}^{(k)} + 4\eta_j^{(k)} + \eta_{j+1}^{(k)}\right)
\end{aligned} \tag{14.21}$$

Eqns (14.18),(14.19), and (14.20) along with eqn (14.21) can be written as

$$\begin{aligned}
&l_1\omega_{-1}^{(k+1)} + y_3\eta_{-1}^{(k+1)} + l_2\omega_0^{(k+1)} + y_5\eta_0^{(k+1)} + l_3\omega_1^{(k+1)} + y_3\eta_1^{(k+1)} \\
&+l_4\omega_2^{(k+1)} + l_5\omega_3^{(k+1)} - y_1\omega_4^{(k+1)} = l_6\omega_{-1}^{(k)} + y_3\eta_{-1}^{(k)} + l_7\omega_0^{(k)} + y_5\eta_0^{(k)} \\
&\qquad +l_8\omega_1^{(k)} + y_3\eta_1^{(k)} - l_4\omega_2^{(k)} - l_5\omega_3^{(k)} + y_1\omega_4^{(k)}, \\
&y_3\omega_{-1}^{(k+1)} + y_8\eta_{-1}^{(k+1)} + y_5\omega_0^{(k+1)} + y_9\eta_0^{(k+1)} + y_3\omega_1^{(k+1)} + y_8\eta_1^{(k+1)} \\
&\quad = y_3\omega_{-1}^{(k)} - y_8\eta_{-1}^{(k)} + y_5\omega_0^{(k)} - y_9\eta_0^{(k)} + y_3\omega_1^{(k)} - y_8\eta_1^{(k)},
\end{aligned}$$

$$\begin{aligned}
&y_1\omega_{j-2}^{(k+1)} + y_2\omega_{j-1}^{(k+1)} + y_3\eta_{j-1}^{(k+1)} + y_4\omega_j^{(k+1)} + y_5\eta_j^{(k+1)} + y_2\omega_{j+1}^{(k+1)} \\
&+y_3\eta_{j+1}^{(k+1)} + y_1\omega_{j+2}^{(k+1)} = -y_1\omega_{j-2}^{(k)} + y_6\omega_{j-1}^{(k)} + y_3\eta_{j-1}^{(k)} + y_7\omega_j^{(k)} \\
&\qquad +y_5\eta_j^{(k)} + y_6\omega_{j+1}^{(k)} + y_3\eta_{j+1}^{(k)} - y_1\omega_{j+2}^{(k)}, \\
&y_3\omega_{j-1}^{(k+1)} + y_8\eta_{j-1}^{(k+1)} + y_5\omega_j^{(k+1)} + y_9\eta_j^{(k+1)} + y_3\omega_{j+1}^{(k+1)} + y_8\eta_{j+1}^{(k+1)} \\
&\quad = y_3\omega_{j-1}^{(k)} - y_8\eta_{j-1}^{(k)} + y_5\omega_j^{(k)} - y_9\eta_j^{(k)} + y_3\omega_{j+1}^{(k)} - y_8\eta_{j+1}^{(k)},
\end{aligned}$$

$$
\begin{aligned}
&-y_1\omega_{n-4}^{(k+1)} + l_5\omega_{n-3}^{(k+1)} + l_4\omega_{n-2}^{(k+1)} + l_3\omega_{n-1}^{(k+1)} + y_3\eta_{n-1}^{(k+1)} + l_2\omega_n^{(k+1)} \\
&+y_5\eta_n^{(k+1)} + l_1\omega_{n+1}^{(k+1)} + y_3\eta_{n+1}^{(k+1)} = y_1\omega_{n-4}^{(k)} - l_5\omega_{n-3}^{(k)} - l_4\omega_{n-2}^{(k)} + l_8\omega_{n-1}^{(k)} \\
&\qquad +y_3\eta_{n-1}^{(k)} + l_7\omega_n^{(k)} + y_5\eta_n^{(k)} + l_6\omega_{n+1}^{(k)} + y_3\eta_{n+1}^{(k)}, \\
&y_3\omega_{n-1}^{(k+1)} + y_8\eta_{n-1}^{(k+1)} + y_5\omega_n^{(k+1)} + y_9\eta_n^{(k+1)} + y_3\omega_{n+1}^{(k+1)} + y_8\eta_{n+1}^{(k+1)} \\
&\quad = y_3\omega_{n-1}^{(k)} - y_8\eta_{n-1}^{(k)} + y_5\omega_n^{(k)} - y_9\eta_n^{(k)} + y_3\omega_{n+1}^{(k)} - y_8\eta_{n+1}^{(k)}.
\end{aligned}
$$

These equations can be arranged into the following matrix representation as

$$\mathcal{A}\Im^{(k+1)} = \mathcal{C}\Im^{(k)}, \tag{14.22}$$

here $\mathcal{A}$, $\mathcal{C}$, and $\Im$ are given as

$$
\mathcal{A} = \left[\begin{array}{cccccccccccccccc}
l_1 & y_3 & l_2 & y_5 & l_3 & y_3 & l_4 & 0 & l_5 & 0 & -y_1 & & & & & \\
y_3 & y_8 & y_5 & y_9 & y_3 & y_8 & & & & & & & & & & \\
y_1 & 0 & y_2 & y_3 & y_4 & y_5 & y_2 & y_3 & y_1 & & & & & & & \\
 & y_3 & y_8 & y_5 & y_9 & y_3 & y_8 & & & & & & & & & \\
 & y_1 & 0 & y_2 & y_3 & y_4 & y_5 & y_2 & y_3 & y_1 & & & & & & \\
 & & y_3 & y_8 & y_5 & y_9 & y_3 & y_8 & & & & & & & & \\
 & & y_1 & 0 & y_2 & y_3 & y_4 & y_5 & y_2 & y_3 & y_1 & & & & & \\
 & & & y_3 & y_8 & y_5 & y_9 & y_3 & y_8 & & & & & & & \\
 & & & \ddots & & & & & \ddots & & & & & & & \\
 & & & & \ddots & & & & & \ddots & & & & & & \\
 & & & & & & y_1 & 0 & y_2 & y_3 & y_4 & y_5 & y_2 & y_3 & y_1 & \\
 & & & & & & & y_3 & y_8 & y_5 & y_9 & y_3 & y_8 & & & \\
 & & & & & & & y_1 & 0 & y_2 & y_3 & y_4 & y_5 & y_2 & y_3 & y_1 \\
 & & & & & & & & y_3 & y_8 & y_5 & y_9 & y_3 & y_8 & & \\
 & & & & -y_1 & 0 & l_5 & 0 & l_1 & 0 & l_3 & y_3 & l_2 & y_5 & l_1 & y_3 \\
 & & & & & & & & & & y_3 & y_8 & y_5 & y_9 & y_3 & y_8
\end{array}\right],
$$

$$
\mathcal{C} = \begin{bmatrix}
l_6 & y_3 & l_7 & y_5 & l_8 & y_3 & -l_4 & 0 & -l_5 & 0 & y_1 & & & & & \\
y_3 & -y_8 & y_5 & -y_9 & y_3 & -y_8 & & & & & & & & & & \\
-y_1 & y_6 & y_3 & y_7 & y_5 & y_6 & y_3 & -y_1 & & & & & & & & \\
 & y_3 & -y_8 & y_5 & -y_9 & y_3 & -y_8 & & & & & & & & & \\
 & -y_1 & y_6 & y_3 & y_7 & y_5 & y_6 & y_3 & -y_1 & & & & & & & \\
 & & y_3 & -y_8 & y_5 & -y_9 & y_3 & -y_8 & & & & & & & & \\
 & & -y_1 & y_6 & y_3 & y_7 & y_5 & y_6 & y_3 & -y_1 & & & & & & \\
 & & & y_3 & -y_8 & y_5 & -y_9 & y_3 & -y_8 & & & & & & & \\
 & & & \ddots & & & & & \ddots & & & & & & & \\
 & & & & \ddots & & & & & \ddots & & & & & & \\
 & & & & & & & -y_1 & y_6 & y_3 & y_7 & y_5 & y_6 & y_3 & -y_1 & \\
 & & & & & & & & y_3 & -y_8 & y_5 & -y_9 & y_3 & -y_8 & & \\
 & & & & & & & & -y_1 & y_6 & y_3 & y_7 & y_5 & y_6 & y_3 & -y_1 \\
 & & & & & & & & & y_3 & -y_8 & y_5 & -y_9 & y_3 & -y_8 & \\
 & & & & y_1 & 0 & -l_5 & 0 & -l_4 & 0 & l_8 & y_3 & l_7 & y_5 & l_6 & y_3 \\
 & & & & & & & & & & y_3 & -y_8 & y_5 & -y_9 & y_3 & -y_8
\end{bmatrix}
$$

and $\Im = [\omega_{-1}, \eta_{-1}, \omega_0, \eta_0, \ldots, \eta_n, \eta_n, \omega_{n+1}, \eta_{n+1}]^T$. Also, here:

$$l_1 = 3\gamma \mathrm{K}^2 + \beta + \frac{7\alpha}{h^2}, \quad l_2 = 4(3\gamma \mathrm{K}^2 + \beta) - \frac{33\alpha}{2h^2}, \quad l_3 = 3\gamma \mathrm{K}^2 + \beta + \frac{14\alpha}{h^2},$$
$$l_4 = -\frac{7\alpha}{h^2}, \quad l_5 = \frac{3\alpha}{h^2}, \quad l_6 = \gamma \mathrm{K}^2 - \beta - \frac{7\alpha}{h^2}, \quad l_7 = 4(\gamma \mathrm{K}^2 - \beta) + \frac{33\alpha}{2h^2},$$
$$l_8 = \gamma \mathrm{K}^2 - \beta - \frac{14\alpha}{h^2}, \quad y_2 = 3\gamma \mathrm{K}^2 + \beta + \frac{4\alpha}{h^2}, \quad y_4 = 4(3\gamma \mathrm{K}^2 + \beta) - \frac{9\alpha}{h^2},$$
$$y_1 = \frac{\alpha}{2h^2}, \quad y_3 = \frac{2}{\Delta t}, \quad y_5 = \frac{8}{\Delta t}, \quad y_6 = \gamma \mathrm{K}^2 - \beta - \frac{4\alpha}{h^2},$$
$$y_7 = 4(\gamma \mathrm{K}^2 - \beta) + \frac{9\alpha}{h^2}, \quad y_8 = -1, \quad y_9 = -4.$$

It is noted that the resulting discretized system consists of $2n + 2$ linear equations involving a total of $2n + 6$ unknowns. This discrepancy indicates that the system is underdetermined and, therefore, additional conditions are necessary to ensure a unique solution. To close the system, four supplementary conditions are derived from the boundary conditions of the problem, specifically the Neumann-type conditions:

$$\phi_\xi(L, t) = \phi_\xi(R, t) = 0, \quad \psi_\xi(L, t) = \psi_\xi(R, t) = 0.$$

These conditions allow for the elimination of the ghost point variables ω_{-1}, η_{-1}, and ω_{n+1}, η_{n+1}, which lie outside the physical computational domain. After applying these conditions, the system reduces to a solvable set of equations with an equal number of unknowns and equations. Once the initial approximation is available, the resulting system at each time level is efficiently solved using the classical Gaussian elimination method. The construction of the initial approximation is described in detail in Section 14.3.1.

14.3.1 The initial solution

The initial value $(\omega_0^{(0)}, \omega_1^{(0)}, \ldots, \omega_{n-1}^{(0)}, \omega_n^{(0)})$ and $(\eta_0^{(0)}, \eta_0^{(0)}, \ldots, \eta_{n-1}^{(0)}, \eta_n^{(0)})$ can be determined by the initial values of ϕ and ψ and boundary conditions for ϕ_ξ and ψ_ξ:

$$\phi(\xi_j, 0) = f_1(\xi_j), \quad 0 \le j \le n.$$

B-spline approximation of these equations give rise to:

$$\omega_{j-1}^{(0)} + 4\omega_j^{(0)} + \omega_{j+1}^{(0)} = f_1(\xi_j), \quad 0 \le j \le n. \tag{14.23}$$

Using boundary conditions $\phi_\xi(a,t) = \phi_\xi(b,t) = 0$, we have:

$$\omega_{-1}^{(0)} = \omega_1^{(0)} \text{ and } \omega_{n-1}^{(0)} = \omega_{n+1}^{(0)}.$$

Thus eqn (14.23) to following n×n system:

$$\begin{pmatrix} 4 & 2 & & & \\ 1 & 4 & 1 & & \\ & \cdots & \cdots & & \\ & \cdots & \cdots & & \\ & & 1 & 4 & 1 \\ & & & 2 & 4 \end{pmatrix} \begin{pmatrix} \omega_0^{(0)} \\ \omega_1^{(0)} \\ \vdots \\ \omega_{n-1}^{(0)} \\ \omega_n^{(0)} \end{pmatrix} = \begin{pmatrix} f_1(\xi_0) \\ f_1(\xi_1) \\ \vdots \\ f_1(\xi_{n-1}) \\ f_1(\xi_n) \end{pmatrix}.$$

The solution of this system can be easily obtained by Thomas algorithm. In the same manner $(\eta_0^{(0)}, \eta_0^{(0)}, \ldots, \eta_{n-1}^{(0)}, \eta_n^{(0)})$ can also be evaluated.

14.4 Numerical Simulation

This section reports the approximate solutions of the nonlinear K–G eqn (14.1) approximated via implementing the proposed technique, to test and assess its performance, accuracy, and efficiency. The efficiency and accuracy of the proposed technique is assessed by the evaluation of the errors (L_2, & L_∞) in the approximated solutions. L_2, & L_∞ error norms are given as

$$L_2 = \left(\sum_{i=0}^{n} (\phi_i - \Phi_i)^2 \right)^{1/2} \quad \text{and } L_\infty = max\{|\phi_i - \Phi_i| : i = 0, 1, \ldots, n\},$$

where ϕ_i and Φ_i represents the numerical solutions and exact solutions at the grid ξ_i, respectively.

Also, the spatial convergence rate of proposed scheme is assessed by

$$r = \frac{\ln(E_2/E_1)}{\ln(2)},$$

where E_1 and E_2 are L_∞ errors with nodal points n and $2n$, respectively.

Analytic solution of K–G eqn (14.1) for the considered case, that it $P'(\phi) = \beta\phi + \gamma\phi^3$ is given as following:

$$\phi(\xi, t) = \sqrt{\frac{\beta}{\gamma}} \tanh\left(\sqrt{\frac{-\beta}{2(\alpha + c^2)}} (\xi - c\,t) \right). \tag{14.24}$$

The ICs and BCs for this test problem can be extracted from the analytical solution.

Case 1: To assess the efficiency and accuracy of the proposed numerical scheme, we begin by considering the nonlinear Klein–Gordon eqn (14.1) with parameters $\gamma = 1$, $\beta = -1$, and $\alpha = -1$, over the spatial domain $[-30, 30]$. The numerical results are presented in Table 14.1, where the L_2 and L_∞ error norms are computed for different values of the wave speed parameter c ($c = 0.5, 0.05$) at time levels $t = 1, 2, 3$, and 4.

Table 14.2 provides a comparison between the results obtained using the present method and those reported in [12], which employed the Exponential Cubic B-Spline Collocation Method (ExCBCM). Despite the application of optimized parameters in [12], the present scheme demonstrates superior accuracy in terms of both L_2 and L_∞ norms.

Moreover, Table 14.3 confirms the fourth-order spatial convergence of the method, in alignment with the theoretical predictions. The distribution of absolute errors at various spatial nodes for $c = 0.5$ and $c = 0.05$ at $t = 1$ is illustrated in Figure 14.1, while the corresponding surface profiles of the numerical solution are shown in Figure 14.2. These graphical representations further support the robustness and reliability of the proposed numerical approach.

Table 14.1 L_2 and L_∞ errors with $h = 0.3$ and $\Delta t = 0.01$

t	L_2	L_∞	L_2	L_∞
	$c = 0.5$		$c = 0.05$	
1	6.493121E-05	6.213846E-05	3.898110E-05	3.393853E-05
2	9.373511E-05	1.063123E-04	4.739763E-05	3.521336E-05
3	1.154384E-04	1.182828E-04	3.991347E-05	2.823775E-05
4	1.353793E-04	1.162402E-04	3.318105E-05	2.619113E-05

Table 14.2 For $c = 0.5$ at $t = 10$, comparing L_∞ error with that reported in [12].

		Present	ExCBCM[12]
h	Δt		
0.2	0.05	7.02292E-04	3.6155E-03($\rho = 0.0000012$)
0.1	0.02	1.12451E-04	2.5250E-04($\rho = 0.0000053$)
0.05	0.01	2.83142E-05	5.8300E-05($\rho = 0.0000151$)
0.02	0.005	7.07765E-06	7.6000E-06($\rho = 0.0000589$)

Case 2: Next, we consider the nonlinear Klein–Gordon eqn (14.1) with the parameter values $\gamma = 1.5$, $\beta = -1$, and $\alpha = -2.5$, defined over the spatial

Table 14.3 L_∞ error and rate of convergence for case 1 at $t = 1$.

n	c=0.5		c=0.05	
	L_∞	Rate of convergence	L_∞	Rate of convergence
40	2.249052E-02	–	1.332197E-02	–
80	1.506761E-03	3.8998	8.878046E-04	3.9074
160	1.543119E-04	3.2875	7.415165E-05	3.5817
320	7.774637E-06	4.3109	5.315957E-06	3.8021

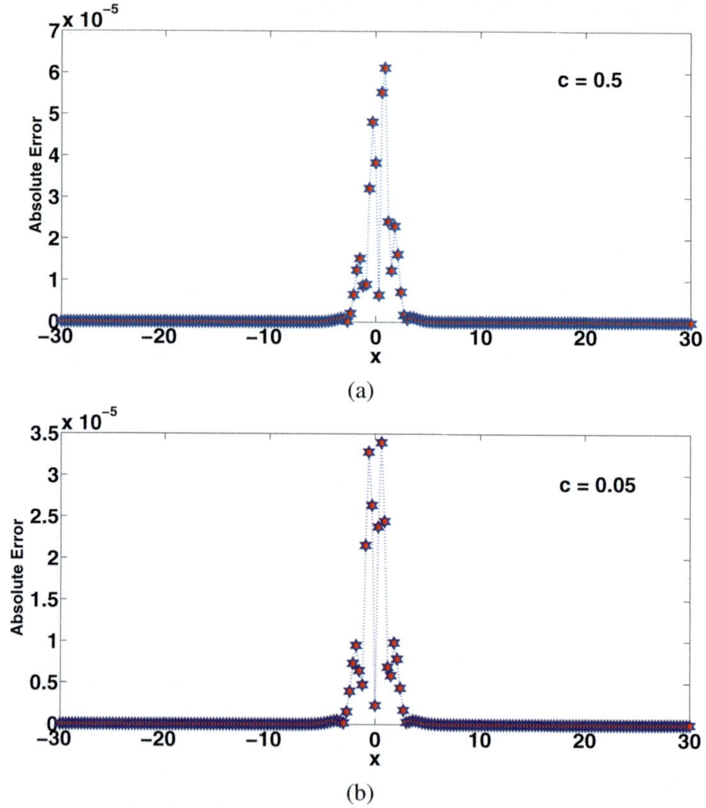

Figure 14.1 Absolute error in case 1 for $t = 1$ and $h = 0.3, \Delta t = 0.01$.

domain $[-30, 30]$. The initial and boundary conditions are constructed using the exact analytical solution of eqn. (14.1).

Table 14.4 presents the computed L_2 and L_∞ error norms for different values of the wave speed parameter c ($c = 0.5, 0.05$), with spatial step size $h = 0.3$ and time step $\Delta t = 0.01$, at time levels $t = 1, 2, 3$, and 4. The spatial convergence behavior of the method is analyzed in Table 14.5, where

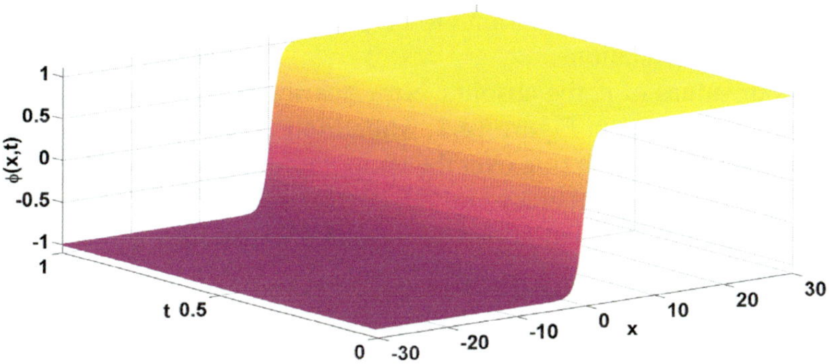

Figure 14.2 Surface behaviour in case 1 for $t = 1$ and $h = 0.3, \Delta t = 0.01$ for c = 0.5.

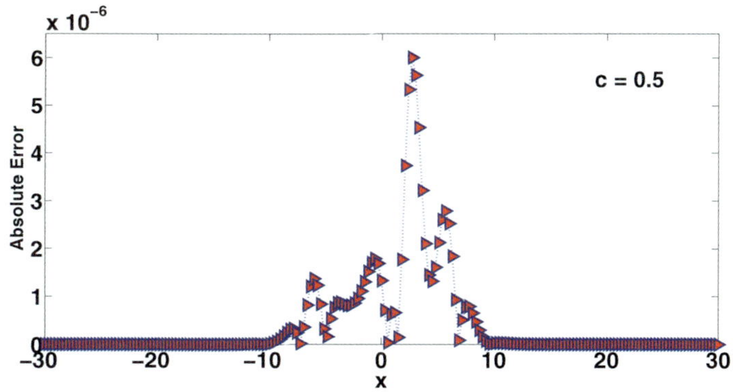

(a)

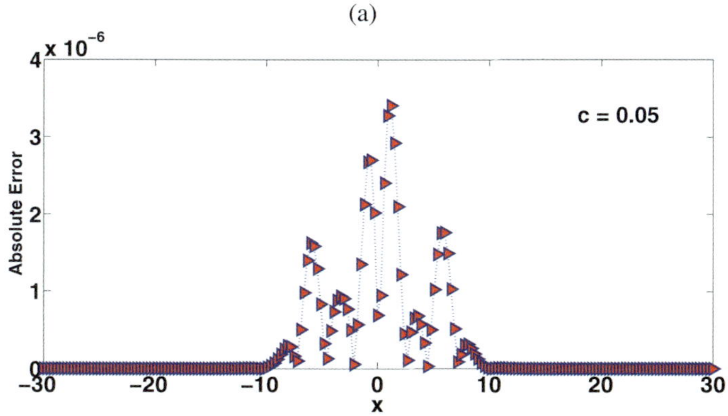

(b)

Figure 14.3 Absolute error in case 2 for $t = 4$ with $h = 0.3, \Delta t = 0.01$.

L_∞ errors are reported for varying numbers of spatial nodes. The observed spatial rate of convergence is found to be of fourth order, in agreement with the theoretical expectations.

For both values of c, the absolute error distributions across spatial nodes at $t = 1$ are illustrated in Figure 14.3, while Figure 14.4 displays the surface plots of the numerical solution over the temporal interval $t \leq 4$. These results further substantiate the accuracy and effectiveness of the proposed numerical approach in solving the nonlinear Klein–Gordon equation.

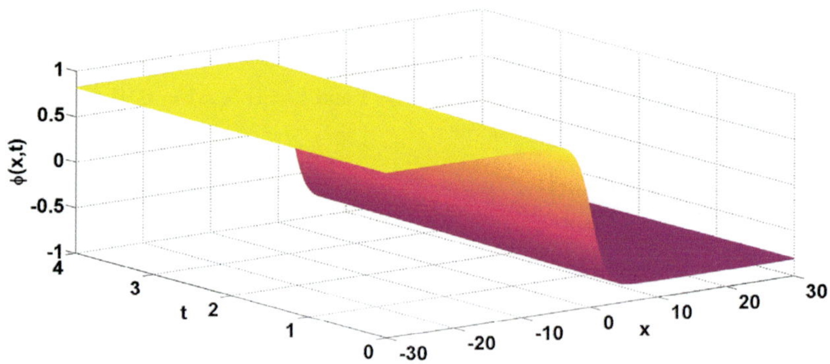

Figure 14.4 Surface behavior in case 2 for $t = 4$ and $h = 0.3$, $\Delta t = 0.01$ for c $= 0.5$.

Table 14.4 L_2 and L_∞ errors with $h = 0.3$ and $\Delta t = 0.01$.

t	L_2	L_∞	L_2	L_∞
	$c = 0.5$		$c = 0.05$	
1	7.178672E-06	5.197219E-06	6.542177E-06	4.532523e-06
2	8.211931E-06	6.398575E-06	7.859561E-06	4.536581E-06
3	8.015366E-06	5.559144E-06	6.602833E-06	3.735268E-06
4	8.085038E-06	6.000669E-06	5.474979E-06	3.407842E-06

Table 14.5 L_∞ error and rate of convergence for case 2 at $t = 1$.

	c=0.5		c=0.05	
n	L_∞	Rate of convergence	L_∞	Rate of convergence
40	2.477704E-03	-	1.752064E-03	-
80	1.772969E-04	3.8048	1.486281E-04	3.5593
160	1.387049E-05	3.6761	1.059827E-05	3.8098
320	5.972200E-07	4.5376	6.929436E-07	3.9349

14.5 Conclusion

In this chapter, the Klein–Gordon (K–G) equation with cubic nonlinearity, an equation that arises in various branches of physics, has been investigated using a collocation method based on cubic B-spline basis functions, offering fourth-order spatial accuracy. The results, presented in Tables 14.1–14.5, report the L_2 and L_∞ error norms along with the spatial rates of convergence for the selected test problems. These results clearly demonstrate that the proposed numerical scheme is both accurate and efficient in approximating the solution of the K–G equation. Furthermore, the fourth-order convergence of the method has been validated numerically. Figures 14.1 and 14.3 illustrate the distribution of absolute errors across various spatial nodes, while Figures 14.2 and 14.4 depict the surface profiles of the numerical solutions, further confirming the accuracy and reliability of the scheme. Based on the outcomes, the proposed approach proves to be a robust and efficient numerical tool for the study of a wide class of linear and nonlinear physical models governed by differential equations.

References

[1] J. K. Perring and T. H. Skyrme. A model unified field equation. *Nucl. Phys.*, 31:550–555, 1962.

[2] A Biswas, A Yildirim, T Hayat, Omar M Aldossary, and R Sassaman. Soliton perturbation theory for the generalized klein-gordon equation with full nonlinearity. *Proceedings of the Romanian Academy, Series A*, 13(1):32–41, 2012.

[3] Anjan Biswas, Chenwi Zony, and Essaid Zerrad. Soliton perturbation theory for the quadratic nonlinear klein–gordon equation. *Applied Mathematics and Computation*, 203(1):153–156, 2008.

[4] PJ Caudrey, JC Eilbeck, and JD Gibbon. The sine-gordon equation as a model classical field theory. *Il Nuovo Cimento B (1971-1996)*, 25(2): 497–512, 1975.

[5] B Bulbul, M Sezer, and W Greiner. Relativistic quantum mechanicsâĂŤwave equations, 2000.

[6] M. Dehghan and A. Shokri. Numerical solution of the nonlinear klein-gordon equation using radial basis functions. *J Comput Appl Math*, 230: 400–410, 2009.

[7] S.M. El-Sayed. The decomposition method for studying the klein-gordon equation. *Chaos Solitions Fractals*, 18:1025–1030, 2003.

[8] N. Ikot Akpan, P. Obong Hillary, and M. Abbey tamunoimi. Approximate analytical solution of klein-gordon equation with an exponential-type potential. *New physics: Sae Mulli*, 65(8):825–836, 2015.

[9] M.G. Hafez and M. Ali Akbar Md. Nur Alam. Exact traveling wave solutions to the klein-gordon equation using novel (g'/g)-expansion method. *Results in Physics*, 14:177–184, 2014.

[10] J. Rashidinia, M. Ghasemi, and Jalilian. Numerical solution of nonlinear klein-gordon equation. *J Comput Appl Math*, 233:1866–1878, 2010.

[11] Sirendaoreji. Auxilary equation method and new solutions of klein-gordon equation. *Chaos Solitions Fractals*, 31(4):943–950, 2007.

[12] O.E. Hepson, A. Korkmaz, and I. Dag. On the numerical solution of the klein-gordon equation by exponential cubic b-spline collocation method. *Commun. Fac. Sci. Univ. Ank. Ser. A1 Math. Stat.*, 68(1): 412–421, 2019.

[13] J. Rashidinia and R. Mohammadi. Tension spline approach for the numerical solution of nonlinear klein-gordon equation. *Comput. Phys. Commun.*, 181:78–91, 2010.

[14] N. Raza, A. Rashid Butt, and A. Javid. Approximate solution of nonlinear klein-gordon equation using sobolev gradients. *Hindawi Publ. Corpor. J. Funct. Sp.*, 2016:1391594, 2016.

[15] F. Yin, T. Tian, J. Song, and M. Zhu. Spectral methods using legendre wavelets for nonlinear klein/sine-gordon equations. *J. Comput. Appl. Math.*, 275:321–334, 2015.

[16] M.S.H. Chowdhury and I. Hashim. Application of homotopy-perturbation method to klein-gordon and sine-gordon equations. *Chaos Solitons Fract.*, 39:1928–1935, 2009.

[17] J. Rashidinia and M. Jokar. Numerical solution of nonlinear klein-gordon equation using polynomial wavelets. *Advances in Intelligent Systems and Computing*, pages 199–214, 2016.

[18] A.R. Kanth and K. Aruna. Differential transform method for solving the linear nonlinear klein-gordon equation. *Comput. Phys. Commun.*, 180: 708–711, 2009.

[19] S Kumbinarasaiah, HS Ramane, KS Pise, and G Hariharan. Numerical-solution-for-nonlinear-klein–gordon equation via operational-matrix by clique polynomial of complete graphs. *International Journal of Applied and Computational Mathematics*, 7:1–19, 2021.

[20] Anas AM Arafa. Analytical solutions for nonlinear fractional physical problems via natural homotopy perturbation method. *International Journal of Applied and Computational Mathematics*, 7(5):179, 2021.

[21] AM Nagy. Numerical solution of time fractional nonlinear klein–gordon equation using sinc–chebyshev collocation method. *Applied Mathematics and Computation*, 310:139–148, 2017.

[22] G Hariharan. Wavelet method for a class of fractional klein-gordon equations. *Journal of Computational and Nonlinear Dynamics*, 8(2): 021008, 2013.

[23] M Ghasemi. An efficient algorithm based on extrapolation for the solution of nonlinear parabolic equations. *International Journal of Nonlinear Sciences and Numerical Simulation*, 19(1):37–51, 2018.

[24] Vijay Kumar Kukreja. An improvised collocation algorithm with specific end conditions for solving modified burgers equation. *Numerical Methods for Partial Differential Equations*, 37(1):874–896, 2021.

[25] Rajni Rohila and R. C. Mittal. Numerical study of reaction diffusion fisher's equation by fourth order cubic b-spline collocation method. *Mathematical Sciences*, 12(2):79–89, 2018.

[26] G Arora, BK Singh, Neetu Singh, and M Gupta. A comparative study: Modified cubic b-spline-based dqm and sixth-order cfds for the klein–gordon equation. In *Advance Numerical Techniques to Solve Linear and Nonlinear Differential Equations*, pages 85–101. River Publishers, 2024.

[27] Mukesh Gupta, Jai Prakash Shukla, and BK Singh. Study of dynamical behavior of burgers' equation in quantum modeling. *Mathematical Methods in the Applied Sciences*, 2023.

[28] Aditi Singh, Sumita Dahiya, and SP Singh. A fourth-order b-spline collocation method for nonlinear burgers–fisher equation. *Mathematical Sciences*, 14(1):75–85, 2020.

[29] Brajesh Kumar Singh and Mukesh Gupta. A new efficient fourth order collocation scheme for solving burgers' equation. *Applied Mathematics and Computation*, 399:126011, 2021.

[30] Brajesh Kumar Singh and Mukesh Gupta. A new efficient fourth order collocation scheme for solving sine–gordon equation. *International Journal of Applied and Computational Mathematics*, 7:1–18, 2021.

[31] Brajesh Kumar Singh and Mukesh Gupta. Numerical solution of fisher's equation by using fourth-order collocation scheme based on modified cubic b-splines. In *Computing and Simulation for Engineers*, pages 209–220. CRC Press, 2022.

15

Numerical Approximation of the Sine-Gordon Equation by using the Hybrid B-spline Differential Quadrature Method

Mohammad Tamsir[1] and Neeraj Dhiman [2]

[1]Department of Mathematics, College of Science, Jazan University, Jazan, Saudi Arabia
[2]Department of Mathematics, Graphic Era Deemed to be University, India
E-mail: tamsiriitm@gmail.com; mtamsir@jazanu.edu.sa; neeraj.dhiman1@gmail.com

Abstract

In this work, we propose a modified cubic hybrid B-Spline (HB-Spline) differential quadrature method (DQM) for the numerical approximation of the sine-Gordon (S-G) equation along with the Neumann boundary conditions. The cubic HB-Spline functions (combinations of CB-Spline and CTB-Spline) are reformed and then applied in the DQM to get the weighting coefficients of spatial derivatives. After applying weighting coefficients, we get systems of ODEs. The SSP-RK(5,4) scheme is used to solve the obtained systems of ODEs. Numerical simulations for the undamped and damped conditions of S-G equation are considered. It commences an excellent agreement with the analytical solution.

Keywords: S-G equation; DQM; HB-spline functions; SSP-RK(5,4) scheme

15.1 Introduction

The classical S-G equation has been employed to model several wave phenomena, such as the phase differences among Josephson junctions,

DOI: 10.1201/9788743806691-15

dislocations in crystals, propagation, torsion waves in pendula and strings, and waves alongside lipid membranes [1]. The authors of [1, 2] proposed cubic B-spline-based DQM and cubic exponential B-spline-based Crank–Nicolson-DQM to solve the 1D S-G equation. Mirzaee et al. [3] used RBFs together with FDM to solve the 2D time-fractional S-G equation.

The 2D sine-Gordon equation has been numerically investigated by different methods like meshless Kriging, RDT, split cosine, Finite Difference, modified predictor-corrector, continuous linear element, LBIE, MLPG, meshless local RPIM, polynomial DQM, semi-implicit FDM, singular boundary and method of approximate particular solutions by many researchers [4–21].

In [22, 23], the authors applied different spline functions with DQM to solve differential equations. Gottlieb [24] developed a SSP-RK (5,4) scheme for solving hyperbolic equations. Shukla and Tamsir [25], and Tamsir et al. [26] presented DQM based on cubic and exponential B-spline functions. In [27], the DQM based on cubic hyperbolic B-spline was used for solving 3D wave equations.

In this work, we propose a DQM based on hybrid B-spline (HB-spline) functions for the S-G equation. The 2D S-G equation is given by

$$\frac{\partial^2 v}{\partial \iota^2} = -\alpha \frac{\partial v}{\partial \iota} + \frac{\partial^2 v}{\partial x^2} + \frac{\partial^2 v}{\partial y^2} - g \sin v(x, y, \iota), (x, y) \in \widehat{D}, \iota > 0, \quad (15.1)$$

where $\widehat{D} = \{(x, y) : \bar{\rho}_1 \leq x \leq \bar{\rho}_2, \hat{\rho}_1 \leq y \leq \hat{\rho}_2\}$, subject to:

$$v(x, y, 0) = \Phi_1(x, y) \text{ and } \frac{\partial v}{\partial \iota}(x, y, 0) = \Phi_2(x, y), \quad (15.2)$$

and Neumann boundary conditions (NBCs):

$$\frac{\partial v}{\partial x}(\bar{\rho}_1, y, \iota) = \varphi_1(y, \iota) \text{ and } \frac{\partial v}{\partial x}(\bar{\rho}_2, y, \iota) = \varphi_2(y, \iota), y \in [\hat{\rho}_1, \hat{\rho}_2], \quad (15.3)$$

$$\frac{\partial v}{\partial y}(x, \hat{\rho}_1, t) = \varphi_3(x, \iota) \text{ and } \frac{\partial v}{\partial y}(x, \hat{\rho}_2, \iota) = \varphi_4(x, \iota), x \in [\bar{\rho}_1, \bar{\rho}_2]. \quad (15.4)$$

In eqn (15.1) g is a current density depending on x and y. The S-G eqn (15.1) becomes an undamped S-G equation when $\alpha = 0$ and acts as a damped when $\alpha > 0$.

The subsequent sections of the article are structured as follows. Section 15.2 emphasizes the implementation of the approach on the sine-Gordon model. Section 15.3 presents the results and discussions pertaining to the applied method, while Section 15.4 provides a conclusion regarding the analysis of the method.

15.2 Hybrid B-spline DQM

The domain $\overline{D}$ is equivalently distributed for the considered problem, with the step size of $h = \frac{\bar{\rho}_2 - \bar{\rho}_1}{N_x - 1}$ and $k = \frac{\hat{\rho}_2 - \hat{\rho}_1}{N_y - 1}$ in x and y-directions, respectively. With the grid points N_x and N_y.

The HB-spline functions $Hb_i(x)$ define for $i = -1, 0, 1, 2, ..., N_x + 1$ as

$$Hb_i(x) = \begin{cases} -\frac{\sigma}{6h^3}(x_{j-2} - x)^3 + \frac{(1-\sigma)}{\varsigma}\hat{\xi}^3(x_{j-2}), \text{if} x \in [x_{j-2}, x_{j-1}), \\ \frac{\sigma}{6h^3}\left(h^3 - 3h^2(x_{j-1} - x) + 3h(x_{j-1} - x)^2 + 3(x_{j-1} - x)^3\right) + \\ \frac{(1-\sigma)}{\varsigma}\left(\hat{\xi}(x_{j-2})\left(\hat{\xi}(x_{j-2})\bar{\xi}(x_i) + \bar{\xi}(x_{j+1})\hat{\xi}(x_{j-1})\right) + \bar{\xi}(x_{j+2})\hat{\xi}^2(x_{j-1})\right), \\ \text{if} x \in [x_{j-1}, x_i), \\ \frac{\sigma}{6h^3}\left(h^3 + 3(x - x_{j+1})^3 + 3h(x - x_{j+1})^2 - 3h^2(x - x_{j+1})\right) + \\ \frac{(1-\sigma)}{\varsigma}\left(\bar{\xi}(x_{j+2})\left(\hat{\xi}(x_{j-1})\bar{\xi}(x_{j+1}) + \bar{\xi}(x_{j+2})\hat{\xi}(x_i)\right) + \bar{\xi}^2(x_{j+1})\hat{\xi}(x_{j-2})\right), \\ \text{if} x \in [x_i, x_{j+1}), \\ -\frac{\sigma}{6h^3}(x - x_{j+2})^3 + \frac{(1-\sigma)}{\varsigma}\bar{\xi}^3(x_{j+2}), \text{if} x \in [x_{j+1}, x_{j+2}) \\ 0, \text{else}. \end{cases} \tag{15.5}$$

The $\hat{\xi}(x_i) = \sin\frac{(x - x_i)}{2}, \bar{\xi}(x_i) = -\hat{\xi}(x_i), \varsigma = \sin\left(\frac{h}{2}\right)\sin(h)\sin\left(\frac{3h}{2}\right)$ and σ is a hybrid parameter. Eqn (15.5) is obtained as

$$Hb_i(x) = \sigma B_i(x) + (1 - \sigma)TB_i(x), \tag{15.6}$$

where, $B_i(x)$ and $TB_i(x)$ are CB-spline functions and CTB-spline functions [28, 29] given by

$$B_i(x) = \begin{cases} -\frac{1}{6h^3}(x_{j-2} - x)^3, \text{if} x \in [x_{j-2}, x_{j-1}), \\ \frac{1}{6h^3}\left(h^3 - 3h^2(x_{j-1} - x) + 3h(x_{j-1} - x)^2 + 3(x_{j-1} - x)^3\right), \\ \text{if} x \in [x_{j-1}, x_i), \\ \frac{1}{6h^3}\left(h^3 - 3h^2(x - x_{j+1}) + 3h(x - x_{j+1})^2 + 3(x - x_{j+1})^3\right), \\ \text{if} x \in [x_i, x_{j+1}), \\ -\frac{1}{6h^3}(x - x_{j+2})^3, \text{if} x \in [x_{j+1}, x_{j+2}), \\ 0, \text{else}, \end{cases}$$

and

$$TB_i(x) = \begin{cases} \frac{1}{\varsigma}\hat{\xi}^3(x_{j-2}), \text{if} x \in [x_{j-2}, x_{j-1}), \\ \frac{1}{\varsigma}\left(\hat{\xi}(x_{j-2})\left(\hat{\xi}(x_{j-2})\bar{\xi}(x_i) + \bar{\xi}(x_{j+1})\hat{\xi}(x_{j-1})\right) + \bar{\xi}(x_{j+2})\hat{\xi}^2(x_{j-1})\right), \\ \text{if} x \in [x_{j-1}, x_i), \\ \frac{1}{\varsigma}\left(\bar{\xi}(x_{j+2})\left(\hat{\xi}(x_{j-1})\bar{\xi}(x_{j+1}) + \bar{\xi}(x_{j+2})\hat{\xi}(x_i)\right) + \bar{\xi}^2(x_{j+1})\hat{\xi}(x_{j-2})\right), \\ \text{if} x \in [x_i, x_{j+1}), \\ \frac{1}{\varsigma}\bar{\xi}^3(x_{j+2}), \text{if} x \in [x_{j+1}, x_{j+2}), \\ 0, \text{else}. \end{cases}$$

Table 15.1 The values of $Hb_i(x)$, $Hb'_i(x)$ and $Hb''_i(x)$.

	x_{j-2}	x_{j-1}	x_j	x_{j+1}	x_{j+2}
$Hb_i(x)$	0	τ_1	τ_2	τ_1	0
$Hb'_i(x)$	0	τ_3	0	τ_4	0
$Hb''_i(x)$	0	τ_5	τ_6	τ_5	0

The HB-spline shape functions form a basis over the domain ($\widehat{D}$). The derivatives of the functions are presented in Table 15.1.

Now, we calculate $\tau_1, \tau_2, \tau_3, \tau_4, \tau_5$ and τ_6 using eqn (15.5) as at the knots x_{j-1}, using $x = x_{j-1}$ in the interval $[x_{j-1}, x_i)$, we get:

$$\tau_1 = \frac{\sigma}{6} + (1-\sigma)\sin^2\left(\frac{h}{2}\right)\csc(h)\csc\left(\frac{3h}{2}\right).$$

At the knots x_j, using $x = x_j$ in the interval $[x_j, x_{j+1})$, we get:

$$\tau_2 = \frac{2\sigma}{3} + \frac{2(1-\sigma)}{1+2\csc(h)}.$$

At the knots x_{j+1}, using $x = x_{j+1}$ in the interval $[x_{j+1}, x_{j+2})$, we get:

$$\tau_1 = \frac{\sigma}{6} + (1-\sigma)\sin^2\left(\frac{h}{2}\right)\csc(h)\csc\left(\frac{3h}{2}\right).$$

Similarly, at the knots x_{j-1},x_j and x_{j+1} in the first and second derivatives of $Hb_i(x)$, we get:

$$\tau_3 = -\left(\frac{\sigma}{2h} + \frac{3}{4}(1-\sigma)\csc\left(\frac{3h}{2}\right)\right),$$

$$\tau_4 = -\tau_3 = \left(\frac{\sigma}{2h} + \frac{3}{4}(1-\sigma)\csc\left(\frac{3h}{2}\right)\right),$$

$$\tau_5 = \frac{\sigma}{h^2} + (1-\sigma)\frac{3(1+3\cos(h))\csc^2\left(\frac{h}{2}\right)}{16\left(2\cos\left(\frac{h}{2}\right)+\cos\left(\frac{3h}{2}\right)\right)},$$

and

$$\tau_6 = -\left(\frac{2\sigma}{h^2} + (1-\sigma)\frac{\left(3\cos^2\left(\frac{h}{2}\right)\csc^2\left(\frac{h}{2}\right)\right)}{(2+4\cos(h))}\right).$$

Now, we approximate $\frac{\partial^r v}{\partial x^r}$ and $\frac{\partial^r v}{\partial y^r}$ at $(x_{\mathrm{i}}, y_{\mathrm{j}})$ as follows:

$$\frac{\partial^r v_{\mathrm{jj}}}{\partial x^r} = \sum_{s=1}^{N_{\mathrm{x}}} a_{\mathrm{js}}^{(r)} u_{sj}. \tag{15.7}$$

$$\frac{\partial^r v_{\mathrm{jj}}}{\partial y^r} = \sum_{s=1}^{N_{\mathrm{y}}} \tilde{a}_{\mathrm{js}}^{(r)} u_{\mathrm{js}}, \tag{15.8}$$

where $i = 1, 2, 3....N_{\mathrm{x}}; j = 1, 2, 3....Ny$ and $v_{\mathrm{jj}} = v(x_{\mathrm{i}}, y_{\mathrm{j}}, \iota)$. Now, the HB-splines are modified as [4, 5, 2, 30, 31, 32, 33, 34]

$$\left.\begin{array}{l} \tilde{H}b_1(x) = Hb_1(x) + 2Hb_0(x) \\ \tilde{H}b_2(x) = Hb_2(x) - Hb_0(x) \\ \tilde{H}b_{m=3,4,.....,N_{\mathrm{x}}-2}(x) = Hb_m(x) \\ \tilde{H}b_{N_{\mathrm{x}}-1}(x) = Hb_{N_{\mathrm{x}}-1}(x) - Hb_{N_{\mathrm{x}}+1}(x) \\ \tilde{H}b_{N_{\mathrm{x}}}(x) = Hb_{N_{\mathrm{x}}}(x) + 2Hb_{N_{\mathrm{x}}+1}(x) \end{array}\right\}, \tag{15.9}$$

such that, $\left\{\tilde{H}b_0, \tilde{H}b_1, ..., \tilde{H}b_{N_{\mathrm{x}}}, \tilde{H}b_{N_{\mathrm{x}}+1}\right\}$ forms a basis under the concerned domain. Making use of $r = 1$ in eqn (15.7) and substituting $\tilde{H}b_{\mathrm{j}}(x)$, $j = 1, 2, ...N_{\mathrm{x}}$, we get a linear system:

$$\tilde{H}b'_{\mathrm{k}}(x_{\mathrm{i}}) = \sum_{j=1}^{N_{\mathrm{x}}} a_{\mathrm{jj}}^{(1)} \tilde{H}b_{\mathrm{k}}(x_{\mathrm{j}}), \quad \text{for } i, k = 1, 2, ..., N_{\mathrm{x}}, \tag{15.10}$$

which can be written as

$$\begin{aligned} \tilde{H}b'_{\mathrm{k}}(x_{\mathrm{i}}) &= a_{\mathrm{j1}}^{(1)} \tilde{H}b_{\mathrm{k}}(x_1) + a_{\mathrm{j2}}^{(1)} \tilde{H}b_{\mathrm{k}}(x_2) + a_{\mathrm{j3}}^{(1)} \tilde{H}b_{\mathrm{k}}(x_3) + ... \\ &+ a_{\mathrm{jN_x}}^{(1)} \tilde{H}b_{\mathrm{k}}(x_{N_{\mathrm{x}}}), i, k = 1, 2, ..., N_{\mathrm{x}}. \end{aligned} \tag{15.11}$$

For $i = 1$ and $k = 1$, using eqn (15.9) in eqn (15.11), we get:

$$\begin{aligned} Hb'_1(x_1) + 2Hb'_0(x_1) &= (Hb_1(x_1) + 2Hb_0(x_1))\, a_{11}^{(1)} + (Hb_1(x_2) \\ &+2Hb_0(x_2))\, a_{12}^{(1)} + (Hb_1(x_3) + 2Hb_0(x_3))\, a_{13}^{(1)} \\ &+ ... + (Hb_1(x_{N_{\mathrm{x}}}) + 2Hb_0(x_{N_{\mathrm{x}}}))\, a_{1N_{\mathrm{x}}}^{(1)}. \end{aligned}$$

Using the values from Table 15.1 in the above equation, we get:

$$2\tau_4 = (\tau_2 + 2\tau_1)\, a_{11}^{(1)} + \tau_1 a_{12}^{(1)}. \tag{15.12}$$

Next, for $i = 1$ and $k = 2$, using eqn (15.9) in eqn (15.11), we get:

$$Hb_1'(x_1) + 2Hb_0'(x_1) = (Hb_1(x_1) + 2Hb_0(x_1))\, a_{11}^{(1)} + (Hb_1(x_2) \\ +2Hb_0(x_2))\, a_{12}^{(1)} + (Hb_1(x_3) + 2Hb_0(x_3))\, a_{13}^{(1)} + ... \\ + (Hb_1(x_{N_x}) + 2Hb_0(x_{N_x}))\, a_{1N_x}^{(1)}.$$

Using the values from Table 15.1 in the above equation, we get:

$$2\tau_3 = \tau_2 a_{12}^{(1)} + \tau_1 a_{12}^{(1)}. \tag{15.13}$$

Next, for $i = 1$ and $k = 3, 4, ..., N_x - 2$, using eqn (15.9) in eqn (15.11), we get:

$$Hb_k'(x_1) = a_{11}^{(1)} Hb_k(x_1) + a_{12}^{(1)} Hb_k(x_2) + a_{13}^{(1)} Hb_k(x_3) + ... \\ + a_{1N_x}^{(1)} Hb_k(x_{N_x}), k = 3, 4, ..., N_x - 2.$$

Using the values from Table 15.1 in the above equation, we get:

$$\begin{cases} 0 = \tau_1 a_{13}^{(1)} + \tau_2 a_{14}^{(1)} + \tau_1 a_{15}^{(1)}, \\ 0 = \tau_1 a_{14}^{(1)} + \tau_2 a_{15}^{(1)} + \tau_1 a_{16}^{(1)}, \\ \vdots \\ 0 = \tau_1 a_{1(N_x-4)}^{(1)} + \tau_2 a_{1(N_x-3)}^{(1)} + \tau_1 a_{1(N_x-2)}^{(1)}. \end{cases} \tag{15.14}$$

Next, for $i = 1$ and $k = N_x - 1$, using eqn (15.9) in eqn (15.11), we get:

$$Hb_{N_x-1}'(x_1) - Hb_{N_x+1}'(x_1) = (Hb_{N_x-1}(x_1) - Hb_{N_x+1}(x_1))\, a_{11}^{(1)} + ... \\ + (Hb_{N_x-1}(x_{N_x-2}) - Hb_{N_x+1}(x_{N_x-2})) \times a_{1(N_x-2)}^{(1)} \\ + (Hb_{N_x-1}(x_{N_x-1}) - Hb_{N_x+1}(x_{N_x-1}))\, a_{1(N_x-1)}^{(1)} \\ + (Hb_{N_x-1}(x_{N_x}) - Hb_{N_x+1}(x_{N_x}))\, a_{1N_x}^{(1)}$$

Using the values from Table 15.1 in the above equation, we get:

$$0 = \tau_1 a_{1(N_x-2)}^{(1)} + \tau_2 a_{1(N_x-1)}^{(1)}. \tag{15.15}$$

Finally, for $i = 1$ and $k = N_x$, using eqn (15.9) in eqn (15.11), we get:

$$Hb_{N_x}'(x_1) + 2Hb_{N_x+1}'(x_1) = (Hb_{N_x}(x_1) + 2Hb_{N_x+1}(x_1))\, a_{11}^{(1)} + \cdots \\ + (Hb_{N_x}(x_{N_x-1}) + 2Hb_{N_x+1}(x_{N_x-1}))\, a_{1(N_x-1)}^{(1)} \\ + (Hb_{N_x}(x_{N_x}) + 2Hb_{N_x+1}(x_{N_x}))\, a_{1N_x}^{(1)}.$$

Now, using the values from Table 15.1 in above equation, we get:

$$0 = \tau_1 a^{(1)}_{1(N_x-1)} + (2\tau_1 + \tau_2)\, a^{(1)}_{1N_x}. \tag{15.16}$$

Eqn (15.12)–(15.16) lead to a system of linear equations as follows

$$\begin{bmatrix} 2\tau_1+\tau_2 & \tau_1 & & & & & \\ 0 & \tau_2 & \tau_1 & & & & \\ & \tau_1 & \tau_2 & \tau_1 & & & \\ & & \ddots & \ddots & \ddots & & \\ & & & \tau_1 & \tau_2 & \tau_1 & \\ & & & & \tau_1 & \tau_2 & 0 \\ & & & & & \tau_1 & 2\tau_1+\tau_2 \end{bmatrix} \begin{bmatrix} a^{(1)}_{11} \\ a^{(1)}_{12} \\ a^{(1)}_{13} \\ a^{(1)}_{14} \\ \vdots \\ a^{(1)}_{1(N_x-1)} \\ a^{(1)}_{1N_x} \end{bmatrix} = \begin{bmatrix} 2\tau_0 \\ 2\tau_3 \\ 0 \\ 0 \\ 0 \\ \vdots \\ 0 \end{bmatrix}. \tag{15.17}$$

In the same way, for $i = 2, 3, \ldots\ldots, N_x$ from eqn (15.11), we have:

$$A\vec{a}^{(1)}[i] = R[i], \text{for} i = 2, 3, \cdots, N_x, \tag{15.18}$$

where $A_{N_x \times N_x}$ is given by

$$A = \begin{bmatrix} 2\tau_1+\tau_2 & \tau_1 & & & & & \\ 0 & \tau_2 & \tau_1 & & & & \\ & \tau_1 & \tau_2 & \tau_1 & & & \\ & & \ddots & \ddots & \ddots & & \\ & & & \tau_1 & \tau_2 & \tau_1 & \\ & & & & \tau_1 & \tau_2 & 0 \\ & & & & & \tau_1 & 2\tau_1+\tau_2 \end{bmatrix}$$

$$\vec{a}^{(1)}[i] = [a^{(1)}_{j1}, a^{(1)}_{j2}, a^{(1)}_{j3}, \cdots, a^{(1)}_{jN_x}]^{tr}, i = 2, 3, \cdots, N_x,$$

and

$$\begin{aligned} &\vec{R}[2] = \begin{bmatrix} \tau_4 & 0 & \tau_3 & 0 & \cdots & 0 & 0 \end{bmatrix}, \\ &\vec{R}[3] = \begin{bmatrix} 0 & \tau_4 & 0 & \tau_3 & \cdots & 0 & 0 \end{bmatrix}, \\ &\vdots \\ &\vec{R}[N_x - 1] = \begin{bmatrix} 0 & 0 & \cdots & 0 & \tau_4 & 0 & \tau_3 \end{bmatrix}, \\ &\vec{R}[N_x] = \begin{bmatrix} 0 & 0 & \cdots & 0 & 0 & 2\tau_4 & 2\tau_3 \end{bmatrix}. \end{aligned}$$

Next, we have to find coefficients $a_{\rm jj}^{(2)}$, i, j $=$ $1, 2, ..., N_{\rm x}$, where the calculation is done by using the formula of Shu [23] and shown as follows:

$$\begin{cases} a_{\rm jj}^{(2)} = 2\left(a_{\rm jj}^{(1)} a_{\rm ji}^{(1)} - \frac{a_{\rm jj}^{(1)}}{x_{\rm i}-x_{\rm j}}\right), \text{for} j \neq i \text{and} i = 1, 2, ..., N_{\rm x}, \\ a_{\rm ji}^{(2)} = -\sum\limits_{j=1, j\neq i}^{N_{\rm x}} a_{\rm jj}^{(2)}, \text{for} j = i. \end{cases} \tag{15.19}$$

By using modified form of HB-spline shape function in DQM, eqn (15.1) becomes:

$$\frac{d^2 v_{\rm jj}}{d\iota^2} + \alpha \frac{d v_{\rm jj}}{d\iota} = \sum_{s=1}^{N_{\rm x}} a_{\rm js}^{(2)} v_{sj} + \sum_{s=1}^{N_{\rm y}} \tilde{a}_{js}^{(2)} v_{\rm js} - g_{\rm jj} \sin v_{\rm jj}, \\ i = 1, 2, 3......., N_{\rm x}; j = 1, 2, 3......, N_{\rm y}, \tag{15.20}$$

along with the initial conditions:

$$v_{\rm jj} = \Phi_1 (x_{\rm i}, y_{\rm j}) \text{ and } \frac{d v_{\rm jj}}{d\iota} = \Phi_2 (x_{\rm i}, y_{\rm j}) \, \text{at} \iota = 0. \tag{15.21}$$

Now let:

$$\frac{d v_{\rm jj}}{d\iota} = \hat{v}_{\rm jj}. \tag{15.22}$$

By using eqn (15.20) to eqn (15.22), we get:

$$\frac{d\hat{v}_{\rm jj}}{d\iota} + \alpha \hat{v}_{\rm jj} = \sum_{s=1}^{N_{\rm x}} a_{\rm js}^{(2)} v_{sj} + \sum_{s=1}^{N_{\rm y}} \tilde{a}_{js}^{(2)} v_{\rm js} - g_{\rm jj} \sin v_{\rm jj}. \tag{15.23}$$

From the NBCs (3) with $x = \bar{\rho}_1$ and $x = \bar{\rho}_2$, we have:

$$\sum_{s=1}^{N_{\rm x}} a_{1s}^{(1)} v_{sj} = \varsigma_1(\bar{\rho}_1, y, \iota), \tag{15.24}$$

and

$$\sum_{s=1}^{N_{\rm x}} a_{N_{\rm x}s}^{(1)} v_{sj} = \varsigma_2(\bar{\rho}_2, y, \iota). \tag{15.25}$$

Eqn (15.24) and (15.25) produce:

$$v_{1j} = \frac{a_{1N_{\rm x}}^{(1)} (\varsigma_2 - A_2) - a_{N_{\rm x}N_{\rm x}}^{(1)} (\varsigma_1 - A_1)}{\left(a_{1N_{\rm x}}^{(1)} a_{N_{\rm x}1}^{(1)} - a_{11}^{(1)} a_{N_{\rm x}N_{\rm x}}^{(1)}\right)}, \tag{15.26}$$

$$v_{N_x j} = \frac{a^{(1)}_{N_x 1}\left(\varsigma_1 - A_1\right) - a^{(1)}_{11}\left(\varsigma_2 - A_2\right)}{\left(a^{(1)}_{1N_x} a^{(1)}_{N_x 1} - a^{(1)}_{11} a^{(1)}_{N_x N_x}\right)}, \tag{15.27}$$

where $A_1 = \sum_{s=2}^{N_x-2} a^{(1)}_{1s} v_{sj}$, $A_2 = \sum_{s=2}^{N_x-2} a^{(1)}_{N_x s} v_{sj}$, $\varsigma_1 = \varsigma_1\left(\bar{\rho}_1, y, \iota\right)$, and $\varsigma_2 = \varsigma_2\left(\bar{\rho}_2, y, \iota\right)$.

Similarly, from the remaining boundary conditions, eqn (15.3) at $y = \hat{\rho}_1$ and $y = \hat{\rho}_2$, we get:

$$\sum_{s=1}^{N_y} \tilde{a}^{(1)}_{1s} v_{\mathrm{js}} = \varsigma_3, \tag{15.28}$$

$$\sum_{s=1}^{N_y} \tilde{a}^{(1)}_{N_y s} v_{\mathrm{js}} = \varsigma_4. \tag{15.29}$$

Eqn (15.28) and (15.29) give:

$$v_{\mathrm{j}1} = \frac{\tilde{a}^{(1)}_{1N_y}\left(\varsigma_4 - A_4\right) - \tilde{a}^{(1)}_{N_y N_y}\left(\varsigma_3 - A_3\right)}{\left(\tilde{a}^{(1)}_{1N_y} \tilde{a}^{(1)}_{N_y 1} - \tilde{a}^{(1)}_{11} \tilde{a}^{(1)}_{N_y N_y}\right)}, \tag{15.30}$$

$$v_{\mathrm{j}N_y} = \frac{\tilde{a}^{(1)}_{N_y 1}\left(\varsigma_3 - A_3\right) - \tilde{a}^{(1)}_{11}\left(\varsigma_4 - A_4\right)}{\left(\tilde{a}^{(1)}_{1N_y} \tilde{a}^{(1)}_{N_y 1} - \tilde{a}^{(1)}_{11} \tilde{a}^{(1)}_{N_y N_y}\right)}, \tag{15.31}$$

where $A_3 = \sum_{s=2}^{N_y-2} \tilde{a}^{(1)}_{1s} v_{\mathrm{js}}$, $A_4 = \sum_{s=2}^{N_y-2} \tilde{a}^{(1)}_{N_y s} v_{\mathrm{js}}$, $\varsigma_3 = \varsigma_3\left(x, \hat{\rho}_1, \iota\right)$, $\varsigma_4 = \varsigma_4\left(x, \hat{\rho}_2, \iota\right)$, and $i = 1, 2, ..., N_x$.

The obtain system from eqn (15.22) and (15.23) solved using SSP-RK(5,4) [24].

15.3 Findings and Analysis

In this section, four distinct examples have been examined for the purpose of numerical approximation. First, Example 15.3.1 is approximated to validate the accuracy of the anticipated method. The accuracy is measured in terms of different error norms like L_∞ and RMS. These error norms and ROC have been calculated by using the following formulas:

$$L_\infty = \max_{\mathrm{j,j}} \left|v^{ex}_{\mathrm{jj}} - v^{nu}_{\mathrm{jj}}\right|,$$

$$\mathrm{RMS} = \sqrt{\frac{1}{N_{\mathrm{x}}N_{\mathrm{y}}}\sum_{\mathrm{j}=1}^{N_{\mathrm{x}}}\sum_{j=1}^{N_{\mathrm{y}}}(v_{\mathrm{jj}}^{ex} - v_{\mathrm{jj}}^{nu})^2},$$

$$\mathrm{ROC} = \frac{\log(er^{h_1}/er^{h_2})}{\log(h_1/h_2)},$$

where $i = 1,2,3....N_{\mathrm{x}}$ and $j = 1,2,3.....N_{\mathrm{y}}$. The terms v_{jj}^{ex} and v_{jj}^{nu} represents exact and numerical solutions while er representing the error at the grid h_1 and h_2.

Example 15.3.1: First, we solve problem (1) for validating the accuracy with $g(x,y) = -1$ and $\alpha = 0$ in [-7,7] [-7,7] as in [12, 17, 20, 21, 25, 29] along with:

$$v\,(x,y,0) = 4\tan^{-1}\left(e^{x+y}\right) \text{ and } \frac{\partial v}{\partial \iota}\,(x,y,0) = -\frac{4e^{x+y}}{1+e^{2(x+y)}}, \quad (15.32)$$

and NBCs

$$\left.\begin{array}{l} \frac{\partial v}{\partial x} = -\frac{4\mathrm{e}^{x+y+t}}{\mathrm{e}^{(2\iota)}+\mathrm{e}^{2(x+y)}}, x = -7, x = 7 \text{ and } y \in [-7,7], \\ \frac{\partial v}{\partial y} = -\frac{4\mathrm{e}^{x+y+t}}{e^{2\iota}+e^{2(x+y)}}, y = -7, y = 7, \text{ and } x \in [-7,7], \end{array}\right\} \quad (15.33)$$

and with $v\,(x,y,\iota) = 4\tan^{-1}\left(\mathrm{e}^{x+y-\iota}\right)$.

The results we obtain for Example 15.3.1 by considering the parameters: $\Delta\iota = 1.0e-02$, $h = 1e-01$, and $\sigma = \frac{1}{2}$. The L_∞ and RMS errors norms for the method applied in the study is numerically compared with the method High order compact ADI method (HADIM) [21] and LMAPS [20] at various times. According to Table 15.2, the numerical results obtained in this study are superior to those found in [20]. In Table 15.3, the obtained results are better than the results in literature [12, 17, 25, 33] for the parameters $\sigma = 1/2, \Delta\iota = 1.0e-03$, and $h = 1/4$ at different times ι. The analysis results are also compared with the obtained results with considering the same parameters $\sigma = 1/2\ \Delta\iota = 1.0e-03$, at different times ι (see Table 15.4). The findings illustrated in Table 15.4 demonstrate a superiority over the results documented in the litrature. Table 15.5 illustrates the root mean square (RMS) values and their corresponding errors norms for the parameters $t = 0.1, 0.3, 0.5$, and 1, which are strikingly minimal, thus validating the accuracy of the proposed method. It is evident that the performance of the proposed approach is commendable. Figure 15.1 shows the errors against $\Delta\iota = 0.05$ at $t = 1$.

Table 15.2 The comparison of RMS and L_∞ with refs. [20, 21] for Example 15.3.1

ι	LMAPS [20] $L_\infty \times 10^{-4}$	RMS $\times 10^{-5}$	Present $L_\infty \times 10^{-4}$	RMS $\times 10^{-4}$
1	4.02	2.18	0.385	0.211
3	2.52	2.82	0.923	0.573
5	3.95	3.62	1.22	1.00
7	5.85	4.01	1.81	1.53

Table 15.3 The results obtained from Example 15.3.1 are being compared.

ι	L_∞ (10^{-4})					RMS (10^{-4})			
	[17] $\times 10^{-3}$	[12] $\times 10^{-2}$	[25] $\times 10^{-4}$	[33] $\times 10^{-2}$	Current Method $\times 10^{-4}$	[17] $\times 10^{-4}$	[12] $\times 10^{-2}$	[25] $\times 10^{-4}$	Current $\times 10^{-4}$
1	2.7	6.70	3.0	3.50	2.53	5.0	0.50	2.0	1.54
3	2.0	8.34	6.0	4.31	5.81	0.5	1.03	4.0	3.80
5	3.3	10.1	8.0	4.04	7.76	7.0	1.45	7.0	6.64
7	5.9	15.2	12.0	3.53	11.5	11	1.87	10	9.95

Table 15.4 The L_∞ error at different times for Example 15.3.1.

ι	1	3	5	7
Current method $\times 10^{-4}$	2.22	5.03	6.75	10.0
Available results $\times 10^{-3}$	3.4	3.8	4.5	4.9

Table 15.5 The errors at various times ι for Example 15.3.1.

ι	0.1	0.3	0.5	0.8	1.0
RMS $\times 10^{-5}$	2.463	4.588	7.754	12.69	15.97
$L_\infty \times 10^{-4}$	0.4536	0.8104	1.331	2.124	2.637

Example 15.3.2:

This example is about studying the circular ring solitons of the problem (1) for $g(x, y) = -1$ and $\alpha = 0$ in [-7,7] [-7,7]:

$$v(x, y, 0) = 4\tan^{-1}\left(e^{\left(3-\sqrt{x^2+y^2}\right)}\right) \text{ and } \frac{\partial v}{\partial \iota}(x, y, 0) = 0, \tag{15.34}$$

and NBCs

$$\frac{\partial v}{\partial x} = 0 \text{ for } x = \pm 7 \text{ and } y \in [-7, 7], \frac{\partial v}{\partial y} = 0, \text{ for } y = \pm 7 \text{ and } x \in [-7, 7]. \tag{15.35}$$

In this example, we obtained circular ring soliton waves when $h = 0.4$ and $\Delta\iota = 0.2$ with the changes in time ι values as shown in Figure 15.2.

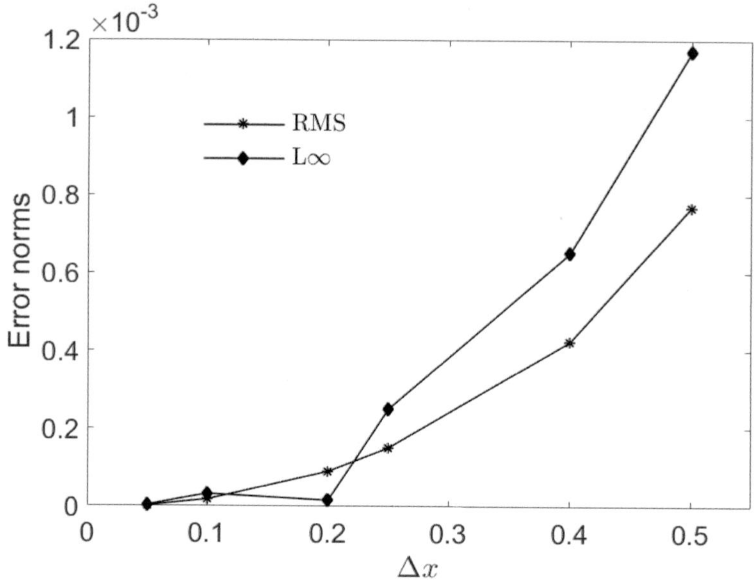

Figure 15.1 The graphical representation of error norms L_∞ and RMS errors with $\Delta\iota = 0.05$ at $t = 1$ for Example 15.3.1.

We can see from these plots that the ring solitons shrink from their primary position until $\iota = 2.8$ and appear like a single soliton. The radiation can be seen when time $\iota = 5.6$ and the expansion phase starts from this point and remains until $\iota = 11.2$. The ring solitons are almost reconstructed at $\iota = 11.2$. The surface plots shown in Figure 15.2 have excellent agreement with those given in [5, 2, 8, 16, 19, 25].

Example 15.3.3:

We show that the elliptical breather is seen for the S-G eqn (15.1) when $g(x, y) = -1$ and $\alpha = 0$ in $[-7, 7] \times [-7, 7]$ with:

$$\left.\begin{aligned} &v(x,y,0) = 4\tan^{-1}\left(2\text{sech}\left(0.866\sqrt{\tfrac{(x-y)^2}{3} + \tfrac{(x+y)^2}{2}}\right)\right), \\ &\tfrac{\partial v}{\partial \iota}(x,y,0) = 0. \end{aligned}\right\} \quad (15.36)$$

We obtained the elliptical breather solitons with $h = 0.4$ and $\Delta\iota = 0.2$ at $\iota = 0, 1.6, 8.0, 9.6, 11.2, 12.8$, and presented them in Figure 15.3. Figure 15.3 shows that the curves of the elliptical breather solitons turn to be clockwise, shrinking from the initial direction $y - x$ up to $\iota = 1.6$. A refection phase happens at $\iota = 8$ and $\iota = 9.6$. The initial direction $y = -x$ has been almost

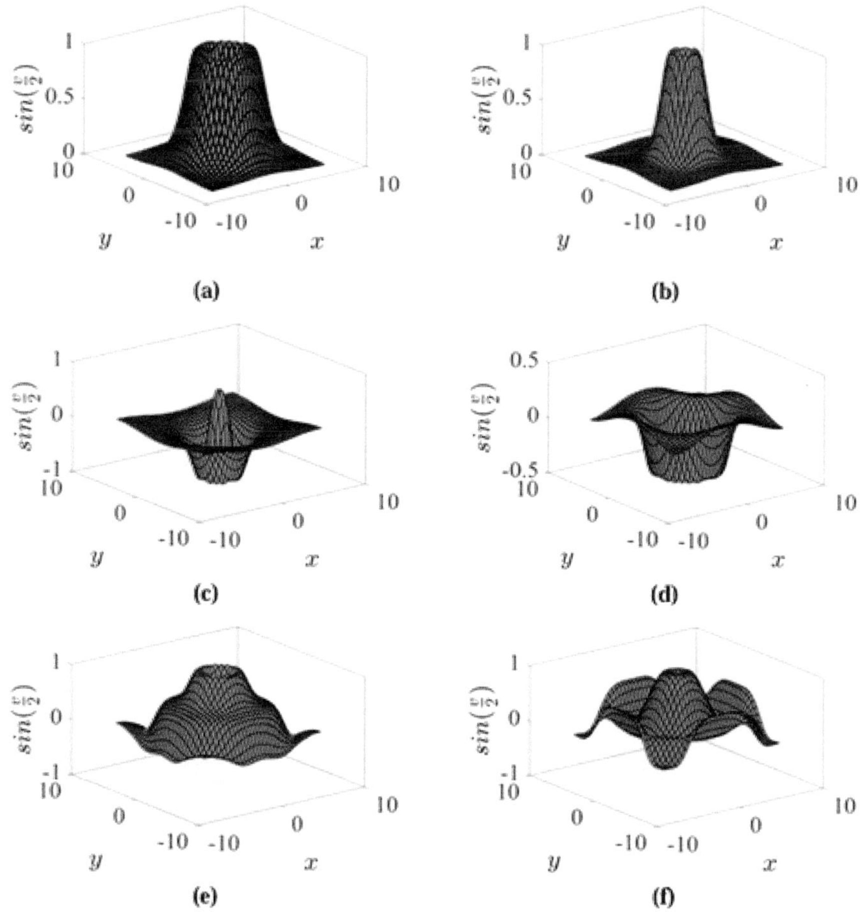

Figure 15.2 The surface plots of circular ring solitons at $\iota =$ (a) 0, (b) 2.8, (c) 5.6, (d) 8.4, (e) 11.2, and (f) 12.6 for Example 15.3.2.

recovered at $\iota = 11.2$ and we note robust oscillations. The graphs presented in Figure 15.2 have excellent agreement with those given in [5, 2, 9, 16, 17, 25].

Example 15.3.4:

Next, we studied the line solitons for when $g(x, y) = -1$ and $\alpha = 0.05$ in $[-7, 7] \times [-7, 7]$ with:

$$\begin{array}{l} v\,(x, y, 0) = 4\tan^{-1}\left(\exp\left\{\frac{x-3.5}{0.954}\right\}\right) \\ \text{and}\frac{\partial v}{\partial \iota}\,(x, y, 0) = 0.629\sec h\left(\exp\left\{\frac{x-3.5}{0.954}\right\}\right) \end{array}. \qquad (15.37)$$

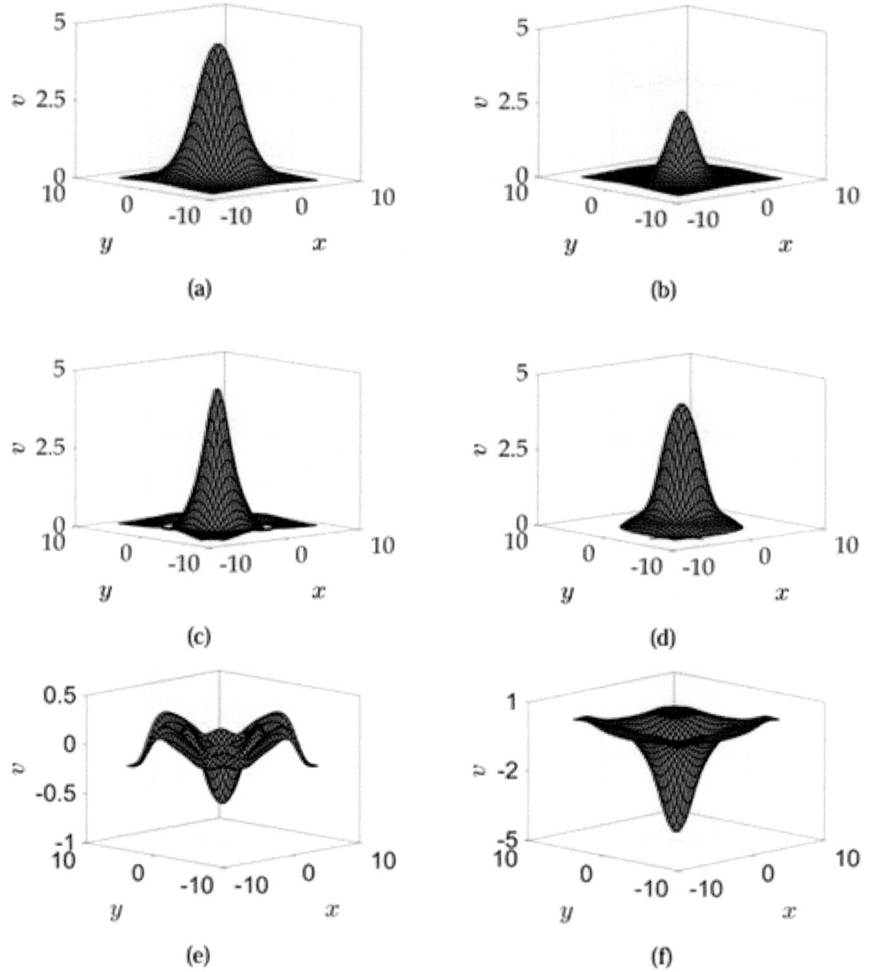

Figure 15.3 The surface plots of an elliptical breather at $\iota =$ (a) 0, (b) 1.6, (c) 8.0, (d) 9.6, (e) 11.2, and (f) 12.8 for Example 15.3.3.

This example is estimated numerically with $h = 0.5$ and $\Delta\iota = 0.001$ at $\iota = 0, 5, 10$, and 15. The surface plots are shown in Figure 15.3. We can see from the graph that the line soliton goes just about in a straight away for the period of diffusion throughout the inhomogeneous medium. As $\iota =$ increases from 5 to 10, a twist appears and a movement seems to be prevented until $\iota = 15$. Subsequently, the solitons recover their straightness at $\iota = 20$.

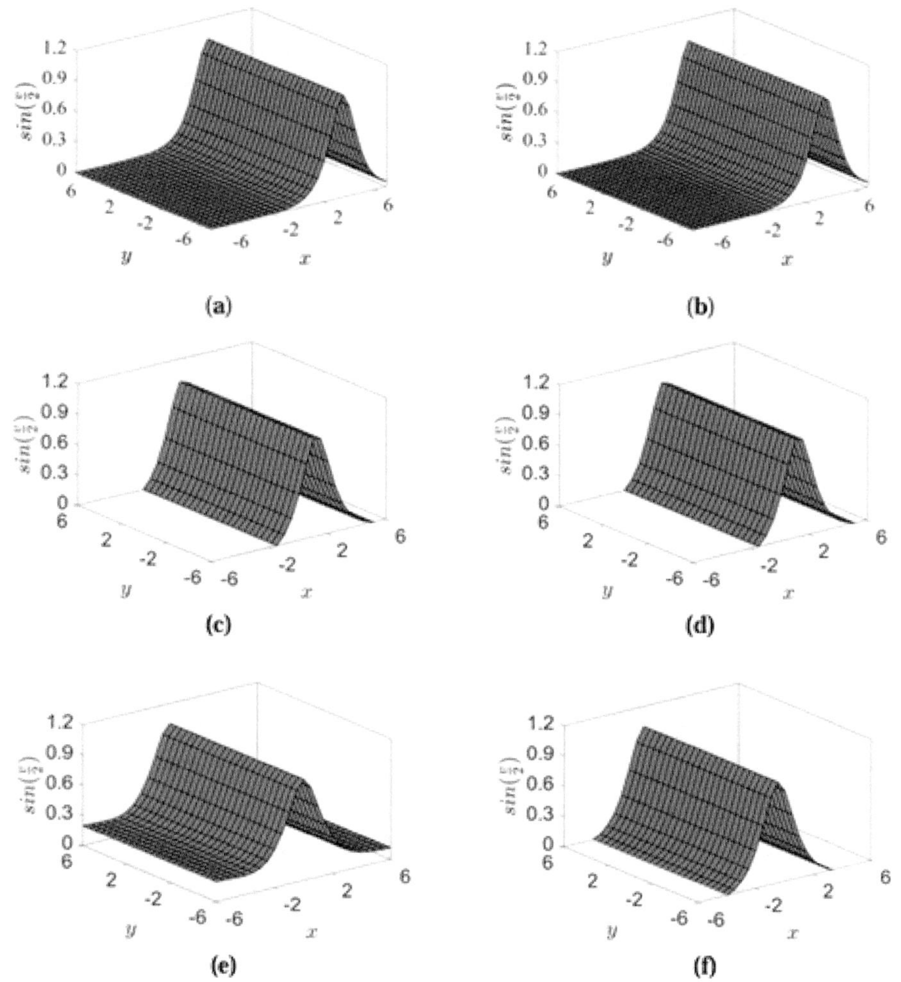

Figure 15.4 The surface plots of line solitons at ι = (a) 0, (b) 5, (c) 10, (d) 15, (e) 20, and (f) 25 for Example 15.3.4.

15.4 Conclusions

In this work, the 2D undamped, as well as damped S-G equation, has been approximated numerically using the proposed method. We have discussed various circumstances of the line and ring solitons containing circular and elliptical forms. The solutions that have been obtained through analysis and computational methodologies have been subjected to a comprehensive comparison with the solutions that are currently available in the existing literature. The method that has been proposed not only delivers results that are significantly superior in quality but also provides exceptionally accurate approximations when evaluated against the established analytical solution. The figures of surface curves and contours show that the solutions are very similar to the existing solutions. Moreover, our method is reasonably easy from an implementation point of view and takes less storage.

Data Availability Statements

All data generated or analyzed during this study are included in this published article.

References

[1] S.N.M. Ruijsenaars. Sine-gordon equation. *Encyclopedia of Mathematical Physics*, pages 576–583, 2006.

[2] A.H. Msmali, M. Tamsir, and A. A. H. Ahmadini. Crank-nicolson-dqm based on cubic exponential b-splines for the approximation of nonlinear sine-gordon equation. *Ain Shams Engineering Journal*, 12(4):4091–4097, 2021.

[3] F. Mirzaee, S. Rezaei, and N. Samadyar. Numerical solution of two-dimensional stochastic time-fractional sine–gordon equation on non-rectangular domains using finite difference and meshfree methods. *Engineering Analysis with Boundary Elements*, 127:53–63, 2021.

[4] P. Guo, A. Boldbaatar, D. Yi, and P. Dai. Numerical solution of sine-gordon equation with the local kriging meshless method. *Mathematical Problems in Engineering*, page 9057387, 2020.

[5] A.T. Deresse, Y.O. Mussa, and A.K. Gizaw. Analytical solution of two-dimensional sine-gordon equation. *Advances in Mathematical Physics*, page 6610021, 2021.

[6] Q. Sheng, A.Q.M Khaliq, and D.A Voss. Numerical simulation of two-dimensional sine-gordon solitons via a split cosine scheme. *Math. Comput. Simulation*, 68:355–373, 2005.

[7] A.G. Bratsos. An explicit numerical scheme for the sine-gordon equation in 2+1 dimensions. *Appl. Numer. Anal. Comput. Math*, 2:189–211, 2005.

[8] A.G. Bratsos. A modified predictor-corrector scheme for the two-dimensional sine-gordon equation. *Numer. Algor.*, 43:295–308, 2006.

[9] A.G. Bratsos. A third order numerical scheme for the two-dimensional sine-gordon equation. *Math. Comput. Simulation*, 76:271–282, 2007.

[10] A.G. Bratsos. The solution of two-dimensional sine-gordon equation using the method of lines. *J. Comput. Appl. Math.*, 206:251–277, 2007.

[11] M. Dehghan and D. Mirzaei. The dual reciprocity boundary element method (drbem) for two-dimensional sine-gordon equation. *Comput. Methods Appl. Mech. Engrg.*, 197:476–486, 2008.

[12] M. Dehghan and A. Shokri. A numerical method for solution of the two-dimensional sine-gordon equation using the radial basis functions. *Math. Comput. Simulation*, 79:700–715, 2008.

[13] M. Mirzaei and M. Dehghan. Boundary element solution of the two-dimensional sine-gordon equation using continuous linear elements. *Eng. Anal. Boundary Elements*, 33:12–24, 2009.

[14] D. Mirzaei and M. Dehghan. Implementation of meshless lbie method to the 2d non-linear sg problem. *Int. J. Numer. Methods Engrg.*, 79:1662–1682, 2009.

[15] D. Mirzaei and D. Dehghan. Meshless local petrov–galerkin (mlpg) approximation to the two-dimensional sine-gordon equation. *J. Comput. Appl. Math.*, 233:2737–2754, 2010.

[16] M. Dehghan and A. Ghesmati. Numerical simulation of two-dimensional sine-gordon solitons via a local weak meshless technique based on the radial point interpolation method (rpim). *Comput. Phys. Comm.*, 181:772–786, 2010.

[17] R. Jiwari, S. Pandit, and R.C. Mittal. Numerical simulation of two-dimensional sine-gordon solitons by differential quadrature method. *Comput. Phy. Comm.*, 183:600–616, 2012.

[18] X. Kanga, W. Fengb, C. Chenga, and K. Guoc. An efficient finite difference scheme for the 2d sine-gordon equation. *J. Nonlinear Sci. Appl.*, 10:2998–3012, 2017.

[19] Y. Ji. Meshless singular boundary method for nonlinear sine-gordon equation. *Math. Problems Eng.*, pages 1–11, 2018.

[20] L. Su. Numerical solution of two-dimensional nonlinear sine-gordon equation using localized method of approximate particular solutions. *Eng. Anal. Boundary Elements*, 108:95–107, 2019.

[21] M. Cui. High order compact alternating direction implicit method for the generalized sine-gordon equation. *J. Comput. Appl. Math.*, 235:837–849, 2010.

[22] R. Bellman, B.G. Kashef, and J. Casti. Differential quadrature: a technique for the rapid solution of nonlinear differential equations. *J. Comput. Phy.*, 10:40–52, 1972.

[23] C. Shu. *Differential Quadrature and its Application in Engineering.* Springer, London, 2000.

[24] S. Gottlieb, D.I. Ketcheson, and C.W. Shu. High order strong stability preserving time discretizations. *J. Sci. Comput.*, 38:251–289, 2009.

[25] H.S Shukla, M. Tamsir, and V.K Srivastava. Numerical simulation of two dimensional sine-gordon solitons using modified cubic b-spline differential quadrature method. *AIP Advances*, 5:017121, 2015.

[26] M. Tamsir, V.K Srivastava, and R. Jiwari. An algorithm based on exponential modified cubic b-spline differential quadrature method for nonlinear burgers' equation. *Appl. Math. Comput.*, 290:111–124, 2016.

[27] M. Tamsir, M.Z. Meetei, and H.M. Ahmed. Hyperbolic b-spline function-based differential quadrature method for approximation of 3d wave equations. *axioms*, 11:597, 2022.

[28] M. Tamsir and N. Dhiman. Solution of convection diffusion equation via a collocation method based on modified cubic trigonometric b-splines. *Nonlinear Studies*, 26(2):303–311, 2019.

[29] M. Tamsir, N. Dhiman, F.S. Gill, and Robin. Approximation of 3d convection diffusion equation using dqm based on modified cubic trigonometric b-splines. *Journal of Computational Methods in Sciences and Engineering*, 20(4):1357–1366, 2020.

[30] M. Tamsir, V.K. Srivastava, and N. Dhiman. Numerical computation of nonlinear fisher's reaction–diffusion equation with exponential modified cubic b-spline differential quadrature method. *Int. J. Appl. Comput. Math*, 4(6), 2018.

[31] N. Dhiman, M.J. Huntul, and M. Tamsir. A modified trigonometric cubic b-spline collocation technique for solving the time-fractional diffusion equation. *Engineering Computations*, 38(7):2921–2936, 2021.

[32] M. Tamsir and N. Dhiman. Dqm based on the modified form of ctb shape functions for coupled burgers' equation in 2d and 3d. *IJMEMS*, 4(4):1051–1067, 2019.

[33] S.R. Thottoli, M. Tamsir, N. Dhiman, and G. Souadi. Computational modeling of the balitsky–kovchegov equation and its numerical solution using hybrid b-spline collocation technique. *Partial Differential Equations in Applied Mathematics*, 5:100348, 2022.

[34] H.S. Shukla and M. Tamsir. An exponential cubic b-spline algorithm for multi-dimensional convection-diffusion equations. *Alexandria Engineering Journal*, 57(3):1999–2006, 2018.

16

An Overview of High-resolution Schemes in the OpenFOAM Toolkit

Tushar Chourushi[1,*], Saptarshi Mandal [2], Chandrakant Sonawane [3], Sunil Dingare[1], and Sudarshan B Sanap [1]

[1]Department of Aerospace Engineering, MIT Art, Design and Technology University, India
[2]Department of Aerospace Engineering, Amity University Mumbai, India
[3]Symbiosis Institute of Technology Pune, Symbiosis International Deemed University, India
E-mail: tchourushi@gmail.com; tushar.chourushi@mituniversity.edu.in; smandal@mum.amity.edu; chandrakant.sonawane@sitpune.edu.in; sunil.dingare@mituniversity.edu.in; sudarshan.sanap@mituniversity.edu.in

Abstract

All numerical solutions contain certain numerical inaccuracies due to the limitations of interpolation schemes. These shortcomings become noticeable as the flow regime transforms into high-speed flows. To elude these drawbacks, high-resolution schemes (HRS) are used for convection-dominated flows. The functional relationship for these schemes relies on the composite function of lower/higher-order schemes. The existing open-source computational fluid flow solver "OpenFOAM" comes with a few built-in HRS, and some of the well-known HRS were implemented in this work. The present work compares these HRS for varied convective flow problems to assess their numerical performance. Results suggest that only those HRS that rely on the QUICK and MUSCL schemes in their functional relationship deliver comparatively better solutions than other HRS.

Keywords: Convective flows, high-resolution schemes, OpenFOAM

DOI: 10.1201/9788743806691-16

16.1 Introduction

With the advancement in computer technologies, numerical studies are often considered a major alternative to experimental and analytical approaches to solving the fluid dynamics problem. On the other hand, the experimental approach requires a substantial monetary investment, and the analytical approach is limited to simple flow problems. In general, numerical studies provide a cost-effective solution compared to other approaches. However, in the case of convection-dominated flows, numerical works often remain distant from the actual solution due to the limitations of numerical methodologies [1]. These numerical methodologies rely on discretization techniques, interpolation schemes, algorithms, solvers, etc., to solve the governing flow equations on the discretized cell [2].

In the present context, the finite volume method (FVM) is used to study convection-dominated flows. In FVM, the computational domain is decomposed into a finite number of control volumes, and fluid flow equations are integrated over a control volume [3]. These flow equations constitute convective and diffusive terms, and the fluxes are the rate at which these terms change across a control volume. The diffusive fluxes are isotropic and are equally spread in all directions. On the other hand, in convective flows, the flow field's direction significantly impacts the flux transfer via the control volume face. Thus, a higher-order (HO) numerical scheme is needed to accurately assess the convective flux due to its dependence on the far-off neighboring points of the cell [4, 5]. However, these HO schemes give rise to unwanted numerical oscillations in convective flow problems [1, 6]. These shortcomings are overcome by incorporating high-resolution schemes (HRS) for convective fluxes [7].

These HRS suppresses unphysical oscillations by incorporating any of these approaches: (1) the total variation diminishing (TVD) approach [8]; or (2) the normalized variable formulations (NVF) approach [7, 9]. Over the years, several HRS have been presented using these approaches [9–17]. In this work, HRS based on the NVF approach, such as SMART [9], WACEB [13], CLAM [14], GAMMA [15], CUBISTA [16], and HREG [17], are only being considered. Moreover, some of these HRS satisfy any of these constraints: convection boundedness criteria (CBC), boundedness, accuracy, and interpolative reasonableness (BAIR), and refined convective boundedness criterion (RCBC) conditions, to attain boundedness [9, 18]. These criteria ensure boundedness and second-order accuracy for the numerical scheme.

Furthermore, implementation of the HRS is not straightforward as it requires special numerical procedures to implement far-off control volume grid points. Several implementation methods have been presented to implement these HRS [9, 19, 20, 21, 22]. Depending upon their formulation, these implementation methods generate either an implicit tri-diagonal matrix [7, 19] or a penta-diagonal matrix [20]. The existing open-source CFD software toolkit "OpenFOAM" employs the blending factor method (BFM) for implementing HRS over an arbitrarily unstructured control volume cell [15, 17, 23, 24]. This way of treatment is inexpensive as it removes the necessity of additional storage of the far-off nodes, thereby generating an implicit tri-diagonal matrix. Because it eliminates the need for extra storage of the distant nodes, this treatment method is inexpensive and creates an implicit tri-diagonal matrix. Predominantly, minimal literature is available on HRS and the BFM in the context of OpenFOAM [17, 23].

This work thus presents a comprehensive study of varied HRS against stringent test cases using OpenFOAM. These tests include a pure advection study of different profiles: (1) step; (2) sine-square; (3) semi-ellipse, to assess the numerical artifacts present in the HRS due to numerical limitations. Additionally, the benchmark test of Sod's shock tube is performed. The sections of the chapter are as follows: Section 2 outlines the governing flow equations. Section 16.3 details the numerical methodology, high-resolution schemes, and numerical solver. Section 16.4 discusses the numerical assessment of HRS against varied test cases. Finally, in Section 16.5, some concluding remarks are made.

16.2 Governing Flow Equations

Two categories are used to group the governing equations in this work: The scalar variable's pure advection equation (ϕ), and compressible fluid flow equations for an ideal gas.

16.2.1 Scalar variable's (ϕ) pure advection:

The governing equation for any scalar variable (ϕ) in vector form [17] can be represented as follows:

$$\nabla \cdot (\rho \mathbf{u} \phi) = Q, \tag{16.1}$$

where ρ, $\mathbf{u}$, and Q represent the mass density, fluid velocity, and source term, respectively.

Compressible fluid flow: The governing equations for the conservation of mass, momentum, and energy for compressible gas flow in vector form are given by:

$$\frac{\partial}{\partial t}\begin{bmatrix}\rho \\ \rho\mathbf{u} \\ \rho E\end{bmatrix} + \nabla\cdot\begin{bmatrix}\rho\mathbf{u} \\ \rho\mathbf{u}\mathbf{u}+p\mathbf{I} \\ (\rho E+p)\,\mathbf{u}\end{bmatrix} = 0. \tag{16.2}$$

Here, $\rho, \mathbf{u}, p$, and E are the mass density, fluid velocity, pressure, and total energy, respectively. Further, the total energy density formulation is simplified by considering the working gas as a calorific perfect gas [25, 26]:

$$E = C_{\mathrm{v}}T + \frac{1}{2}\mathbf{u}\cdot\mathbf{u}, \tag{16.3}$$

where C_{v} is the specific heat at constant volume and T denotes the temperature [27].

16.3 Numerical Discretization

A set of discretized equations is produced by integrating and discretizing the governing flow eqn (16.1) and (16.2) over a control volume [28, 29]. Usually, the OpenFOAM framework adopts the collocated grid arrangement, wherein all flow variables are stored at the centroid of the control volume [24, 29, 30]. The discretized governing equations' explicit, first-order Euler approach is used to solve all time-dependent terms [17]. The central difference scheme is used to discretize the diffusive terms, whereas the high-resolution schemes (HRS) [9, 18] are used to discretize the convective terms. Furthermore, in the case of compressible gas flows, flow variable properties are transported by both convective forces and propagation of waves [31, 32]. Therefore, the convective terms in the discretized governing equations can be represented as

$$\int_V \nabla\cdot[\mathbf{u}\phi]\,dV = \sum_{\mathrm{f}}(S\cdot\mathbf{u})_{\mathrm{f}}\,\phi_{\mathrm{f}} = \sum_{\mathrm{f}} F_{\mathrm{f}}\phi_{\mathrm{f}}. \tag{16.4}$$

The volumetric flow and the convected variable crossing the face f are represented by the expressions F_{f} and ϕ_{f} , respectively, for any arbitrary control volume V. These convected variables ϕ_{f} are computed using the localized normalized variable approach [7, 9], as shown in Figure 16.1. While the term $\sum_{\mathrm{f}}$ denotes the summation of all control volume faces. The

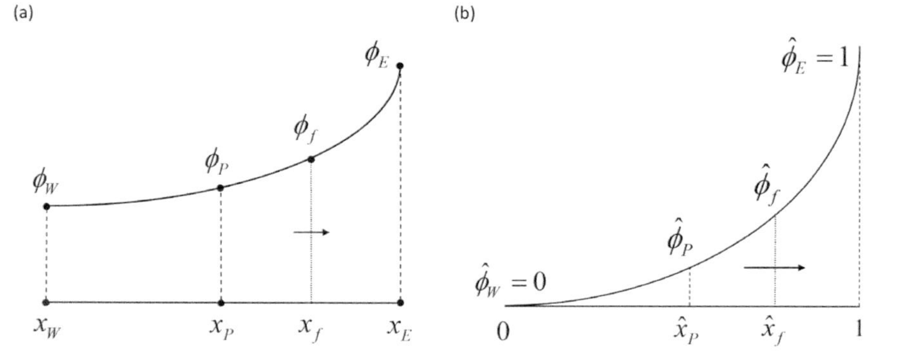

Figure 16.1 Plot of convected variables: (a) actual flow variable; (b) normalized flow variables on the normalized variable diagram, for the case of flow along the positive $x-$direction.

convected variable ϕ_f for the central grid point P, is normalized as follows:

$$\hat{\phi}_\mathrm{P} = 1 - \frac{\phi_\mathrm{f+} - \phi_\mathrm{P}}{\phi_\mathrm{f+} - \phi_\mathrm{f-}}. \tag{16.5}$$

Here $\phi_\mathrm{f+}$ and $\phi_\mathrm{f-}$ refer to the interpolated convected face variables. Furthermore, in the case of an arbitrary control volume mesh, the interpolation factor f_x is as under:

$$f^+ = (fE)/(PE). \tag{16.6}$$

Using the above equation, the numerator in eqn (16.5) can be reformulated as follows [23]:

$$\phi_\mathrm{f+} - \phi_\mathrm{P} = f^+ \left(\nabla\phi\right)_\mathrm{f} \cdot \hat{\mathbf{d}} \left(x_\mathrm{E} - x_\mathrm{P}\right), \tag{16.7}$$

where $\hat{\mathbf{d}}, x_\mathrm{E}$ and x_P denotes the unit vector, and distances of central and downwind grid points, respectively. Similarly, the denominator term can be represented as

$$\phi_\mathrm{f+} - \phi_\mathrm{f-} = \left(\nabla\phi\right)_\mathrm{P} \cdot \hat{\mathbf{d}} \left(x_\mathrm{f+} - x_\mathrm{f-}\right). \tag{16.8}$$

By substituting, in eqn (16.5) gives [17, 23]:

$$\hat{\phi}_\mathrm{P} = 1 - \frac{\left(\nabla\phi\right)_\mathrm{f} \cdot \mathbf{d}}{2\left(\nabla\phi\right)_\mathrm{P} \cdot \mathbf{d}} = 1 - \frac{\phi_\mathrm{E} - \phi_\mathrm{P}}{2\left(\nabla\phi\right)_\mathrm{P} \cdot \mathbf{d}}. \tag{16.9}$$

The normalized convected variables for the central grid point $\hat{\phi}_{\mathrm{P}}$ along with the normalized distances $(\hat{x}_{\mathrm{P}}, \hat{x}_{\mathrm{f}})$ is then used to formulate the localized normalized functional relationship for any HRS as

$$\hat{\phi}_{\mathrm{f}} = f\left(\hat{\phi}_{\mathrm{P}}, \hat{x}_{\mathrm{P}}, \hat{x}_{\mathrm{f}}\right). \tag{16.10}$$

These normalized convected face variables are then bounded between 0 and 1 values to attain numerical boundedness. Subsequently, the blending factor method of OpenFOAM [15] is applied to calculate the un-normalized variables:

$$\phi_{\mathrm{f}} = \left(1 - \gamma\left(1 - f^{+}\right)\right)\phi_{\mathrm{P}} + \gamma\left(1 - f^{+}\right)\phi_{\mathrm{E}}. \tag{16.11}$$

The term "γ" refers to the blending factor which depends on the localized value of $\hat{\phi}_{\mathrm{P}}$ and f^{+}[23].

16.3.1 High-resolution schemes

Table 16.1 lists the normalized functional relationship of some HRS [7, 17] considered in this study. A few of these HRS, such as GAMMA [15] and MINMOD [23], come with the built-in OpenFOAM package, and the remaining HRS are implemented into the OpenFOAM code using the blender factor method [23]. Furthermore, to attain better numerical convergence, the GAMMA scheme assumes that the user's preference factor, b, is 0.5 [15].

16.3.2 Numerical solver of OpenFOAM (scalarTransportFoam and rhoCentralFoam)

Scalar Transport FOAM: The OpenFOAM basic solver, scalarTransport-FOAM, is incorporated to discretize eqn (16.1). In this equation, the velocity field is set to be uniform for the 2D classical pure advection tests [17], that is, $u_{\mathrm{x}} = u_{\mathrm{y}} = u$, at an angle of 45^{0}. Furthermore, the source term Q is ignored for mathematical simplicity. Thus, the governing scalar transport equation can be rewritten as

$$\frac{\partial\left(u_{\mathrm{x}}\phi\right)}{\partial x} + \frac{\partial\left(u_{\mathrm{y}}\phi\right)}{\partial y} = 0. \tag{16.12}$$

Here the scalar variable ϕ is discretized using HRS, for different test profiles. Consecutively, eqn (16.12) transforms to the algebraic equations [17, 22]:

$$a_{\mathrm{P}}\phi_{\mathrm{P}} = \sum_{\mathrm{NB}} a_{\mathrm{NB}}\phi_{\mathrm{NB}} + B, \tag{16.13}$$

Table 16.1 Normalized functional relationships based on the non-uniform and uniform grids for high-resolution schemes [7, 17].

HRS	Non-uniform grid	Uniform grid
GAMMA	$\hat{\phi}_{\mathrm{f}} = \begin{cases} \left[1+\frac{1}{\beta}\frac{\hat{x}_{\mathrm{f}}-\hat{x}_{\mathrm{P}}}{1-\hat{x}_{\mathrm{P}}}\left(1-\hat{\phi}_{\mathrm{P}}\right)\right]\hat{\phi}_{\mathrm{P}} & 0<\hat{\phi}_{\mathrm{P}}<\beta \\ \frac{1-\hat{x}_{\mathrm{f}}}{(1-\hat{x}_{\mathrm{P}})}\hat{\phi}_{\mathrm{P}}+\frac{\hat{x}_{\mathrm{f}}-\hat{x}_{\mathrm{P}}}{(1-\hat{x}_{\mathrm{P}})} & \beta\le\hat{\phi}_{\mathrm{P}}<1 \\ \hat{\phi}_{\mathrm{P}} & \text{elsewhere} \end{cases}$	$\hat{\phi}_{\mathrm{f}} = \begin{cases} \left[1+\frac{0.5}{\beta}\left(1-\hat{\phi}_{\mathrm{P}}\right)\right]\hat{\phi}_{\mathrm{P}} & 0<\hat{\phi}_{\mathrm{P}}<\beta \\ \frac{1}{2}\hat{\phi}_{\mathrm{P}}+\frac{1}{2} & \beta\le\hat{\phi}_{\mathrm{P}}<1 \\ \hat{\phi}_{\mathrm{P}} & \text{elsewhere} \end{cases}$
CLAM	$\hat{\phi}_{\mathrm{f}} = \begin{cases} \frac{\hat{x}_{\mathrm{f}}-\hat{x}_{\mathrm{P}}^2}{\hat{x}_{\mathrm{P}}(1-\hat{x}_{\mathrm{P}})}\hat{\phi}_{\mathrm{P}}+\frac{\hat{x}_{\mathrm{f}}-\hat{x}_{\mathrm{P}}}{\hat{x}_{\mathrm{P}}(1-\hat{x}_{\mathrm{P}})}\hat{\phi}_{\mathrm{P}}^2 & 0\le\hat{\phi}_{\mathrm{P}}<1 \\ \hat{\phi}_{\mathrm{P}} & \text{elsewhere} \end{cases}$	$\hat{\phi}_{\mathrm{f}} = \begin{cases} \hat{\phi}_{\mathrm{P}}\left(2-\hat{\phi}_{\mathrm{P}}\right) & 0<\hat{\phi}_{\mathrm{P}}<1 \\ \hat{\phi}_{\mathrm{P}} & \text{elsewhere} \end{cases}$
WACEB	$\hat{\phi}_{\mathrm{f}} = \begin{cases} 2\hat{\phi}_{\mathrm{P}} & 0<\hat{\phi}_{\mathrm{P}}<\frac{\hat{x}_{\mathrm{P}}\hat{x}_{\mathrm{f}}(\hat{x}_{\mathrm{f}}-\hat{x}_{\mathrm{P}})}{2\hat{x}_{\mathrm{P}}(1-\hat{x}_{\mathrm{P}})-\hat{x}_{\mathrm{f}}(1-\hat{x}_{\mathrm{f}})} \\ \frac{\hat{x}_{\mathrm{f}}(1-\hat{x}_{\mathrm{f}})}{\hat{x}_{\mathrm{P}}(1-\hat{x}_{\mathrm{P}})}\hat{\phi}_{\mathrm{P}}+\frac{\hat{x}_{\mathrm{f}}(\hat{x}_{\mathrm{f}}-\hat{x}_{\mathrm{P}})}{(1-\hat{x}_{\mathrm{P}})} & \frac{\hat{x}_{\mathrm{P}}\hat{x}_{\mathrm{f}}(\hat{x}_{\mathrm{f}}-\hat{x}_{\mathrm{P}})}{2\hat{x}_{\mathrm{P}}(1-\hat{x}_{\mathrm{P}})-\hat{x}_{\mathrm{f}}(1-\hat{x}_{\mathrm{f}})}\le\hat{\phi}_{\mathrm{P}}\le\frac{\hat{x}_{\mathrm{P}}}{\hat{x}_{\mathrm{f}}}(1+\hat{x}_{\mathrm{f}}-\hat{x}_{\mathrm{P}}) \\ 1 & \frac{\hat{x}_{\mathrm{P}}}{\hat{x}_{\mathrm{f}}}(1+\hat{x}_{\mathrm{f}}-\hat{x}_{\mathrm{P}})<\hat{\phi}_{\mathrm{P}}<1 \\ \hat{\phi}_{\mathrm{P}} & \text{elsewhere} \end{cases}$	$\hat{\phi}_{\mathrm{f}} = \begin{cases} 2\hat{\phi}_{\mathrm{P}} & 0<\hat{\phi}_{\mathrm{P}}<\frac{\hat{x}_{\mathrm{P}}\hat{x}_{\mathrm{f}}(\hat{x}_{\mathrm{f}}-\hat{x}_{\mathrm{P}})}{2\hat{x}_{\mathrm{P}}(1-\hat{x}_{\mathrm{P}})-\hat{x}_{\mathrm{f}}(1-\hat{x}_{\mathrm{f}})} \\ \frac{\hat{x}_{\mathrm{f}}(1-\hat{x}_{\mathrm{f}})}{\hat{x}_{\mathrm{P}}(1-\hat{x}_{\mathrm{P}})}\hat{\phi}_{\mathrm{P}}+\frac{\hat{x}_{\mathrm{f}}(\hat{x}_{\mathrm{f}}-\hat{x}_{\mathrm{P}})}{(1-\hat{x}_{\mathrm{P}})} & \frac{\hat{x}_{\mathrm{P}}\hat{x}_{\mathrm{f}}(\hat{x}_{\mathrm{f}}-\hat{x}_{\mathrm{P}})}{2\hat{x}_{\mathrm{P}}(1-\hat{x}_{\mathrm{P}})-\hat{x}_{\mathrm{f}}(1-\hat{x}_{\mathrm{f}})}\le\hat{\phi}_{\mathrm{P}}\le\frac{\hat{x}_{\mathrm{P}}}{\hat{x}_{\mathrm{f}}}(1+\hat{x}_{\mathrm{f}}-\hat{x}_{\mathrm{P}}) \\ 1 & \frac{\hat{x}_{\mathrm{P}}}{\hat{x}_{\mathrm{f}}}(1+\hat{x}_{\mathrm{f}}-\hat{x}_{\mathrm{P}})<\hat{\phi}_{\mathrm{P}}<1 \\ \hat{\phi}_{\mathrm{P}} & \text{elsewhere} \end{cases}$
SMART	$\hat{\phi}_{\mathrm{f}} = \begin{cases} \frac{\hat{x}_{\mathrm{f}}(1-3\hat{x}_{\mathrm{P}}+2\hat{x}_{\mathrm{f}})}{\hat{x}_{\mathrm{P}}(1-\hat{x}_{\mathrm{P}})}\hat{\phi}_{\mathrm{P}} & 0<\hat{\phi}_{\mathrm{P}}<\frac{\hat{x}_{\mathrm{P}}}{3} \\ \frac{\hat{x}_{\mathrm{f}}(1-\hat{x}_{\mathrm{f}})}{\hat{x}_{\mathrm{P}}(1-\hat{x}_{\mathrm{P}})}\hat{\phi}_{\mathrm{P}}+\frac{\hat{x}_{\mathrm{f}}(\hat{x}_{\mathrm{f}}-\hat{x}_{\mathrm{P}})}{(1-\hat{x}_{\mathrm{P}})} & \frac{\hat{x}_{\mathrm{P}}}{3}\le\hat{\phi}_{\mathrm{P}}\le\frac{\hat{x}_{\mathrm{P}}}{\hat{x}_{\mathrm{f}}}(1+\hat{x}_{\mathrm{f}}-\hat{x}_{\mathrm{P}}) \\ 1 & \frac{\hat{x}_{\mathrm{P}}}{\hat{x}_{\mathrm{f}}}(1+\hat{x}_{\mathrm{f}}-\hat{x}_{\mathrm{P}})<\hat{\phi}_{\mathrm{P}}<1 \\ \hat{\phi}_{\mathrm{P}} & \text{elsewhere} \end{cases}$	$\hat{\phi}_{\mathrm{f}} = \begin{cases} 3\hat{\phi}_{\mathrm{P}} & 0<\hat{\phi}_{\mathrm{P}}<\frac{1}{6} \\ \frac{3}{4}\hat{\phi}_{\mathrm{P}}+\frac{3}{8} & \frac{1}{6}\le\hat{\phi}_{\mathrm{P}}\le\frac{5}{6} \\ 1 & \frac{5}{6}<\hat{\phi}_{\mathrm{P}}<1 \\ \hat{\phi}_{\mathrm{P}} & \text{elsewhere} \end{cases}$
CUBISTA	$\hat{\phi}_{\mathrm{f}} = \begin{cases} \left[1+\frac{\hat{x}_{\mathrm{f}}-\hat{x}_{\mathrm{P}}}{3(1-\hat{x}_{\mathrm{P}})}\right]\frac{\hat{x}_{\mathrm{P}}}{\hat{x}_{\mathrm{f}}}\hat{\phi}_{\mathrm{P}} & 0<\hat{\phi}_{\mathrm{P}}<\frac{3}{4}\hat{x}_{\mathrm{P}} \\ \frac{\hat{x}_{\mathrm{f}}(1-\hat{x}_{\mathrm{f}})}{\hat{x}_{\mathrm{P}}(1-\hat{x}_{\mathrm{P}})}\hat{\phi}_{\mathrm{P}}+\frac{\hat{x}_{\mathrm{f}}(\hat{x}_{\mathrm{f}}-\hat{x}_{\mathrm{P}})}{(1-\hat{x}_{\mathrm{P}})} & \frac{3}{4}\hat{x}_{\mathrm{P}}\le\hat{\phi}_{\mathrm{P}}\le\frac{1+2(\hat{x}_{\mathrm{f}}-\hat{x}_{\mathrm{P}})}{2\hat{x}_{\mathrm{f}}-\hat{x}_{\mathrm{P}}}\hat{x}_{\mathrm{P}} \\ 1-\frac{(1-\hat{x}_{\mathrm{f}})}{2(1-\hat{x}_{\mathrm{P}})}\left(1-\hat{\phi}_{\mathrm{P}}\right) & \frac{1+2(\hat{x}_{\mathrm{f}}-\hat{x}_{\mathrm{P}})}{2\hat{x}_{\mathrm{f}}-\hat{x}_{\mathrm{P}}}\hat{x}_{\mathrm{P}}<\hat{\phi}_{\mathrm{P}}<1 \\ \hat{\phi}_{\mathrm{P}} & \text{elsewhere} \end{cases}$	$\hat{\phi}_{\mathrm{f}} = \begin{cases} \frac{7}{4}\hat{\phi}_{\mathrm{P}} & 0<\hat{\phi}_{\mathrm{P}}<\frac{3}{8} \\ \frac{3}{4}\hat{\phi}_{\mathrm{P}}+\frac{3}{8} & \frac{3}{8}\le\hat{\phi}_{\mathrm{P}}\le\frac{3}{4} \\ 1 & \frac{3}{4}<\hat{\phi}_{\mathrm{P}}<1 \\ \hat{\phi}_{\mathrm{P}} & \text{elsewhere} \end{cases}$
HREG	$\hat{\phi}_{\mathrm{f}} = \begin{cases} \left[1+\frac{\hat{x}_{\mathrm{f}}-\hat{x}_{\mathrm{P}}}{3(1-\hat{x}_{\mathrm{P}})}\right]\frac{\hat{x}_{\mathrm{f}}}{\hat{x}_{\mathrm{P}}}\hat{\phi}_{\mathrm{P}} & 0<\hat{\phi}_{\mathrm{P}}<\frac{((\hat{x}_{\mathrm{f}}-\hat{x}_{\mathrm{P}})\hat{x}_{\mathrm{P}})3(1-\hat{x}_{\mathrm{P}})}{3\hat{x}_{\mathrm{f}}-4\hat{x}_{\mathrm{P}}\hat{x}_{\mathrm{f}}+\hat{x}_{\mathrm{f}}^2-3\hat{x}_{\mathrm{P}}+3\hat{x}_{\mathrm{P}}^2} \\ \hat{\phi}_{\mathrm{P}}+(\hat{x}_{\mathrm{f}}-\hat{x}_{\mathrm{P}}) & \frac{((\hat{x}_{\mathrm{f}}-\hat{x}_{\mathrm{P}})\hat{x}_{\mathrm{P}})3(1-\hat{x}_{\mathrm{P}})}{3\hat{x}_{\mathrm{f}}-4\hat{x}_{\mathrm{P}}\hat{x}_{\mathrm{f}}+\hat{x}_{\mathrm{f}}^2-3\hat{x}_{\mathrm{P}}+3\hat{x}_{\mathrm{P}}^2}\le\hat{\phi}_{\mathrm{P}}\le\frac{1-\hat{x}_{\mathrm{f}}-2\hat{x}_{\mathrm{P}}\hat{x}_{\mathrm{f}}-2\hat{x}_{\mathrm{P}}^2}{1-2\hat{x}_{\mathrm{P}}+\hat{x}_{\mathrm{f}}} \\ 1-\frac{1-\hat{x}_{\mathrm{f}}}{2(1-\hat{x}_{\mathrm{P}})}\left(1-\hat{\phi}_{\mathrm{P}}\right) & \frac{1-\hat{x}_{\mathrm{f}}-2\hat{x}_{\mathrm{P}}\hat{x}_{\mathrm{f}}-2\hat{x}_{\mathrm{P}}^2}{1-2\hat{x}_{\mathrm{P}}+\hat{x}_{\mathrm{f}}}<\hat{\phi}_{\mathrm{P}}<1 \\ \hat{\phi}_{\mathrm{P}} & \text{elsewhere} \end{cases}$	$\hat{\phi}_{\mathrm{f}} = \begin{cases} \frac{7}{4}\hat{\phi}_{\mathrm{P}} & 0<\hat{\phi}_{\mathrm{P}}<\frac{1}{3} \\ \hat{\phi}_{\mathrm{P}}+\frac{1}{4} & \frac{1}{3}\le\hat{\phi}_{\mathrm{P}}\le\frac{2}{3} \\ \frac{1}{4}\hat{\phi}_{\mathrm{P}}+\frac{3}{4} & \frac{2}{3}<\hat{\phi}_{\mathrm{P}}<1 \\ \hat{\phi}_{\mathrm{P}} & \text{elsewhere} \end{cases}$

where ϕ_{P} and ϕ_{NB} denote the scalar variables at central grid point “P” and nearby grid points “NB” of the control volume, respectively. The a_{P} and a_{NB} coefficients are the contributions due to the central node and neighboring nodes, respectively. The term B denotes the source term in the governing equation, if any.

rho CentralFOAM: In the case of compressible gas flows, the rhoCentralFOAM solver is considered to discretize the eqn (16.2), by following the flux-splitting approach of Kurganov and Tadmor [31]. Thereby, the convective flux (eqn (16.4)) is rewritten as

$$\sum_{\mathrm{f}} F_{\mathrm{f}}\phi_{\mathrm{f}} = \sum_{\mathrm{f}} \{\alpha\left(F_{\mathrm{f}+}\phi_{\mathrm{f}+} + F_{\mathrm{f}-}\phi_{\mathrm{f}-}\right) + \omega_{\mathrm{f}}\left(\phi_{\mathrm{f}-} - \phi_{\mathrm{f}+}\right)\}. \tag{16.14}$$

Here $\alpha = 0.5$ is the weighted coefficient and ω_{f} is the volumetric flux evaluated based on the maximum speed of propagation of discontinuities [31, 32] as

$$\omega_{\mathrm{f}} = 0.5 \max \begin{pmatrix} \max\left(a_{\mathrm{f}+}\left|\mathbf{S}_{\mathrm{f}}\right| + \phi_{\mathrm{f}+}, a_{\mathrm{f}-}\left|\mathbf{S}_{\mathrm{f}}\right| + \phi_{\mathrm{f}-}, 0\right) \\ \max\left(a_{\mathrm{f}+}\left|\mathbf{S}_{\mathrm{f}}\right| - \phi_{\mathrm{f}+}, a_{\mathrm{f}-}\left|\mathbf{S}_{\mathrm{f}}\right| - \phi_{\mathrm{f}-}, 0\right) \end{pmatrix}. \tag{16.15}$$

where a_{f} refers to the speed of sound of the gas at the respective control volume face. While the transport variables are evaluated using the HRS. Upon substitution, the discretized governing flow equations transcend into a similar set of algebraic eqn (16.13). These algebraic equations are further solved using the preconditioned bi-conjugate gradient (PbiCG) solver with diagonal incomplete LU (DILU) for asymmetric matrices and the preconditioned conjugate gradient (PCG) solver with diagonal incomplete-Cholesky (DIC) preconditioner [24].

16.4 Results and Discussions

In this section, numerical solutions obtained using the varied high-resolution schemes (HRS) for scalarTransportFOAM and rhoCentralFOAM solvers were discussed, respectively. At first, all HRS were assessed against the pure advection test profiles: (i) step; (ii) sine-square; and (3) semi-ellipse [16, 33]. In these tests, the 2D computational domain was decomposed into 60×60 uniform grid cells in each coordinate direction. Later, the Sod’s shock tube problem was studied. The Sod’s shock tube [34] is a classical 1D problem with an imposed discontinuity at the initial time step. The computational domain was discretized into 50 uniform grid points. All numerical simulations were considered to be converged when the initial residual error for transport equations falls under 10^{-7}.

16.4.1 Pure advection profiles

Numerical simulations were performed for pure advection of scalar profiles (step, sine-square, semi-ellipse) using the HRS to provide an assessment against the numerical accuracy, unphysical oscillations, and symmetricity [16, 17]. The detailed boundary conditions are further briefed for the reader's understanding:

- *Step profile* (size of the domain: $0 \leq x \leq 1$ and $0 \leq y \leq 1$)

$$\begin{aligned} \phi(y) &= 1 \quad \text{for } 0 \leq y \leq 1, \\ \phi(x) &= 0 \quad \text{for } 0 < x \leq 1. \end{aligned} \tag{16.16}$$

- *Sine-square profile* (size of the domain:$0 \leq x \leq 1$ and $0 \leq y \leq 1$)

$$\begin{aligned} &\phi(y) = \begin{cases} \sin^2\left(\frac{10}{3}\pi y\right) & \text{for } 0 \leq y \leq 3/20 \\ 1 & \text{for } 3/20 \leq y \leq 1 \end{cases}, \\ &\phi(x) = 0 \quad \text{for } 0 < x \leq 1. \end{aligned} \tag{16.17}$$

- *Semi-ellipse profile* (Size of the domain: $0 \leq x \leq 6$ and $0 \leq y \leq 6$)

$$\begin{aligned} \phi(y) &= \begin{cases} \left\{1 - \left(\frac{y}{1/6}\right)^2\right\} & \text{for } |y| < 1, \\ 0 & \text{elsewhere.} \end{cases} \\ \phi(x) &= \begin{cases} \left\{1 - \left(\frac{x}{1/6}\right)^2\right\} & \text{for } |x| < 1, \\ 0 & \text{elsewhere.} \end{cases} \end{aligned} \tag{16.18}$$

Figures 16.2, 16.3, and 16.4 detail the sketch of the computational domain with imposed boundary conditions for the pure advection profiles, at a mesh-to-flow angle of 45^0. Further comparison of the step profile for the upwind (UDS), GAMMA [15], CLAM [14], SMART [9], WACEB [13], CUBISTA [16], and HREG [17] schemes against the analytical solution are presented in Figure 16.2. It was observed that the lower-order UDS captured an inaccurate step profile in comparison with other HRS, followed by the GAMMA and CLAM schemes. On the other hand, the WACEB, SMART, CUBISTA, and HREG schemes predicted far better discontinuity.

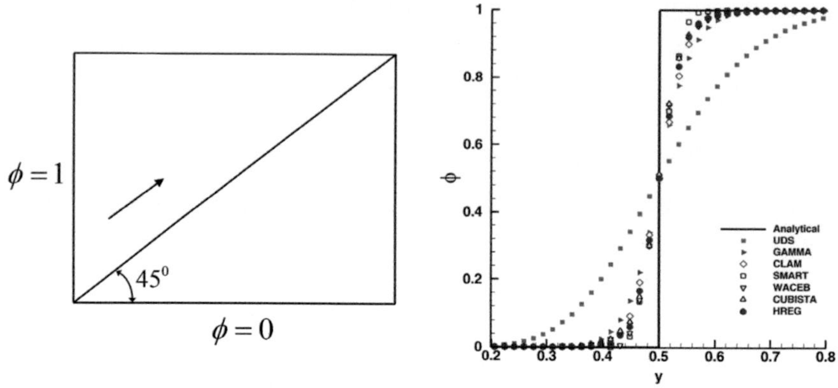

Figure 16.2 Assessment of high-resolution schemes with the analytical solution [16, 17] at $x = 0.5$ line, for the pure advection of step-profile.

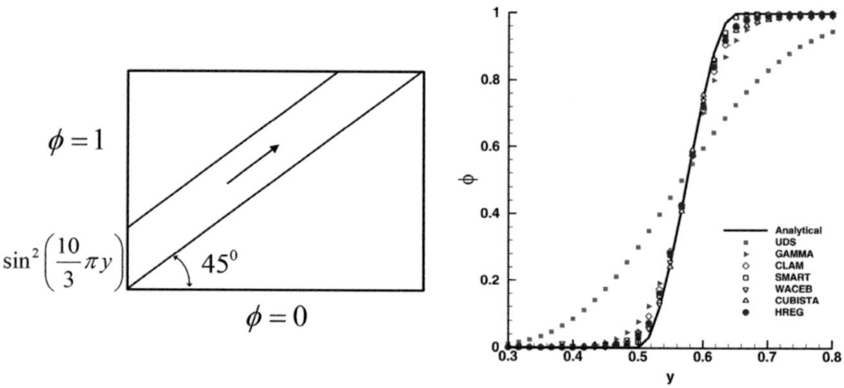

Figure 16.3 Assessment of high-resolution schemes with the analytical solution [16, 17] at $x = 0.5$ line, for the pure advection of sine-square profile.

In the next sine-square profile test, HRS were assessed against the smoothly varying gradient of the scalar variable ϕ, as shown in Figure 16.3. It was noticed that the lower-order UDS scheme convects an unacceptable profile due to significant numerical diffusion, followed by the GAMMA scheme. Contrarily, other HRS predict more or less similar features, except the SMART and WACEB, as these schemes show a loss of symmetricity by surpassing the exact solution of the sine-square profile, at a critical position.

Finally, HRS were evaluated against the continuously varying gradient fields with smooth maxima in the middle (semi-ellipse test), as detailed in

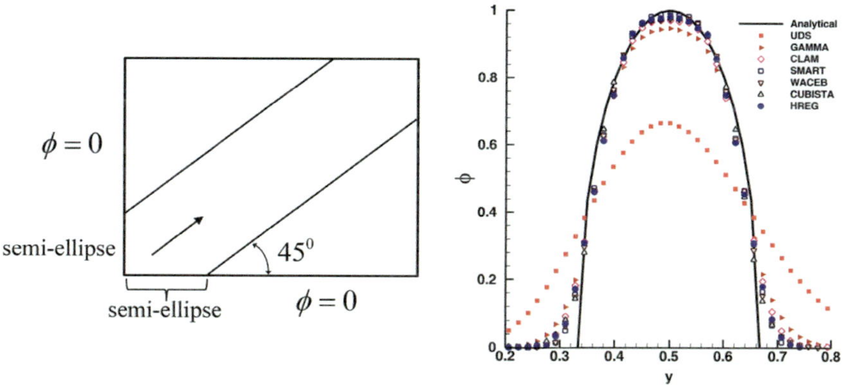

Figure 16.4 Assessment of high-resolution schemes with the analytical solution [16, 17] at $x = 0.5$ line, for the pure advection of a semi-ellipse profile.

Figure 16.4. This test evaluates the steepening/clipping characteristics for the numerical scheme. Because of significant numerical diffusion, the lower-order UDS technique advects the distorted profile and cuts the true peak. While the GAMMA scheme, on the other hand, captured a slightly better semi-ellipse profile. Overall, the CLAM, WACEB, SMART, CUBISTA, and HREG schemes showed a symmetric profile advection.

16.4.2 Sod's shock tube problem

In this test, the classical Sod's shock tube [34] problem was numerically assessed against the accuracy and undue oscillations for varied HRS. The sod's shock tube is a 1D problem with an applied discontinuity at the initial time step. At the time, $t = 0$, an imaginary diaphragm is placed at the tube's center, separating fluids with different densities and pressures. Further, these fluids were presumed to be at rest. The region on the left-hand side of the diaphragm is assumed to be at higher pressure, density, and temperature $p_1 = 10^5$ Pa, $\rho_1 = 1$ Kg/m^3, $\mathrm{T}_1 = 358.4$ K, when compared with the region on the right-hand side $p_2 = 10^4$ Pa, $\rho_1 = 0.125$ Kg/m^3, $\mathrm{T}_2 = 278.7$ K. At time $t > 0$, the diaphragm is instantly removed, leading to a shockwave generation in a low-pressure region; additionally, an expansion wave propagates back into the high-pressure region.

The computational domain comprises a 1D geometry with a domain length of "$-5 \leq x \leq 5$" m, as sketched in Figure 16.5. The computational domain was decomposed into 50 grid points. The absolute tolerance for convergence of the flow variables is set to be 10^{-10} and a run-time modifiable time-step is applied for the maximum CFL number of 0.2. All numerical simulations were simulated up to a time, $t = 7$ ms. At the boundaries, zero-gradient conditions were applied for all flow variables. Figure 16.6 compares the density profiles of UDS, GAMMA, CLAM, SMART, WACEB, CUBISTA, and HREG schemes with the analytical solution [17, 34], at time $(t) = 7$ ms. It was noticed that the lower-order UDS and GAMMA schemes due to their NVF formulations, advected a significant amount of artificial numerical diffusion when with other HRS, for both expansion (Figure 16.6(a)) and shock (Figure 16.6(b)) waves. Comprehensively, the CLAM, SMART, WACEB, CUBISTA, and HREG schemes offered a better resolution of the step discontinuity.

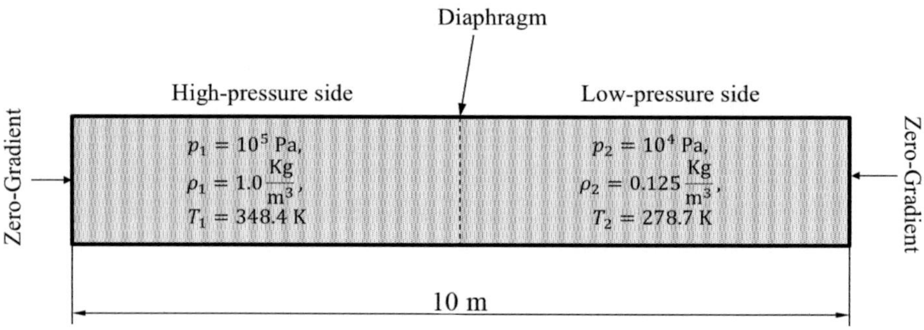

Figure 16.5 Plot of the computational domain for classical Sod's shock tube, at initial time step, $t = 0$.

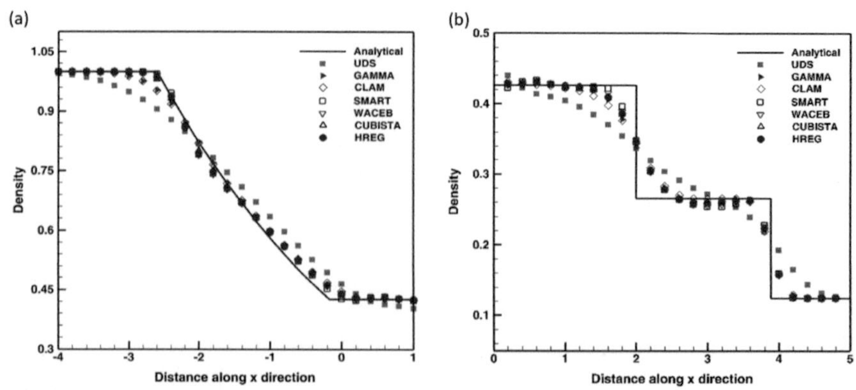

Figure 16.6 Comparison of density profiles obtained using varied high-resolution schemes with the analytical solution: (a) expansion wave; (b) shock wave, at time, $t = 7$ ms.

16.5 Conclusions

A comparative study of various high-resolution schemes (HRS) with the analytical solutions for the classical benchmark test cases was conducted using the open-source toolkit "OpenFOAM." Some of the well-known HRS, like CLAM, SMART, WACEB, CUBISTA, and HREG schemes, which do not come with the existing OpenFOAM solver, were also implemented in this work. Further, to assess the numerical accuracy, unphysical oscillations, and symmetricity, simulations are performed for pure advection tests and sod's shock tube.

In broad terms, lower-order UDS, GAMMA, and CLAM schemes showed inaccurate results and loss of symmetricity for these stringent test cases. However, higher-order HRS such as SMART, WACEB, CUBISTA, and HREG schemes exhibit precise advection of step, sine-square, and semi-ellipse profiles. Moreover, a careful examination of the referred higher-order HRS revealed that these schemes rely on the QUICK and MUSCL schemes in their functional relationships on the normalized variable diagram (NVD).

References

[1] S. V. Patankar. *Numerical heat transfer and fluid flow, 'Hemisphere Series on Computational Methods in Mechanics and Thermal Science'*, volume 1. McGraw Hill books, New York, 1980.

[2] J. H. Ferziger and M. Peric. *Computational methods for fluid dynamics*, volume 3. Springer Verlag, Berlin, 2003.

[3] H. K. Versteeg. *An introduction to computational fluid dynamics the finite volume method*, volume 2. Pearson Education, India, 2007.

[4] R. M. Beam R. F. Warming. Upwind second-order difference schemes and applications in aerodynamic flows. *AIAA Journal*, 14:1241–1249, 1976.

[5] B. P. Leonard. A stable and accurate convective modeling procedure based on quadratic interpolation. *Computer Methods in Applied Mechanics and Engineering*, 19:59–98, 1979.

[6] T. Hayase, J. A. C. Humphrey, and R. Greif. A consistently formulated quick scheme for fast and stable convergence using finite-volume iterative calculation procedures. *Journal of Computational Physics*, 98:108–118, 1992.

[7] M. S. Darwish and F. Moukalled. Normalized variable and space formulation methodology for high-resolution schemes. *Numerical Heat Transfer, Part B*, 26:79–96, 1994.

[8] P. K. Sweby. High-resolution schemes using flux limiters for hyperbolic conservation laws. *SIAM Journal on Numerical Analysis*, 21(5):995–1011, 1984.

[9] P. H. Gaskell and A. K. C. Lau. Curvature compensated convective transport: Smart, a new boundedness preserving transport algorithm. *International Journal for Numerical Methods in Fluids*, 8:617–641, 1988.

[10] B. Van Leer. Towards the ultimate conservative difference scheme ii. monotonicity and conservation combined in a second-order scheme. *Journal of Computational Physics*, 14:361–370, 1974.

[11] G. D. Van Albada, B. Van Leer, and W. W. Roberts. A comparative study of computational methods in cosmic gas dynamics. *Astronomy and Astrophysics*, 108:76–84, 1982.

[12] P. L. Roe. Characteristic based schemes for the euler equations. *Annual Review of Fluid Mechanics*, 18:337–365, 1986.

[13] B. Song, G. R. Liu, K. Y. Lam, and R. S. Amano. On a higher-order bounded discretization scheme. *International Journal for Numerical Methods in Fluids*, 32:881–897, 2000.

[14] J. Zhu. A low-diffusive and oscillation-free convection scheme. *Communications in Applied Numerical Methods*, 7:225–232, 1991.

[15] H. Jasak, H. G. Weller, and A. D. Gosman. High-resolution nvd differencing scheme for arbitrarily unstructured meshes. *International Journal for Numerical Methods in Fluids*, 31:431–449, 1999.

[16] M. A. Alves, P. J. Oliveira, and F. T. Pinho. A convergent and universally bounded interpolation scheme for the treatment of advection. *International Journal for Numerical Methods in Fluids*, 41:47–75, 2003.

[17] T. Chourushi. A high-resolution equi-gradient scheme for convective flows. *Applied Mathematics and Computation*, 338:123–140, 2018.

[18] J.-J. Wei, B. Yu, W.-Q. Tao, Y. Kawaguchi, and H. s. Wang. A new high-order-accurate and bounded scheme for incompressible flow. *Numerical Heat Transfer: Part B: Fundamentals*, 43(1):19–41, 2003.

[19] S. G. Rubin and P. K. Khosla. Polynomial interpolation method for viscous flow calculations. *Journal of Computational Physics*, 27:153–168, 1982.

[20] B. P. Leonard and S. Mokhtari. Beyond first-order upwinding: The ultra-sharp alternative for non-oscillatory steady-state simulation of convection. *International Journal of Numerical Methods in Engineering*, 30:729, 1990.

[21] M. S. Darwish and F. Moukalled. The normalized weighting factor method: A novel technique for accelerating the convergence of high-resolution convective schemes. *Numerical Heat Transfer*, 30:217–237, 1996.
[22] T. Chourushi. Computationally inexpensive and revised normalized weighting factor method for segregated solvers. *International Journal of Computer Mathematics*, 95(8):1622–1653, 2018.
[23] T. Chourushi. Effect of fluid elasticity on the numerical stability of high-resolution schemes for high shearing contraction flows using openfoam. *Theoretical and Applied Mechanics Letters*, 7(1):41–51, 2017.
[24] H. G. Weller, G. Tabor, H. Jasak, and C. Fureby. A tensorial approach to computational continuum mechanics using object-oriented techniques. *Computers in Physics*, 12(6):620–631, 1998.
[25] T. Chourushi, S. Singh, V. A. Sreekala, and R. S. Myong. Computational study of hypersonic rarefied gas flow over re-entry vehicles using the second-order boltzmann-curtiss constitutive model. *International Journal of Computational Fluid Dynamics*, 35(8):566–593, 2021.
[26] S. Singh, A. Karchani, T. Chourushi, and R. S. Myong. A three-dimensional modal discontinuous galerkin method for the second-order boltzmann-curtiss-based constitutive model of rarefied and microscale gas flows. *Journal of Computational Physics*, 457:111052, 2022.
[27] T. Chourushi, A. Rahimi, S. Singh, and R. S. Myong. Computational simulations of near-continuum gas flow using navier–stokes–fourier equations with slip and jump conditions based on the modal discontinuous galerkin method. *Advances in Aerodynamics*, 2:1–37, 2020.
[28] L. F. G. Marcantoni, J. P. Tamagno, and S. A. Elaskar. High-speed flow simulation using openfoam. *MecÃąnica Computacional*, 31(16):2939–2959, 2012.
[29] H. Jasak. *Error analysis and estimation in the finite volume method with applications to fluid flows*. PhD thesis, Imperial College, University of London, 1996.
[30] C. Rhie and W. L. Chow. Numerical study of the turbulent flow past an airfoil with trailing edge separation. *AIAA Journal*, 21(11):1525–1532, 1983.
[31] A. Kurganov and E. Tadmor. New high-resolution central schemes for nonlinear conservation laws and convection-diffusion equations. *Journal of Computational Physics*, 160(1):241–282, 2000.

[32] A. Kurganov, S. Noelle, and G. Petrova. Semidiscrete central-upwind schemes for hyperbolic conservation laws and hamilton–jacobi equations. *SIAM Journal on Scientific Computing*, 23(3):707–740, 2001.

[33] B. P. Leonard. The ultimate conservative difference scheme applied to unsteady one-dimensional advection. *Computer Methods in Applied Mechanics and Engineering*, 88:17–74, 1991.

[34] G. A. Sod. A survey of several finite difference methods for systems of nonlinear hyperbolic conservation laws. *Journal of Computational Physics*, 27(1):1–31, 1978.

17

Trends, Challenges, and Future Directions in Mathematical Solutions for Complex Engineering Systems

Mukesh Kumar Awasthi[1] and Satyvir Singh[2,3]

[1]Department of Mathematics, Babasaheb Bhimrao Ambedkar University, India
[2]Institute for Applied and Computational Mathematics, RWTH Aachen University, Germany;
[3]Department of Mathematics, Graphic Era Deemed to be University, Dehradun, India
E-mail: mukeshiitr.kumar@gmail.com; singh@acom.rwth-aachen.de

Abstract

This concluding chapter provides a synthesis of key themes and challenges discussed in *Mathematical Solutions for Complex Engineering Systems*. It highlights how advanced mathematical tools – such as stochastic processes, computational fluid dynamics, and fractional calculus – enable robust solutions for real-world engineering problems. The chapter distills pivotal contributions across four major domains: mathematical modeling (e.g., eco-epidemiology), analytical tools (e.g., fractional Taylor wavelets and method of fundamental solutions), computational simulations (e.g., rarefied gas flows and OpenFOAM solvers), and real-world engineering applications. It identifies transformative trends such as AI-augmented modeling, hybrid analytic-numeric methods, and interdisciplinary coupling of physical, biological, and ecological systems. Core challenges – like balancing model fidelity and computational cost, multi-scale complexity, and data-driven validation-are critically assessed. Concluding with a forward-looking perspective, the chapter emphasizes the need for adaptive, collaborative, and experimentally

DOI: 10.1201/9788743806691-17

validated modeling strategies to meet the demands of future engineering systems. It positions the book as a foundational reference for researchers and practitioners at the intersection of mathematics and engineering.

Keywords: Mathematical modeling, computational simulations, interdisciplinary approaches, engineering systems, future trends

17.1 Introduction

Objectives and scope of the book: The edited volume *Mathematical Solutions for Complex Engineering Systems* brings together mathematical innovations aimed at solving a diverse array of engineering challenges. The chapters span domains such as fluid mechanics, epidemic modeling, reliability engineering, and applied physics. The objective is to demonstrate how analytical rigor and numerical techniques converge to produce robust, scalable, and interpretable solutions.

The book opens with Chapter 1, which sets the conceptual groundwork by highlighting recent advances in mathematical modeling for engineering systems. It explores contemporary trends in deterministic and stochastic approaches, as well as the integration of multi-physics models. Chapter 2 focuses on stochastic modeling of nuclear power plant systems, utilizing multi-state reliability frameworks to demonstrate the significance of probabilistic approaches in high-risk engineering domains.

Shifting toward nonlinear dynamics, Chapter 3 explores chaos theory through the lens of an epidemic model, offering insights into the complex behavior of engineering systems under nonlinear influences. Chapter 4 investigates eco-epidemiological dynamics, particularly plant-herbivore interactions, by incorporating Allee effects and time-dependent parameters. This theme continues in Chapter 5, which models crop productivity under pesticide exposure and external interventions-providing guidance for sustainable agricultural practices. Chapter 6 bridges biology and mathematics by developing a model of tumor dynamics, showcasing the role of mathematical precision in biomedical systems.

The next section delves into fluid dynamics and rarefied gas behavior. Chapter 7 presents high-order computational models to study shock-induced instabilities in a the chevron-shaped light gas interface. Chapter 8 addresses Rayleigh – Brillouin scattering phenomena in polyatomic rarefied gases using extended hydrodynamic models. Chapter 9 builds on this with a nodal discontinuous Galerkin framework to solve the Grad-17 moment equations in rarefied regimes, offering insights into advanced gas dynamics simulation.

Chapter 10 provides a unified formula for velocity components in incompressible turbulent boundary layers over rotating geometries, relevant to rotating machinery and aerodynamics. Chapter 11 transitions into optimization strategies for heat transfer and fluid dynamics, highlighting evolutionary algorithms and AI-based methods for engineering efficiency.

The final portion of the book is devoted to novel numerical approaches for solving partial differential equations (PDEs). Chapter 12 introduces the method of fundamental solutions (MFS) for tackling nonlinear and nonhomogeneous PDEs. Chapter 13 proposes a fractional-order Taylor wavelet framework for fractional variational problems, expanding the numerical toolbox for non-integer order systems. Chapters 14 and 15 focus on wave equations: the former provides a fourth-order accurate scheme for the Klein – Gordon equation, while the latter presents a hybrid B-spline differential quadrature method for the sine–Gordon equation. Chapter 16 emphasizes the growing role of open-source tools by reviewing high-resolution schemes implemented in OpenFOAM.

Together, these chapters reflect the multi-faceted role of mathematics in modeling, simulating, and optimizing complex engineering systems. The book not only bridges theoretical mathematics with applied engineering but also offers a panoramic view of how interdisciplinary models can yield actionable solutions in fields ranging from nanotechnology and renewable energy to bioengineering and aerospace.

The centrality of mathematical modeling in engineering innovation: As engineering systems become increasingly complex and data-intensive, the demand for precise, scalable, and interpretable mathematical models is growing. This volume illustrates how core techniques – such as stochastic processes, nonlinear dynamical systems, fractional calculus, and high-fidelity simulations-serve as essential tools for analyzing multi-scale, multi-physics environments. Whether forecasting queue delays in stochastic systems or simulating hypersonic shock structures, the mathematical methods showcased in this book enable predictive accuracy and computational efficiency. More importantly, these models form the backbone of engineering decision-making, where reliability, adaptability, and verifiability are paramount.

In this context, the chapters serve as both exemplars and building blocks. They collectively affirm the transformative impact of mathematical modeling on next-generation engineering systems, while inviting future researchers to extend, validate, and integrate these techniques in ever more complex real-world applications.

17.2 Summary of Key Contributions

Innovations in mathematical modeling: This volume presents a diverse array of mathematical modeling strategies tailored to address complex challenges in modern engineering systems. Chapter 1 provides a conceptual overview of both deterministic and stochastic approaches, establishing a foundation for subsequent discussions. Chapter 2 advances this by applying stochastic reliability modeling to nuclear power plants, incorporating Markov chains and multi-state logic to evaluate system behavior under uncertainty. Nonlinear dynamical systems take center stage in Chapter 3, which uses an epidemic-style model to examine chaotic transitions in engineering contexts.

Chapters 4 and 5 extend this trajectory by examining ecological and agricultural systems: Chapter 4 focuses on plant-herbivore interactions incorporating the Allee effect and time-dependent factors, while Chapter 5 explores the interplay between pesticide use and crop productivity using nonlinear dynamical models. Chapter 6 shifts to tumor dynamics, blending mathematical precision with biological insight to model tumor growth and response. Together, these chapters illustrate the versatility of mathematical modeling in abstracting real-world complexity across environmental, biological, and engineered domains.

Analytical and approximation techniques: Several chapters are devoted to analytical and semi-analytical strategies that facilitate the solution of complex mathematical models. Chapter 12 introduces the Method of Fundamental Solutions (MFS), a meshless and computationally flexible technique ideal for nonlinear and irregular geometries. Chapter 13 presents a fractional-order Taylor wavelet approach for simulating variational problems with memory effects-an area gaining traction in materials and control theory.

Further advancing numerical approximation, Chapters 14 and 15 employ high-order schemes to solve canonical nonlinear wave equations. Chapter 14 develops efficient fourth-order methods for the Klein–Gordon equation, while Chapter 15 applies a hybrid B-spline differential quadrature technique to the sine–Gordon equation. These methods enhance precision and stability in simulating wave propagation and soliton behavior, thus reinforcing the role of approximation theory in modern applied mathematics.

Advances in computational and numerical simulations: Chapters 7 through 10 highlight the growing prominence of computational modeling in capturing the physics of extreme or multiscale systems. Chapter 7 examines shock-induced instabilities in light gas interfaces using high-resolution numerical solvers. Chapter 8 investigates Rayleigh–Brillouin scattering in polyatomic rarefied gases via extended hydrodynamic models, providing deeper insight into high-frequency flow behavior. Chapter 9 implements a nodal discontinuous Galerkin scheme to solve the Grad-17 moment equations, a cutting-edge tool for modeling nonequilibrium flows in the rarefied regime.

Chapter 10 presents a unifying formula for velocity components in turbulent boundary layers over rotating geometries, streamlining complex flow analysis in rotating systems. These contributions emphasize the indispensable role of computational solvers in domains where analytical methods are insufficient or infeasible.

Optimization Techniques and Practical Applications: Bridging theory and application, Chapters 11 and 16 explore optimization and practical deployment. Chapter 11 provides a comprehensive overview of optimization strategies in heat transfer and fluid dynamics, covering both classical techniques and emerging metaheuristics. Chapter 16 surveys high-resolution convection schemes available within the OpenFOAM framework, underscoring the practical impact of open-source CFD tools for simulating complex fluid flow environments.

These chapters demonstrate how optimization, informed by advanced mathematical techniques and computational infrastructure, enables performance improvement and resource efficiency across engineering sectors.

Integration with Real-World Systems: Many chapters emphasize real-world applicability, reinforcing the engineering relevance of mathematical abstractions. Queueing models and reliability theory (Chapter 2) address safety in nuclear systems, while eco-epidemiological models (Chapters 4-6) provide insight into agriculture, pest management, and tumor therapy. Chapters 7-10 have direct implications for aerospace, microscale flow devices, and high-altitude aerodynamics. Optimization frameworks and high-order solvers (Chapters 11, 14, 15) find use in thermal management, optical systems, and energy conversion technologies.

These applications reveal mathematical modeling not only as a theoretical pursuit but also as a critical engineering tool with measurable impact on sustainability, safety, and design innovation.

Cross-cutting themes: Several themes recur throughout the volume. First, the tension between modeling complexity and computational efficiency is addressed via approximation, hybrid, and meshless methods. Second, the integration of cross-disciplinary domain knowledge-from microfluidics to oncology-underscores the value of collaborative modeling. Third, the promotion of open-source and reproducible tools (e.g., OpenFOAM, MFS) reflects a broader shift toward transparent and accessible scientific computation.

17.3 Emerging Trends and Interdisciplinary Approaches

Contextual overview of emerging modeling paradigms: As engineering problems grow in complexity and scale, the methodologies used to analyze them must evolve accordingly. This book captures several emerging directions in mathematical modeling-ranging from the application of fractional calculus to data-driven hybrid techniques and interdisciplinary system integration. These innovations highlight the trend toward methods that are not only mathematically rigorous but also computationally feasible and broadly applicable.

Analytical and hybrid modeling techniques: Several chapters explore non-classical approaches that extend the boundaries of traditional mathematical modeling. Notably, Chapter 13 employs fractional-order Taylor wavelets to address memory-dependent dynamics common in viscoelastic and anomalously diffusive systems. These techniques allow for compact representations and high accuracy in modeling non-local processes. Stochastic frameworks also continue to gain traction. Chapter 3 utilizes Markov chains and Poisson processes to evaluate reliability in nuclear power plants, emphasizing probabilistic modeling under uncertainty. Hybrid methods, such as those in Chapter 2, blend deterministic and stochastic techniques-offering fluid and diffusion approximations that scale well to large systems while preserving essential random behaviors. This shift toward fractional, stochastic, and hybrid approaches reflects a broader move in the field toward models that are flexible, modular, and capable of capturing subtle system nuances often missed by traditional techniques.

Computational tools and high-fidelity solvers: Advancements in computational infrastructure have enabled the adoption of powerful numerical solvers across a range of engineering domains. Chapter 16 exemplifies this with its OpenFOAM-based high-resolution schemes for convection-dominated flows, democratizing access to cutting-edge CFD techniques through open-source software. Chapter 8 employs the boundary element method (BEM) for solving MHD Stokes flows in microchannels, reducing mesh complexity while preserving accuracy. Further innovations include the method of fundamental solutions (Chapter 12), which solves nonlinear and nonhomogeneous PDEs efficiently without mesh generation. The discontinuous Galerkin method (Chapter 11) applied to Grad-17 equations enables rarefied gas flow simulation beyond classical hydrodynamics. These computational strategies share a common aim: to bridge the gap between analytical accuracy and scalable implementation.

Interdisciplinary system integration: One of the most promising directions emerging from this volume is the fusion of disciplines-particularly biology, ecology, and physics-into unified modeling frameworks. Chapter 4 models rumor spread and pollution using epidemic frameworks, while Chapter 5 integrates ecological dynamics through eco-epidemiological models incorporating Allee effects. Chapter 6 advances this idea further with dynamical models that inform pesticide use in agricultural planning. Chapter 9 demonstrates another form of integration, modeling nanofluid flow over stretched surfaces where magnetic fields and thermal properties interact-linking fluid mechanics and materials science. These contributions underscore the expanding boundary of engineering mathematics, wherein models originally rooted in biological or environmental contexts are used to solve physically driven systems, and vice versa.

Sustainability and societal impact: An implicit yet important trend reflected across the chapters is the increasing alignment of mathematical modeling with sustainability objectives. Whether optimizing resource allocation in queueing systems (Chapter 2), improving agricultural resilience (Chapters 5 and 6), or reducing computational overhead through efficient solvers (Chapters 12 and 16), the methodologies aim to enhance efficiency, minimize waste, and support environmentally informed decision-making. Moreover, tools like uncertainty quantification and real-time adaptability-as touched upon in Chapters 3 and 14-are becoming essential in designing engineering systems that are both resilient and responsive to dynamic operating conditions.

Synthesis and outlook: These emerging trends point to a future where mathematical modeling is deeply interdisciplinary, computationally empowered, and grounded in practical relevance. The shift toward hybrid and fractional frameworks, combined with open-source computational tools, is transforming the accessibility and applicability of mathematical solutions. The integration of biological and environmental modeling further expands the scope of engineering mathematics, preparing it to address complex systems in healthcare, climate, and energy.

Together, the chapters in this book highlight a collective movement toward modeling approaches that are not only scientifically robust but also societally impactful.

17.4 Challenges in Mathematical Modeling of Complex Systems

As engineering systems grow in complexity, traditional modeling approaches often fall short in capturing the nuances of dynamic, nonlinear, and interdisciplinary behaviors. This volume presents a broad spectrum of emerging paradigms in mathematical modeling-ranging from fractional calculus and stochastic systems to hybrid and biologically inspired frameworks. These approaches not only emphasize theoretical rigor but also aim to ensure computational viability and real-world applicability.

Analytical and Hybrid Modeling Techniques: Several chapters in the volume demonstrate a clear departure from classical models by embracing advanced analytical and hybrid strategies. Chapter 13 introduces a fractional-order Taylor wavelets approach for solving fractional variational problems, well-suited for systems with memory effects or non-local dynamics. In contrast, Chapter 14 presents efficient fourth-order numerical schemes for the Klein–Gordon equation, reflecting advances in the numerical treatment of nonlinear wave dynamics.

Stochastic methods receive particular attention in Chapter 2, where Markov chains and Poisson processes are used to analyze nuclear power plant reliability, capturing randomness in system behavior. Hybrid methodologies are showcased in Chapter 15, where a hybrid B-spline differential quadrature method is developed for approximating solutions to the nonlinear sine-Gordon equation, balancing accuracy with computational efficiency. These contributions reflect a clear trend toward modeling approaches that are modular, flexible, and capable of resolving fine-scale system dynamics.

Computational Tools and High-Fidelity Solvers: Recent improvements in computational power and open-source platforms have enabled the deployment of sophisticated solvers across engineering domains. Chapter 16 presents an overview of high-resolution schemes in OpenFOAM, providing access to state-of-the-art CFD tools through open-source infrastructure.

Similarly, Chapter 9 applies a nodal discontinuous Galerkin framework to solve the Grad-17 moment equations in rarefied gas flows-extending simulation capabilities into the non-continuum regime.

Meshless techniques are explored in Chapter 12 via the Method of Fundamental Solutions (MFS), which efficiently handles nonlinear and nonhomogeneous PDEs. Chapter 8 contributes further by addressing extended hydrodynamic models in Rayleigh–Brillouin scattering, underscoring the role of advanced numerical formulations in handling high-fidelity physical scenarios. These chapters collectively highlight a push toward scalable, high-accuracy methods that bridge theoretical modeling with robust computation.

Interdisciplinary System Integration: The integration of mathematical modeling across disciplinary boundaries is a defining feature of this volume. Chapter 3 employs epidemic models to explore nonlinear dynamics and chaos in engineering systems, showcasing a novel crossover between biological modeling and system dynamics. Chapters 4 and 5 investigate eco-epidemiological models in plant–herbivore interactions and pesticide management, incorporating concepts like the Allee effect and external interventions to improve agricultural resilience.

Chapter 6 extends the interdisciplinary approach to tumor dynamics, using mathematical precision to support biological insight. Meanwhile, Chapter 7 models shock-induced instabilities in light gas interfaces, a topic deeply rooted in high-speed fluid mechanics. Chapter 10 develops compact formulations for turbulent boundary layers over rotating bodies. These chapters illustrate the fluid boundaries between physical, biological, and computational sciences in modern engineering modeling.

Sustainability and Societal Impact: An undercurrent of sustainability and real-world relevance runs throughout the volume. In Chapter 2, stochastic queueing models contribute to resource optimization in nuclear systems. Chapters 4-6 address environmental and agricultural planning, proposing models that aim to enhance crop yield, reduce pesticide use, and improve system resilience. Chapter 12 and Chapter 16 emphasize computational efficiency as a means of reducing energy consumption in simulations-contributing to greener computing practices.

Furthermore, the growing importance of uncertainty quantification and adaptive modeling is evident in Chapters 3 and 14, which promote the development of systems capable of real-time response and decision-making under dynamic conditions.

Synthesis and Outlook: Taken together, the contributions in this volume indicate a decisive shift in mathematical modeling practices. The convergence of fractional, stochastic, and hybrid approaches, supported by open-source and high-performance computing platforms, is expanding the boundaries of what engineering mathematics can achieve. The integration of biological, ecological, and physical modeling frameworks reflects the evolving scope of engineering problems-now as likely to involve ecosystems or human health as they are structural or mechanical systems.

Looking ahead, the future of modeling lies in interdisciplinary, adaptive, and accessible methods that can respond to the challenges posed by climate, sustainability, and technological innovation. The chapters compiled here collectively form a forward-looking roadmap for researchers and practitioners working to solve the complex engineering problems of tomorrow.

17.5 Future Research Directions

AI-enhanced and hybrid modeling frameworks: AI-Enhanced and Hybrid Modeling Frameworks The convergence of artificial intelligence (AI), machine learning (ML), and classical physics-based modeling is redefining the landscape of engineering simulation. While traditional solvers offer interpretability grounded in governing equations, they often struggle with scalability and computational cost-particularly for high-dimensional, nonlinear, or multiscale systems discussed throughout this volume. In contrast, AI-based surrogates-trained on high-fidelity simulation data or sensor measurements-can dramatically reduce computational overhead in design optimization, parameter tuning, and uncertainty quantification.

However, black-box models, while efficient, often lack the physical interpretability crucial for safety-critical applications like aerospace, nuclear energy, and climate modeling. A promising research frontier involves hybrid frameworks, such as physics-informed neural networks (PINNs) and operator learning architectures, which embed physical laws into the learning process. These approaches balance data efficiency with theoretical rigor, enabling real-time modeling in domains like fluid dynamics (Chapters 7, 8, 16), rarefied gas flows (Chapters 8, 9), and biological systems (Chapters 4, 6).

Real-time and adaptive simulation architectures: Modern engineering systems-from autonomous vehicles and robotics to smart manufacturing and responsive climate control-operate in dynamically evolving environments. Static models, though accurate under fixed conditions, become inadequate when boundary conditions, loads, or environmental inputs vary in real time. The demand is shifting toward adaptive simulation architectures that integrate online learning, sensor-driven feedback, and edge computing.

Techniques such as reduced-order modeling, model predictive control (MPC), and online filtering algorithms are gaining traction. These enable near-instantaneous decision-making in applications such as turbulence control, thermal optimization, and complex system automation. Challenges remain in ensuring stability, managing latency, and synchronizing diverse data sources. Nevertheless, advances in data assimilation, adaptive meshing, and multi-rate solvers are closing the gap between simulation and control-especially relevant for Chapters 11, 13, and 15, which deal with high-fidelity modeling and optimization.

Toward unified cross-disciplinary modeling frameworks: As the chapters in this volume illustrate, real-world engineering problems are rarely confined to a single domain. From eco-epidemiological modeling (Chapter 4) to coupled biological-chemical interactions (Chapters 5, 6), there is a growing need for cross-disciplinary frameworks capable of handling heterogeneous systems. These may span ordinary differential equations, partial differential equations, stochastic processes, and agent-based models-all under a single computational umbrella.

The development of such unified platforms requires formal ontologies, cross-domain validation protocols, and standardized interfaces to facilitate seamless co-simulation. Efforts toward plug-and-play model interoperability will be vital for the future of digital twins, which aim to mirror complex engineering and ecological systems with real-time fidelity. Modular architectures and cloud-based simulation backends will further enhance scalability and collaboration across disciplinary boundaries.

Rigorous validation and uncertainty quantification: For mathematical models to transition from academic prototypes to industrial standards, validation and credibility are essential. Particularly in high-stakes domains like nuclear power (Chapter 2), medical modeling (Chapter 6), and aerospace (Chapter 14), trust in model predictions must be established through rigorous testing and uncertainty quantification.

Future research should emphasize Bayesian calibration, Monte Carlo simulation, sensitivity analysis, and ensemble modeling to provide probabilistic guarantees of performance. The availability of benchmark datasets, transparent codebases, and reproducible pipelines will be key to achieving methodological comparability. In addition, lab-in-the-loop validation-where experimental setups are coupled with digital models-can accelerate the theory-practice feedback loop.

A forward-looking vision: The trajectory of mathematical modeling for engineering systems must be guided by three foundational pillars: scalability, interpretability, and societal impact. Scalability ensures that models can capture the behavior of large-scale, high-dimensional systems. Interpretability allows domain experts-whether engineers, scientists, or decision-makers-to understand and trust the outputs. Societal relevance ensures that modeling efforts align with global challenges, including sustainability, resilience, and equity.

As highlighted across this volume-from nonlinear epidemic modeling (Chapter 3) to advanced optimization techniques (Chapter 11)-the future of mathematical modeling is integrative, agile, and application-driven. The goal is not just to simulate reality, but to shape it: by optimizing energy systems, predicting biological outcomes, designing efficient materials, and informing robust policy decisions. With continued innovation at the intersection of theory, computation, and data, mathematical sciences are well-positioned to drive the next generation of engineering breakthroughs.

17.6 Conclusion and Outlook

This concluding chapter has synthesized the diverse methodologies and innovations presented throughout the volume *Mathematical Solutions for Complex Engineering Systems*. From stochastic modeling and fractional calculus to high-resolution computational solvers and real-time data integration, the book showcases the evolving role of mathematics as both a lens for understanding complex systems and a toolkit for engineering solutions.

Across the chapters, several cross-cutting themes emerge: the growing reliance on hybrid and data-driven models, the need for scalable and interpretable numerical schemes, and the increasing importance of interdisciplinary integration-linking physical, biological, and ecological models within unified frameworks. These approaches not only enhance computational efficiency and modeling fidelity but also expand the applicability of mathematical tools to sustainability, healthcare, aerospace, agriculture, and beyond. At the same time, the challenges are clear: model validation, uncertainty quantification, and the need for adaptive, real-time simulations demand continued methodological innovation. These frontiers call for deeper collaboration among mathematicians, engineers, computer scientists, and domain experts to ensure that mathematical solutions remain relevant, trustworthy, and transformative.

Looking forward, the future of mathematical modeling lies in its ability to scale with complexity, evolve with data, and align with societal goals. By combining analytical rigor with computational agility and ethical responsibility, the mathematical sciences are uniquely positioned to guide the next generation of engineering breakthroughs and shape a resilient, data-informed, and sustainable world.

Index

About the Editors

Dr. Satyvir Singh is a Research Associate Fellow at the Institute of Applied and Computational Mathematics, RWTH Aachen University, Germany. He obtained his Ph.D. in Computational Fluid Mechanics from Gyeongsang National University, South Korea, where he developed 3D discontinuous Galerkin methods for solving Boltzmann-type gas kinetic equations. He has held research positions at Nanyang Technological University, Singapore, and the Research Center for Aircraft Parts Technology in South Korea. His expertise spans computational fluid dynamics, high-order numerical methods, gas kinetic theory, and hydrodynamic instability. Dr. Singh has authored over 50 research articles with more than 1000 citations and has presented his work globally. He has also received research funding as Co-PI for a project on brain tumor dynamics at Jazan University, Saudi Arabia. Recognized among the top 2% of researchers worldwide by Stanford University in 2025, he continues to make significant contributions to the fields of Physics and Astronomy.

Dr. Mukesh Kumar Awasthi is an Assistant Professor in the Department of Mathematics at Babasaheb Bhimrao Ambedkar University, Lucknow. Specializing in mathematical modeling of flow problems, he has expertise in viscous potential flow, electro-hydrodynamics, magneto-hydrodynamics, and heat and mass transfer. Dr. Awasthi has published over 135 research works, including books and journal articles, and has served as a series editor for CRC Press. He has received multiple research awards and secured funding for his project on nonlinear fluid interfaces. Recognized among the top 2% of researchers worldwide by Stanford University in 2022 and 2023, he continues to contribute significantly to computational mathematics and engineering applications.

For Product Safety Concerns and Information please contact our
EU representative GPSR@taylorandfrancis.com Taylor & Francis
Verlag GmbH, Kaufingerstraße 24, 80331 München, Germany